AF362058

Evolution and Palaeobiology of Flightless Birds

Evolution and Palaeobiology of Flightless Birds

Editors

Eric Buffetaut
Delphine Angst

MDPI • Basel • Beijing • Wuhan • Barcelona • Belgrade • Manchester • Tokyo • Cluj • Tianjin

Editors
Eric Buffetaut
CNRS (UMR 8538),
Laboratoire de Géologie de
l'Ecole Normale Supérieure,
PSL Research University,
24 rue Lhomond
CEDEX 05, 75231 Paris,
France

Delphine Angst
82 rue Pierre Brossolette
92320 Châtillon, France

Editorial Office
MDPI
St. Alban-Anlage 66
4052 Basel, Switzerland

This is a reprint of articles from the Special Issue published online in the open access journal *Diversity* (ISSN 1424-2818) (available at: https://www.mdpi.com/journal/diversity/special_issues/flightless_birds).

For citation purposes, cite each article independently as indicated on the article page online and as indicated below:

LastName, A.A.; LastName, B.B.; LastName, C.C. Article Title. *Journal Name* **Year**, *Volume Number*, Page Range.

ISBN 978-3-0365-4023-8 (Hbk)
ISBN 978-3-0365-4024-5 (PDF)

Cover image courtesy of Agustin Agnolin

Contents

Editorial

An Introduction to Evolution and Palaeobiology of Flightless Birds

Eric Buffetaut [1,2,]* and Delphine Angst [3]

[1] CNRS (UMR 8538), Laboratoire de Géologie de l'Ecole Normale Supérieure, PSL Research University, 24 rue Lhomond, CEDEX 05, 75231 Paris, France
[2] Palaeontological Research and Education Centre, Maha Sarakham University, Maha Sarakham 44150, Thailand
[3] 82 rue Pierre Brossolette, 92320 Châtillon, France; angst.delphine@gmail.com
* Correspondence: eric.buffetaut@sfr.fr

Citation: Buffetaut, E.; Angst, D. An Introduction to Evolution and Palaeobiology of Flightless Birds. *Diversity* 2022, 14, 296. https://doi.org/10.3390/d14040296

Received: 11 April 2022
Accepted: 12 April 2022
Published: 15 April 2022

Although flight is often considered as one of the most salient characteristics of birds, in the course of their evolution various avian lineages have lost the ability to fly. This has happened at different periods of the geological past, beginning in the Cretaceous with such forms as the terrestrial *Patagopteryx* and *Gargantuavis* and the marine Hesperornithiformes, and under very varied circumstances. In some cases, loss of flight is associated with strictly terrestrial habits in usually large forms, as in living and fossil "ratites" and in various extinct groups of giant ground birds (gastornithids, dromornithids, phorusrhacids, etc.). The case of birds deeply adapted to foraging in an aquatic environment, such as penguins and Hesperonthiformes, is a completely different instance of flightlessness. Loss of flight has often taken place in insular environments, where the lack of predators is supposed to have played a crucial part—the dodo is a case in point. However, it also occurred repeatedly on large land masses, as exemplified today by the ostrich and related birds.

This Special Issue explores various aspects of this multi-faceted evolutionary process, from phylogeny to palaeobiology. The nine papers in this collection deal with flightless birds belonging to widely different extinct and extant groups, from many parts of the world and from various time periods, from the Late Cretaceous to the Quaternary. The aim is not to provide a comprehensive review of the evolution and palaeobiology of the many groups that have lost the power of flight over the long time span extending back to the Cretaceous. Rather, the papers published here illustrate both the diversity of flightless birds and the multifarious approaches that can be used to study them, from stratigraphy and functional anatomy to phylogenetic analysis and bone histology.

The paper by Alyssa Bell and Luis Chiappe deals with the Hesperornithiformes, a group of Late Cretaceous diving birds that were among the first avians to become flightless. When first described in the 1870s, these "birds with teeth" attracted much attention, and Charles Darwin considered them as some of the best evidence in favour of his theory of evolution. Although they have been known for a century and a half, however, no recent global review of Hesperornithiforms was available. Bell and Chiappe's timely paper provides such a review, discussing their diversity, geographical and temporal distribution, and ecology. Even though they have been known for a long time, Hesperornithiformes remain a fascinating group of early birds, about which innovative approaches are revealing many new facts.

Federico Agnolin discusses the phylogenetic relationships of *Brontornis burmeisteri*, a giant bird from the Miocene of South America that has puzzled palaeornithologists since its original description in 1891. Although it has often been placed among the cursorial and carnivorous "terror birds" (Phorusrhacoidea), several of its osteological characters rather suggest a graviportal plant-eater. Agnolin's conclusion, based on modified datasets, is that it belongs to Galloanserae and is part of a still poorly known Tertiary radiation of large graviportal birds from South America.

Klara Widrig and Daniel Field provide a comprehensive review of the fossil record and evolution of the Palaeognathae, a large avian group containing, besides the volant tinamous, a large number of flightless forms—including the living ostriches, rheas, emus and cassowaries, as well as many extinct taxa such as the moas and elephant birds. Despite remaining gaps in the fossil record, an evolutionary history emerges, starting with relatively small-sized ground-feeding birds which survived the end-Cretaceous extinction event and diversified considerably during the Paleogene, although the extant sub-clades do not become clearly recognizable until the Neogene. It is increasingly clear that flightlessness and large body size have appeared independently in several lineages.

Anusuya Chinsamy, Aurore Canoville and Delphine Angst present the results of histological studies carried out on a large sample of limb bones from various large flightless birds, including extant and extinct "ratites" as well as the Paleogene giant neognath *Gastornis*. Their results show that bone microanatomy can reflect locomotion type (graviportal versus cursorial), thus providing a useful tool for palaeobiological interpretations. In addition, somewhat unexpectedly, growth marks in the bones of various extant ratites indicate flexible growth patterns (in response to environmental conditions) that may represent the plesiomorphic condition in Palaeognathae and, more widely, Neornithes.

Peter Johnston and Kieren Mitchell's paper on sensory adaptation in flightless birds explores the intriguing topic of the sensory capacities of several extinct and extant forms, including moa, elephant birds, kiwi and the kakapo parrot. On the basis of various cranial skeletal features relating to vision, hearing and olfaction, they show how the different lifestyles of these birds have resulted in contrasting sensory strategies: for instance, the kiwi, the Upland Moa and the aepyornithids apparently were olfactory specialists, but the moa had a well-developed hearing sensitivity range lacking in the other taxa. This approach opens up interesting new directions for palaeobiological investigations and reconstructions.

Warren Handley and Trevor Worthy use morphometric methods to describe in detail the endocranial morphology of the dromornithids, or mihirungs, a group of extinct large flightless birds which flourished in Australia from the Eocene to the Pleistocene. This study has phylogenetic implications, since in terms of endocranial anatomy the mirhirungs appear to be closer to galliforms than to anseriforms, in agreement with a recent interpretation. From a functional point of view, this study supports the conclusion that they were diurnal herbivores with well-developed stereoscopic depth perception—the old myth of the dromornithids as "killer ducks" thus receives an additional blow from endocranial anatomy.

Anusuya Chinsamy and Trevor Worthy describe the bone histology of the Pleistocene dromornithid *Genyornis newtoni*, the last of the mihirungs. This study provides important new evidence about the still poorly known biology of this Australian giant bird. In particular, the growth pattern revealed by this study indicates that *Genyornis newtoni* took more than a single year to become sexually mature, and reached skeletal maturity after sexual maturity. In addition, it apparently retained a plesiomorphic flexible growth strategy, which enabled it to respond to changing environmental conditions.

Eric Buffetaut and Delphine Angst describe a large ostrich femur found in the 1920s in the Lower Pleistocene deposits of the Nihewan basin of northern China, which had hitherto been only very cursorily mentioned. It is referred to *Pachystruthio* and significantly enlarges the geographical distribution of this genus of giant ostriches which was previously known from Hungary, Crimea and Georgia. As shown by a review of the fossil ostriches from China, *Pachystruthio* is an element of the long and apparently complex history of this group of birds in eastern Asia, which extends from the Miocene to the Late Pleistocene.

Eric Buffetaut reviews the stratigraphic distribution of *Psammornis*, an enigmatic egg-based taxon from the Neogene and Quaternary of North Africa and possibly the Middle East. The genus was erected in 1911, on the basis of eggshell fragments indicating a very large bird, from a locality of uncertain geological age in Algeria. Since then, a number of eggshell finds from the Sahara and surrounding areas have been referred to *Psammornis*, but most of them are very poorly dated. Curiously enough, *Psammornis* localities with a reasonably good stratigraphic context, in Tunisia and Mauritania, have often been overlooked, although they are of prime importance for unravelling the obscure history of what were probably giant ostriches.

It is expected that this collection of papers will both provide abundant new information about the evolution and biology of extant and extinct flightless birds and illustrate the wide spectrum of the approaches used to investigate them. Although these approaches have resulted in considerable progress in our understanding of these avian groups, an obvious lesson to be learnt from these contributions is that much remains to be discovered and investigated. Owing to both discoveries of new fossil specimens and the implementation of new, innovative techniques, our current picture of flightless birds is in many respects quite different from what it was a few decades ago. We hope these papers will reflect how fast this branch of ornithology is developing and hint at directions for future studies.

Author Contributions: Writing—original draft preparation, E.B. and D.A.; writing—review and editing, E.B. and D.A. All authors have read and agreed to the published version of the manuscript.

Funding: This research received no external funding.

Conflicts of Interest: The authors declare no conflict of interest.

 *diversity*

Review

The Hesperornithiformes: A Review of the Diversity, Distribution, and Ecology of the Earliest Diving Birds

Alyssa Bell * and Luis M. Chiappe

Dinosaur Institute, Natural History Museum of Los Angeles County, 900 Exposition Boulevard, Los Angeles, CA 90007, USA; lchiappe@nhm.org
* Correspondence: abell@nhm.org

Abstract: The Hesperornithiformes (sometimes referred to as Hesperornithes) are the first known birds to have adapted to a fully aquatic lifestyle, appearing in the fossil record as flightless, foot-propelled divers in the early Late Cretaceous. Their known fossil record—broadly distributed across the Northern Hemisphere—shows a relatively rapid diversification into a wide range of body sizes and degrees of adaptation to the water, from the small *Enaliornis* and *Pasquiaornis* with lesser degrees of diving specialization to the large Hesperornis with extreme morphological specializations. Paleontologists have been studying these birds for over 150 years, dating back to the "Bone Wars" between Marsh and Cope, and as such have a long history of naming, and renaming, taxa. More recent work has focused to varying degrees on the evolutionary relationships, functional morphology, and histology of the group, but there are many opportunities remaining for better understanding these birds. Broad-scale taxonomic evaluations of the more than 20 known species, additional histological work, and the incorporation of digital visualization tools such as computed tomography scans can all add significantly to our understanding of these birds.

Keywords: Hesperornithiformes; Aves; Mesozoic birds; evolution; paleoecology; diving birds

Citation: Bell, A.; Chiappe, L.M. The Hesperornithiformes: A Review of the Diversity, Distribution, and Ecology of the Earliest Diving Birds. *Diversity* 2022, 14, 267. https://doi.org/10.3390/d14040267

Academic Editors: Eric Buffetaut and Delphine Angst

Received: 23 February 2022
Accepted: 25 March 2022
Published: 1 April 2022

Publisher's Note: MDPI stays neutral with regard to jurisdictional claims in published maps and institutional affiliations.

1. Introduction

In the winter of 1870, Othniel Charles Marsh discovered the distal-most end of a tibiotarsus of a large bird in Cretaceous (Coniacian-early Campanian) marine sediments near the Smoky Hill River in western Kansas (specimen 1205 at the Yale Peabody Museum [YPM]) [1]. This unremarkable specimen was the first look at a remarkable group of extinct animals, the first dinosaurs to adapt to a fully aquatic lifestyle and the earliest group of birds to swim away from the ability to fly. On second and third expeditions to western Kansas, in June of 1871 and the fall of 1872, Marsh discovered a more complete specimen (YPM 1200) of the same species as well as fossils of other ancient birds, one of which was nearly complete (YPM 1207) [1]. In 1872, as the infamous "Bone Wars"—an ignominious chapter in American paleontology—were just beginning, Marsh published his first work on these specimens [2]. Marsh designated the material as *Hesperornis regalis*, a large swimming bird that he interpreted as being most closely related to modern loons, albeit with significant differences from "all other known birds, recent and extinct" [3] (p. 361), and later assigned it to the Natatores [4], a paraphyletic group used at the time to unite modern swimming birds that has since been abandoned. Over subsequent years, Marsh sent numerous expeditions back to the Smoky Hill River in Kansas, resulting in the collection of hundreds of specimens of birds belonging to a group termed the Odontornithes, which Marsh erected for Ichthyornis and *Apatornis* [5]. Later, Marsh added the coeval Hesperornis to the Odontornithes on the basis of the presence of teeth in the jaws [6]. Marsh would go on to describe a second species of Hesperornis, *H. gracilus*, and three other related genera, *Baptornis advenus*, *Coniornis altus*, and *Lestornis crassipes* [7], although the latter two are now assigned to Hesperornis

At the same time that Marsh was working on the North American toothed birds, Harry Seeley [8] was describing a group of small fossil birds from the Upper Cretaceous (Cenomanian) Cambridge Greensand in England from material discovered by Lucas Barrett in 1858 and briefly discussed by Lyell a year later [9]. Unlike Marsh's larger birds that included well-preserved, articulated specimens, the two species identified by Seeley—*Enaliornis barretti* and *E. sedgewicki*—were entirely disarticulated and heavily eroded, as part of a reworked deposit [8]. However, like the fossils Marsh was discovering, these British fossils were also abundant, with dozens of isolated bones available for study in the Woodwardian Museum (now the Woodwardian Collection of the Sedgewick Museum) in Cambridge [8]. Seeley did not recover any specimens with teeth. Furthermore, he deemed Marsh's reliance on teeth for designation of the Odontornithes to be unsupported, in light of the variability of teeth across modern mammals and reptiles [8], a view that was upheld by Furbringer [9] in 1888 when he established the order Hesperornithiformes (now phylogenetically defined as all taxa more closely related to *Hesperornis regalis* than to Neornithes or modern birds), removing Hesperornis and *Baptornis* from Odontornithes. Seeley noted numerous similarities between *Enaliornis* and modern loons, and so referred *Enaliornis* to the Natatores [8], as Marsh had originally done with Hesperornis. The placement of *Enaliornis* within the Hesperornithiformes was first proposed by Lydekker [10] a few years after the erection of this clade, which was later supported by Wetmore [11], Storer [12], Martin and Tate [13], and others. Our modern understanding of hesperornithiform phylogenetics places them within the Ornithurae and very close to the divergence of modern birds, Neornithes (Figure 1). Thus, by the end of the 19th century the Hesperornithiformes were the most diverse lineage of Cretaceous birds known, with a wide geographic and stratigraphic distribution and ranging in size from a bird the size of a grebe to birds as much as 1.5 m long.

From these early studies, our modern understanding of the Hesperornithiformes has expanded to over 20 species from across Laurasia, identified from marine, transitional, and continental deposits (Figure 2). Interestingly, the fossil collections amassed at the Peabody Museum in Yale University by Marsh and at the Sedgewick Museum in Cambridge University remain the most abundant in terms of number of specimens, rivaled only by the Carrot River material of *Pasquiaornis* collected in the latter part of the twentieth century [14–16].

Studies over the past 150+ years have explored the evolutionary relationships and trajectories, biomechanics, ecology, life history, and biogeography of these incredible birds, as well as continually identifying new families, genera, and species. This review will provide an overview of this body of research, summarizing both where we are today in our understanding of the evolution and biology of these birds and how we got there, and highlight areas for potential future research.

2. General Anatomy

From the first studies of specimens by both Marsh and Seeley, the highly modified bauplan of these birds was recognized as a significant chapter in avian adaptation. This consists of a streamlined body with an elongated skull and neck, heavily reduced forelimb, and dramatically robust hindlimb (Figure 3). In superficial form, it is easy to see how early researchers, and some not-so-early researchers, identified the similarities to modern foot-propelled diving birds such as loons and concluded that these birds were part of the modern diving lineage (e.g., [3,8,17–19]). However, this view fails to account for the strong convergence found among modern diving bird lineages, as we now recognize foot-propelled diving to have evolved independently at least four times among modern birds (i.e., loons, grebes, diving ducks, and cormorants), and that even such morphologically similar birds as loons and grebes are not closely related at all [20,21]. In fact, the results of a comprehensive morphometric analysis of hesperornithiforms and modern diving birds showed that the former rarely share morphospace with loons and grebes, and that instead, they overlap more in morphospace with cormorants and diving ducks [22].

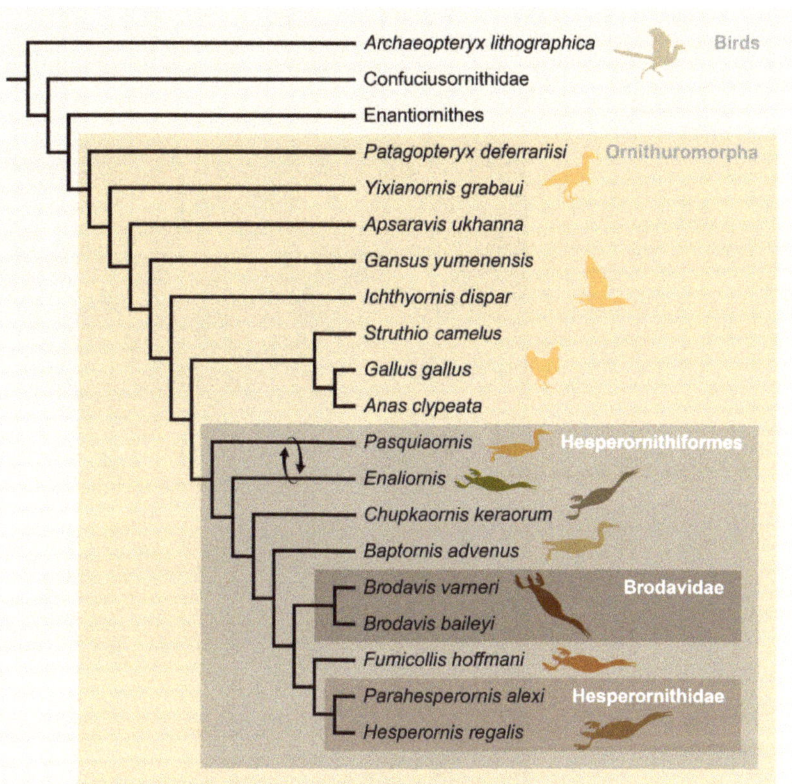

Figure 1. Phylogenetic tree of birds, after Bell and Chiappe [23] and Tanaka et al. [24]. Arrows show the alternative placement of *Enaliornis* and *Pasquiaornis* recovered by Tanaka et al. [24].

The skull of hesperornithiforms is elongate, with a long rostrum similar to that seen in modern foot-propelled diving birds. This elongation is due primarily to the length of the premaxilla, as in modern birds, which makes up nearly half the length of the rostrum in Hesperornis [25]. This is unlike more stemward, longirostrine Mesozoic birds (e.g., *Longipteryx, Rapaxavis, Dingavis*), where elongation of the rostrum is due in part to an extended maxilla [26]. Within hesperornithiforms, there appears to be variation in the degree of elongation of the skull. *Enaliornis*, the most basal hesperornithiform currently known [23], has a proportionally shorter skull than that of hesperornithids (*Parahesperornis* and Hesperornis). This is seen in three different regions of the skull: the parietals and temporal fenestrae, the frontals, and the portion of the rostrum rostral to the nares [25]. Furthermore, these regions of the skull are proportionally shorter in *Parahesperornis* than in Hesperornis, implying potential ecological specializations (i.e., niche partioning) among these likely coeval and sympatric birds [25].

The dentary and maxillae of hesperornithiforms bear small recurved teeth set in a groove (Figure 4). While today we recognize a wide diversity of tooth retention patterns across Mesozoic birds [26,27], when first discovered by Marsh, this feature was striking. The retention of teeth in birds is a conserved character with similar molecular and developmental mechanisms inherited from their nonavian reptilian ancestors [28].

Marsh's [3] first description of the teeth of Hesperornis noted that they were not set in true sockets (i.e., thecodont implantation), but were instead separated by slight projections from the sides of the groove in which they were set in the jaws. While dentary fragments assigned to *Pasquiaornis* have been described as having similar tooth implantation in a groove with incomplete sockets [15], published images of some specimens appear to show much more extensive socket development than the slight projections found in hesperornithids (Figure 4).

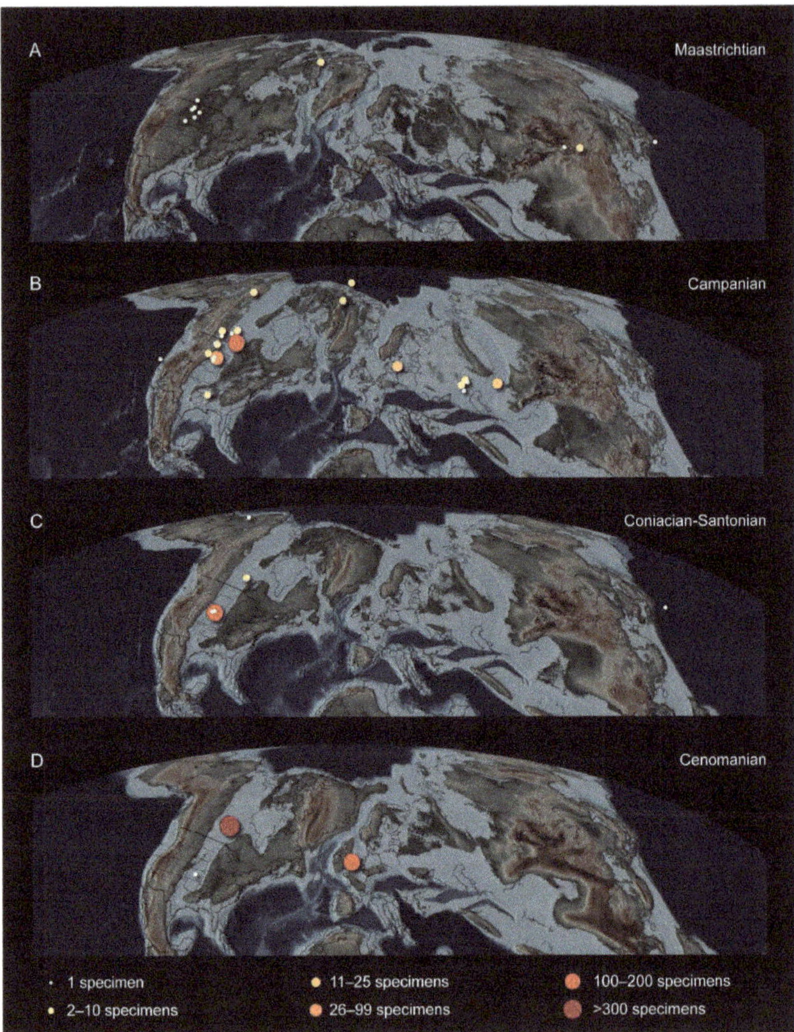

Figure 2. Distribution of hesperornithiform specimens across the Northern Hemisphere, mapped on paleogeographic reconstructions of the Cretaceous (after [29]): (**A**) Maastrichtian; (**B**) Campanian; (**C**) Coniacian to Santonian; (**D**) Cenomanian [19]. After Bell and Chiappe [25].

More recent work exploring the nature of the teeth in Hesperornis via synchrotron imaging found that they have fully thecodont-style root attachments but that secondary loss of periodontal ligaments led to the implantation of the teeth in a groove [30]. While the retention of teeth in the jaw is plesiomorphic in Hesperornis, the emplacement of the teeth is an autapomorphy, uniquely evolved in hesperornithiforms and not seen in other toothed birds such as Archaeopteryx [31], Ichthyornis [32], other toothed ornithuromorphs [33], or toothed enantiornithines [26,34]. The enamel on Hesperornis teeth is thin and simple in structure, with fine fluted ornamentation [35] formed by thickened ridges of enamel [30]. The teeth have a relatively high extension rate (a measure of how fast the tooth growths in height) in the dentine compared to that of nonavian dinosaurs, as calculated from dentine increment lines preserved in the teeth [30]. Tooth replacement involved a resorption pit in the root of the functional tooth, leading to lingual replacement with a calculated mean frequency of 66 days [30]. The teeth of hesperornithiforms are unicuspid and highly recurved, with a hooked shape in side view [20,30]. The teeth exhibit a gradient in curvature, with the mesial teeth more recurved than the distal teeth [1], more than is seen in other Mesozoic birds [36]. Teeth are absent in the premaxillae, as in Ichthyornis and some other early ornithuromorphs (e.g., *Gansus, Iteravis*), while the dentary and maxilla are toothed.

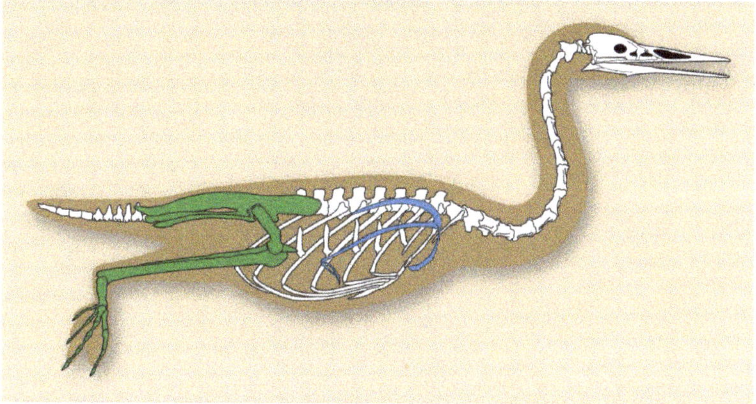

Figure 3. The basic bauplan of a hesperornithiform bird, based off *Hesperornis regalis*. While the degree of reduction of the forelimb and hindlimb proportions vary across taxa, this overall morphology—characterized by an elongated skull and neck, abbreviated forelimbs, and robust hindlimb with a long pelvis—is typical of all hesperornithiforms. Forelimb girdle shown in blue and hindlimb girdle shown in green.

The hesperornithiform skull retains numerous additional ancestral characters that are only well-preserved in two taxa (Hesperornis and *Parahesperornis*). Smaller hesperornithiforms only preserve limited parts of the skull (i.e., *Pasquiaornis, Enaliornis, Potamornis*). Marsh noted a number of similarities of the skull of Hesperornis to those of modern ratites, including palatines and pterygoids that articulate with facets on the basipterygoid process present on the body of the basisphenoid rostrum as well as an undivided quadrate head [1]. Elzanowski and Galton [37] identified additional ancestral features in the skulls of hesperornithiforms, including open frontoparietal and intraparietal sutures, caudal origination of the pseudotemporalis muscle, and the lack of carotid canals, among others.

A key feature contributing to the streamlined body of hesperornithiforms comes from the elongation of the neck. The majority of hesperornithiform specimens do not preserve a complete vertebral column, with the exception of one specimen of *Parahesperornis* (KUVP 2287) that appears to preserve the entirety of the vertebral column, minus the atlas. This specimen was largely collected in articulation, and the fit of articulation between the separated vertebral sections indicates it is likely that only the atlas is missing [25]. Modern foot-propelled diving birds use their elongate necks to increase maneuverability underwater. For example, cormorants have been documented to move their head and neck independently of the body during pursuit diving, thus avoiding limitations imposed by the limited turning radius of the entire body [38]. This contrasts with penguins, wing-propelled diving birds that swim with their necks retracted, thus limiting their range of motion to the turning radius of the entire body [39].

Figure 4. Hesperornithiform teeth and dentaries: (**A**) dentaries of *Hesperornis regalis*, showing the groove with implanted teeth in ventral view (1–KUVP 71012) and a broken specimen in medial view showing the internal projections separating individual teeth (2–YPM 1206) as well as an isolated tooth preserved with KUVP 71012 (3); (**B**) isolated teeth (1, 2) preserved in the roof of the articulated premaxillae of *Parahesperornis alexi* KUVP 2287; (**C**) dentary fragment assigned to *Pasquiaornis tankei* (RSM P2995.5) in lateral (1) and dorsal (2) views showing more extensive socketing (i.e., alveolar configuration) and isolated teeth attributed to *Pasquiaornis* (3, 4) (images in c from [15]).

A third key feature contributing to the overall bauplan of the hesperornithiforms comes from the pelvic girdle. The pelvis is highly elongated, with an expanded preacetabular ilium that varies in degree across hesperornithiforms but is found in all taxa (Figure 5). The pelvis is fused only around the acetabulum, a plesiomorphic feature also found in most other stem birds (e.g., Archaeopteryx, *Sapeornis*, *Confuciusornis*, basal ornithuromorphs, and some enantiornithines). Many enantiornithines have secondarily developed a contact between the ischium and the postacetabular wing of the ilium [40].

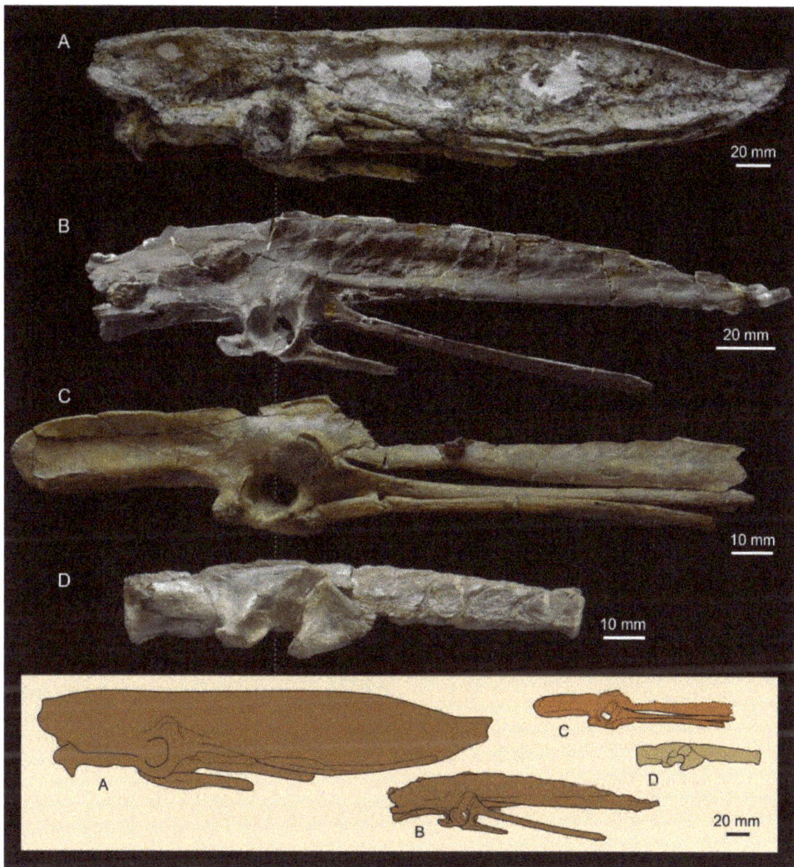

Figure 5. Comparison of hesperornithiform pelves of: (**A**) Hesperornis YPM 1476; (**B**) *Parahesperornis* KUVP 2287; (**C**) *Fumicollis* UNSM 20030; and (**D**) *Baptornis* AMNH 5101 in left lateral view. Elements are scaled to be of a similar acetabular diameter and aligned at the acetabulum. Inset shows silhouettes of the same elements to scale. After Bell and Chiappe [25].

The proportion and, perhaps most importantly, the orientation of the hindlimbs constitute a significant suite of adaptations that underlie interpretations of these birds as foot-propelled divers. In proportion, the femur and tarsometatarsus are reduced in length, and the tibiotarsus is highly elongated, as in modern diving birds (Figure 6). The degree of the shortening of the femur and extension of the tibiotarsus appears to vary dramatically across hesperornithiforms, with more basal taxa such as *Baptornis* and *Fumicollis* having a proportionally longer femur and shorter tibiotarsus than in the more derived *Hesperornis* and *Parahesperornis* [22].

Of particular interest to the mechanics of swimming in these birds is the orientation of the hindlimb. The earliest observations of these birds included reference to the extreme rotation of the femur, orienting the hindlimb behind the body. Marsh noted that the hindlimb orientation was such that the birds must have had difficulty standing and walking on land [3]. The orientation of the hindlimb directly behind the body is important for two reasons. It places the point of propulsion directly in line with the body being propelled, and it reduces the overall surface area of the body in the direction of motion, where resistance from the water is greatest.

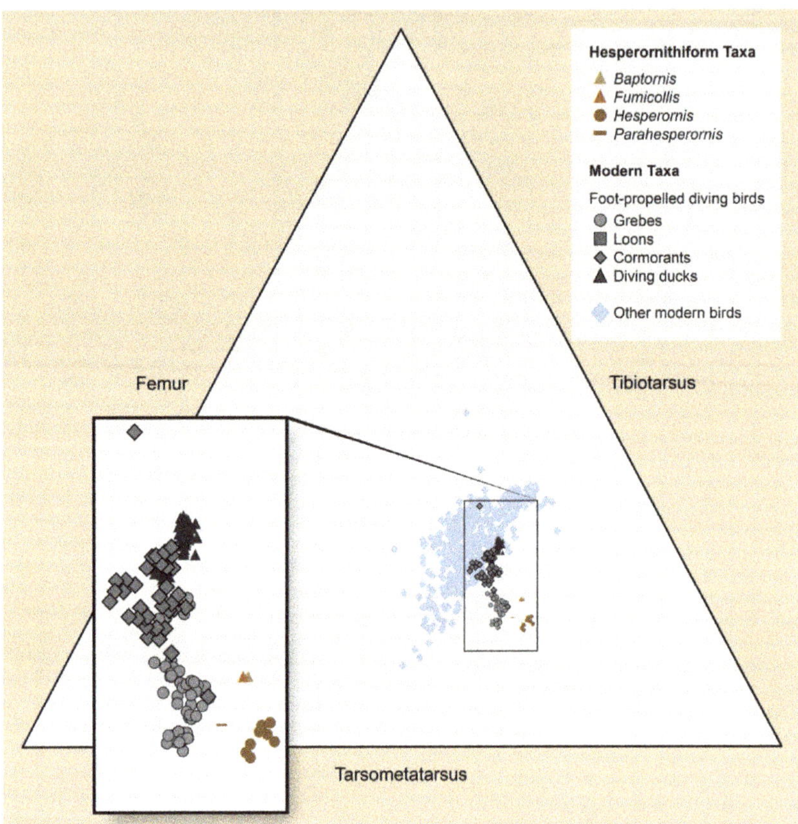

Figure 6. Ternary diagram showing proportions of the tarsometatarsus, tibiotarsus, and femur of hesperornithiform birds and modern foot-propelled diving birds (inset), and other modern birds. After Bell et al. [22].

A final key feature of the hesperornithiform bauplan is the extreme reduction of the forelimb to the point of flightlessness. This is discussed in evolutionary terms in more detail below, but the reduction of the forelimb is seen to varying degrees across the taxa for which forelimb elements are known, with *Pasquiaornis* displaying the least reduction (but see the discussion in Section 4 regarding this problematic taxon) and Hesperornis and *Parahesperornis* displaying the most (Figure 7). While the articular ends of the humerus of *Pasquiaornis* retain easily identifiable morphological landmarks, such as a deltopectoral crest at the proximal end and distinct dorsal and ventral condyles at the distal end, in Hesperornis virtually all such detail is lost, with no discernable deltopectoral crest and only a faint subdivision of the distal end where the condyles would be. Martin and Tate [13] questioned whether the more distal forelimb elements (ulna, radius, carpometatacarpus, and manual digits) developed at all, and proposed that perhaps they had been completely lost in some taxa. The complete loss of flight in hesperornithiforms underscores the success of these birds as foot-propelled divers, the first birds we currently know of to follow this evolutionary path. Among modern foot-propelled divers, flightlessness has evolved occasionally, with one species of flightless cormorant and two of grebes. The loss of flight may indicate that these birds were so well adapted to their foot-propelled lifestyle that they no longer needed flight to be successful for hunting, avoiding predation, and other activities for which birds typically use flight.

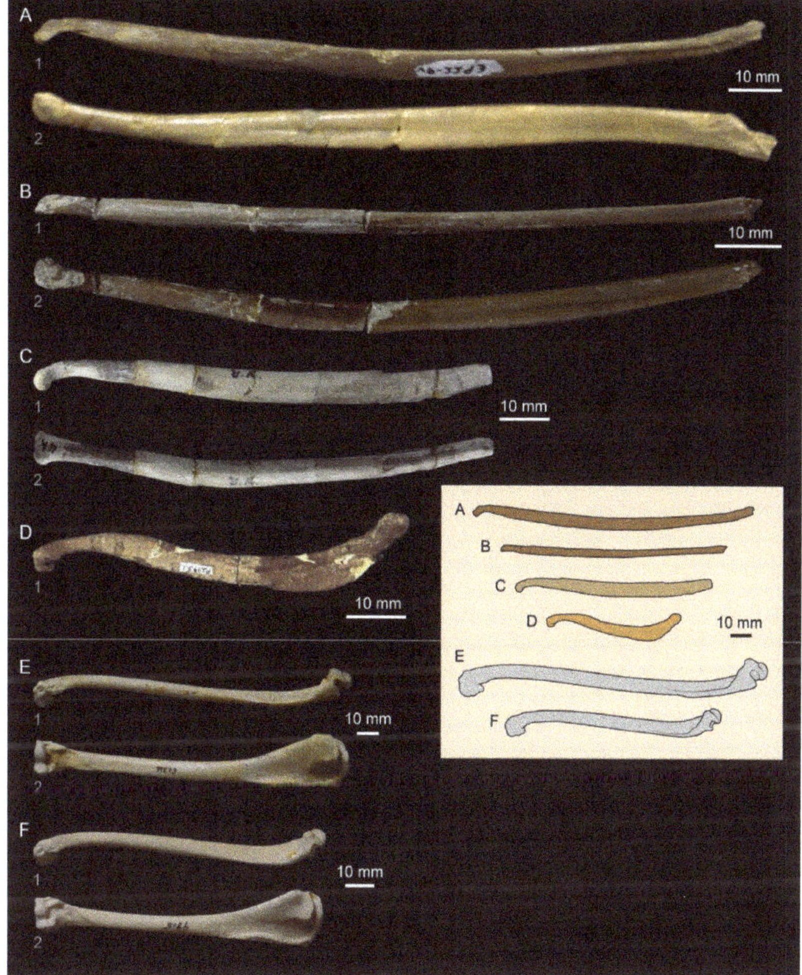

Figure 7. Comparison of the humeri of hesperornithiform taxa: (**A**) Hesperornis FHSM VP-2293; (**B**) *Parahesperornis* KUVP 2287; (**C**) *Baptornis* KUVP 2290; and (**D**) *Pasquiaornis* RSM P2995.1 (from [15]); as well as the modern cormorants: (**E**) *Phalacrocorax penicillatus* (flighted); and (**F**) *Nannopterum harrisi* (flightless) in (1) dorsal; and (2) cranial views. Elements are scaled to be of a similar width across the distal condyles and aligned at the distal ends. Inset shows silhouettes of the same elements to scale. After Bell and Chiappe [25].

3. Taxonomy

The past 150 years of research have resulted in numerous hesperornithiform taxa being named, some of which have been revised or rejected and many of which have never been revisited in light of more recent discoveries. This is undoubtedly one of the main areas for future work on the Hesperornithiformes, as while many taxonomic units are still recognized as valid, they have not undergone robust analysis and lack strong support.

The current taxonomic structure of the Hesperornithiformes recognizes four families: the Enaliornithidae and the Brodavidae, which are both monogeneric; the Baptornithidae, with two genera; and the Hesperornithidae, with five genera (Table 1). Additionally, *Pasquiaornis*, *Potamornis*, and *Fumicollis* are currently unassigned to a family, as discussed in detail below. This overall structure largely predates the use of phylogenetic analyses in developing taxonomic hypotheses, as typified by the initial assignment and later removal of *Pasquiaornis* to the Baptornithidae. These families and intermediate genera are introduced in this section.

Table 1. Current taxonomic framework of the Hesperornithiformes, with taxa that have been previously invalidated shown in red.

Class	Family	Genus	Species
	Enaliornithidae	*Enaliornis*	*barretti*
			sedgewicki
			seeleyi
	Baptornithidae	*Baptornis*	*advenus*
			varneri
		Judinornis	*nogontsavensis*
		Parascaniornis	*stensoei*
	Brodavidae	*Brodavis*	*americanus*
			baileyi
			mongoliensis
			varneriv
	Hesperornithidae	*Asiahesperornis*	*bazhanoviv*
		Canadaga	*arctica*
		Coniornis	*altus*
		Hargeria	*gracilis*
		Hesperornis	*altus*
Hesperornithiformes			*bairdi*
			chowi
			crassipes
			gracilis
			macdonaldi
			mengeli
			montana
			regalis
			rossicus
		Lestornis	*crassipes*
		Parahesperornis	*alexi*
	NA	*Pasquiaornis*	*hardei*
			tankei
	NA	*Chupkaornis*	*keraorum*
	NA	*Fumicollis*	*hoffmani*
	NA	*Potamornis*	*skutchi*

3.1. Enaliornithidae

The Enaliornithidae from the Cambridge Greensand of England are the oldest hesperornithiform family currently known. The Cambridge Greensand is usually interpreted as dating from the early Cenomanian, with reworked late Albian material from the underlying Gault Formation [41], indicating *Enaliornis* is the oldest group of hesperornithiforms. As described above, they were first reported in 1859 by Lyell [42] and then more fully described by Seeley [8].

Originally, two species were identified on the basis of size but without quantification of those differences [8]. As the remains are highly fragmentary and consist entirely of disarticulated and unassociated remains, determining the exact taxonomic diversity of this group has proven difficult (e.g., [8,13,43,44]). Brodkorb [45] identified a lectotype and numerous paralectotypes for each species. Later, extensive evaluation of all known material by Galton and Martin [44] proposed diagnostic features for the genus as well as for each species. The features identified in combination as diagnostic of the genus included a transversely constricted centrum in the preacetabular synsacrum, the presence of an antitrochanter on the ilium, absence of a distinct neck on the femur, tarsometatarsus with a cranioproximal process originating proximally from metatarsal III, a caudomedial ridge leading to trochlea II distally, and the cranial edge of trochlea IV caudal to the cranial edge of trochlea III. They also identified a third species, *E. seeleyi*, as being intermediate in size between the larger *E. barretti* and the smaller *E. sedgewicki*, but it remains unclear how unassociated elements were combined into a single species [44].

3.2. Baptornithidae

The Baptornithidae was erected to place *Baptornis advenus* within the Hesperornithiformes as a unique monogeneric family [13]. Since the establishment of the family, several new genera and species have been added to the group. A small baptornithid, *Judinornis nogontsavensis*, was described from a single thoracic vertebra discovered in Maastrichtian-aged fluvial deposits of Mongolia [46]. The genus *Pasquiaornis*, consisting of two species from Cenomanian-aged marine strata in Canada, was also added to the Baptornithidae [47]. However, more recent phylogenetic analysis has determined this placement to be unsupported (as discussed in Section 4 below).

Baptornis is monotypic and known from several specimens, both partially complete and isolated elements, in North America, an isolated vertebra from Europe, and an isolated tibiotarsus fragment from Mongolia, while *Judinornis* remains known from the isolated vertebrae from Mongolia described above. These specimens all date from the Late Cretaceous, the youngest of which is from the Lincoln Limestone Member of the Greenhorn Formation of Kansas (upper middle Cenomanian) [48] and the latest of which is from the Campanian to Maastrichtian-aged Tsagan-Khusu locality in Mongolia [49].

These studies, along with research into baptornithid specimens from Canada [16] and Kansas [48], have added to the diagnostic features of the family. A suite of features has been used as diagnostic of the Baptornithidae, including a slender coracoid (as compared to hesperornithids [13]), elongate preacetabular illium [13], pyramidal patella [13], and dorsally inclined cotyla of the tarsometatarsus [48].

3.3. Brodavidae

The Brodavidae was erected to unite four fragmentary specimens into a single family [50]. The Brodavidae is the second most taxonomically diverse family of hesperornithiforms following the hesperornithids, with four species that range widely in size (Figure 8). Three of the four species are known from isolated tarsometatarsi. However, the holotype and sole specimen of *B. varneri* is partially complete, preserving a portion of the vertebral column, ribs, pelvis, and most of the hindlimb. This specimen was originally assigned to *Baptornis* [51] but later moved to the Brodavidae [50], an assignment that has since been supported by phylogenetic analysis [23,24].

The brodavids are known primarily from the Maastrichtian (the fluvial Nemegt Formation of Mongolia, the Frenchman Formation of Canada, and the coastal plain Hell Creek Formation of the USA), with the oldest specimen dating to the Campanian (the marine Pierre Shale, USA) [50].

The Brodavidae was defined using features of the tarsometatarsus, known for all taxa, and the rest of the postcranial skeleton, known only from *B. varneri* [50]. Two diagnostic features were identified as separating brodavids from other hesperornithiforms [50]: the shortness compared to breadth of the tarsometatarsus (this value was not quantified) and the proximal displacement of the facet for metatarsal I (described as almost in the middle of the tarsometatarsus).

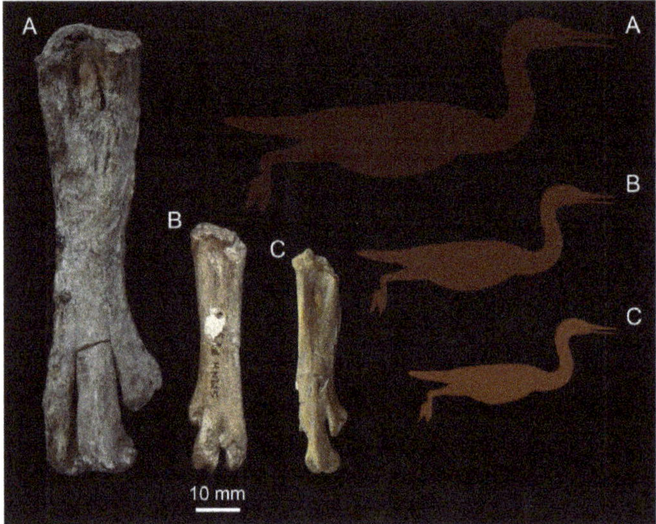

Figure 8. Comparison of body size among the brodavids: (**A**) *Brodavis varneri* SDSM 68430; (**B**) *Brodavis americanus* cast of RSM P2315.6; and (**C**) *Brodavis baileyi* UNSM 50665.

3.4. The Hesperornithidae

The family Hesperornithidae was proposed by Marsh [2] as Hesperornidae, later used as Hesperornithidae [7], and at the time was monogeneric and monospecific. Much later, Clarke [32] provided the first cladistic definition for the group as a stem-based name encompassing all taxa more closely related to *Hesperornis regalis* than to *Baptornis advenus*. Bell and Chiappe [22] later revised this definition to a node-based clade encompassing all taxa descended from the common ancestor of *Hesperornis regalis* and *Parahesperornis alexi*. The Hesperornithidae are the most taxonomically diverse group of hesperornithiforms. In addition to Hesperornis and *Parahesperornis*, *Asiahesperornis* [52], *Canadaga* [53], and eight additional species of Hesperornis [7,54–56] have been added to the family.

The Hesperornithidae are known from North America, Europe, and Asia, ranging in age from the Coniacian/Santonian (Vermillion River Formation, Canada) to the Maastrichtian (Hell Creek Formation, USA and Zhuravlovskaya Svita, Russia). This range may in fact be much older, as a specimen has been reported from the marginal marine Mesa Verde Group, USA [57], but the stratigraphic position is poorly constrained. It is interesting to note that the highest hesperornithid diversity in terms of body size and species richness is known from the oldest deposits (such as the marine Niobrara Formation), implying much information on the evolution and diversification of this group remains unknown.

Unfortunately, Marsh did not diagnose the family separately from the genus Hesperornis, and subsequent work has focused on individual species or genera and not diagnostic features of the entire family. Phylogenetic work by Bell and Chiappe [22] identified 28 unambiguous synapomorphies uniting the monophyletic clade Hesperornithidae, which could be evaluated for expanding the current diagnosis. Typical hesperornithid features include the combination of a dramatically reduced forelimb; robust, blocky coracoid; robust femur with expanded trochanter; expanded proximal tibiotarsus with robust cnemial crests; and enlarged fourth trochlea of the tarsometatarsus and pedal phalanx IV. While often considered the largest of the Hesperornithiformes, it is important to remember that species of hesperornithids range widely in size, with *H. macdonaldi, H. lumgairi, H. mengeli,* and *Parahesperornis* much smaller than the larger species such as *H. regalis* and *H. rossicus* (Figure 9).

Figure 9. Comparison of body size among Hesperornithidae taxa: (**A**) *H. regalis* YPM 1200; (**B**) *H. chowi* YPM PU 17208; (**C**) *H. rossicus* SGU 3442 Ve01; (**D**) *H. gracilis,* YPM 1473; (**E**) *H. lumgairi* CFDC B78.02.07; (**F**) *H. bairdi* YPM PU 17208a; (**G**) *Asiahesperornis bazhanovi* IZASK 5/287/86a; (**H**) *H. mengeli* CFDC 78.01.08; (**I**) *Parahesperornis alexi* KUVP 2287; (**J**) *H. altus* YPM 515; (**K**) *H. macdonaldi* LACM 9727. Silhouettes are approximately scaled to averages of measurements of the corresponding elements preserved in specimens of *H. regalis*.

3.5. Taxa Outside of Recognized Families

3.5.1. Pasquiaornis

Pasquiaornis was erected to unite two species of small hesperornithiforms from the marine Belle Fourche Formation (Cenomanian) of Canada [47]. *Pasquiaornis* was originally assigned to the Baptornithidae [47]. However, subsequent phylogenetic analysis failed to return *Pasquiaornis* + *Baptornis* as a monophyletic clade, and it was suggested that *Pasquiaornis* should not be considered part of the Baptornithidae [22]. Additional analysis has since supported this phylogenetic topography [24]. These taxa are only known in the literature from unassociated and disarticulated elements found in a bone-bed deposit in the Belle Fouche Formation [13].

Tokaryk et al. [47] proposed that a combination of size and select morphological differences could be used to separate the disarticulated elements and assigned *Pasquiaornis* specimens into two species, *P. hardei* and *P. tankei*. While *P. tankei* was described as the larger of the two taxa, specific size differentials used to separate these species have never been quantified. Furthermore, as there are no associated elements known within the genus, it is unclear how different elements are assigned together in one of the two species. *Pasquiaornis* was diagnosed as having a less-expanded trochanter and proximal end of the femur than is seen in *Baptornis*, as well as having the anterior intercotylar eminence on the tarsometatarsus overhanging the shaft and trochlea II of the tarsometatarsus positioned posterior to and near the base of trochlea III [47].

3.5.2. Chupkaornis

Chupkaornis is a small hesperornithiform discovered in the Coniacian to Santonian-aged Kashima Formation in Hokkaido, Japan [24]. The holotype and sole specimen was published as a partial skeleton preserving six vertebrae, distal femora, and a fragment of fibula [24]. However, examination of the photographs published indicates what is described as the distal left femur is actually a distal tibiotarsus. Phylogenetic analysis returned *Chupkaornis* as basal to *Brodavis* within the Hesperornithiformes and derived from *Enaliornis* and *Pasquiaornis*.

Diagnostic features proposed as separating *Chupkaornis* from other hesperornithiforms include the combination of vertebrae that are fully heterocoelous (but see discussion in Section 4 below) with emarginated lateral excavations on the centra and sharp ventral margins; slender ventral process and laterally expanded fibular condyle of the femur with a finger-like projection on the tibiofibular crest [24].

3.5.3. Fumicollis

The holotype of *Fumicollis* was originally identified as *Baptornis* [13] but later recognized as possessing some characters typical of hesperornithids and some typical of *Baptornis* during the course of research for a phylogenetic study of the hesperornithiforms [23]. This specimen was therefore used to erect a new species—*Fumicollis hoffmani*—phylogenetically intermediate between the Baptornithidae and the Hesperornithidae [58]. This placement has since been supported by additional analysis [24]. Two additional specimens, both isolated femora, have also been proposed as belonging to *Fumicollis*. Both of these are known from museum collection studies but are currently unpublished (specimen numbers not available). The holotype and only published specimen preserves a partial vertebral column and nearly complete hindlimb, making it one of the more complete hesperornithiform specimens. *Fumicollis* is known from the marine Smoky Hill Member of the Niobrara Formation (upper Coniacian to lower Campanian [59]) of Kansas (USA) [58].

A combination of features from the vertebrae, pelvis, femur, tibiotarsus, and tarsometatarsus were used to diagnose the genus [58]. These include an elongate preacetabular pelvis, expanded lateral condyle on the femur (defined as midshaft width 75% of lateral condyle width), medial cnemial crest extended to midshaft of the femur, pyramidal patella, a distinct dorsal ridge of the tarsometatarsus formed by the entire length of metatarsal IV, and others. The presence of both baptornithid and hesperornithid characters can be seen in these traits. For example, the degree of expansion of the lateral condyle is also found in *Baptornis*, while the dorsal surface of metatarsal IV forming a prominent ridge along its entire length is typical of hesperornithids.

3.5.4. *Potamornis*

Potamornis skutchi was erected for an isolated quadrate discovered from the fluviodeltaic Lance Formation (late Maastrichtian) in Wyoming (USA) [60]. The element was assigned to the Hesperornithiformes on the basis of an undivided head and an elongate pterygoid condyle, features typical of hesperornithiform quadrates. A unique combination of characters was identified as diagnostic of *Potamornis skutchi*, including: a strongly asymmetrical quadrate head, rostrally open pit near the medial apex of the head, shallow caudomedial depression, small orbital process, a quadratojugal buttress on the lateral process, and medial and lateral mandibular condyles meeting at an angle of 115 degrees [60]. An isolated tarsometatarsus from the same formation was also tentatively assigned to the genus on the basis of size [60]. However, the specimen was not figured or formally described, and no additional work has been done.

3.6. *Taxonomic Challenges*

Our understanding of hesperornithiform taxonomy is plagued by a host of problems common to paleontology, such as the renaming of previously described taxa [59,61]; taxa described from highly fragmentary material [24,50,55,56,62]; elements misidentified [24]; and subjective, unspecific, or incorrect characters used for diagnosis [47,63]; reliance upon which may result in further confusing the assignment of fragmentary taxa [50,64]. The majority of hesperornithiform species have been described from fragmentary material (Figure 10). Of the 25 described species, only six include specimens preserving more than three elements, one is known from two elements, and 18 species were described and remain known from a single bone. Whether or not all of these species are valid taxa has rarely been rigorously examined. For example, debate over the synonymy of *Coniornis altus* [65], *Hesperornis altus* [62], and *Hesperornis montana* [62] has appeared in the literature, with both *Coniornis* and *H. montana* being invalidated without *H. altus* ever being resolved in the form of a concise diagnostic description and justification of the "valid" taxon.

Perhaps due to difficulties arising from the fragmentary nature of the fossil record, a number of taxa have been poorly or inaccurately described. An example of one such recurring error and source of much confusion is the presence or absence of the proximal foramina on the cranial surface of the tarsometatarsi of hesperornithiforms. In his monograph on hesperornithiforms, Marsh did not mention the presence of these foramina in Hesperornis or *Baptornis* [1]. More recent descriptive work has specifically pointed out the lack of these foramina in numerous species of hesperornithiforms [13,44,63]. This has led to the presence of these foramina to be used, in part, as justification for the Cretaceous of Chile from the Hesperornithiformes [64]. Furthermore, the relative degree of development of the foramina has been used as a diagnostic feature of two species of *Brodavis* [50]. However, closer examination of specimens of *H. regalis*, *H. gracilis*, *H. crassipes*, *P. alexi*, *B. advenus*, and other unidentified hesperornithiforms shows that in all cases proximal foramina are present on the cranial surface of the tarsometatarsi [32,48]. Incomplete preparation of the bones may be to blame for the foramina being overlooked by previous authors. Additionally, the appearance of these foramina appears to be closely tied to preservation quality.

As discussed by Bell [65], another problem that is commonly seen in hesperornithiform taxonomy is the reliance on qualitative language to describe quantitative traits. For example, the tarsometatarsus of numerous species has been described as being "slender". This is essentially a qualitative way of describing the length to width ratio of the element. Using this sort of language in the diagnosis of numerous species instead of presenting morphometric data to precisely define important aspects of morphology creates uncertainty and confusion when making comparisons across dozens of species.

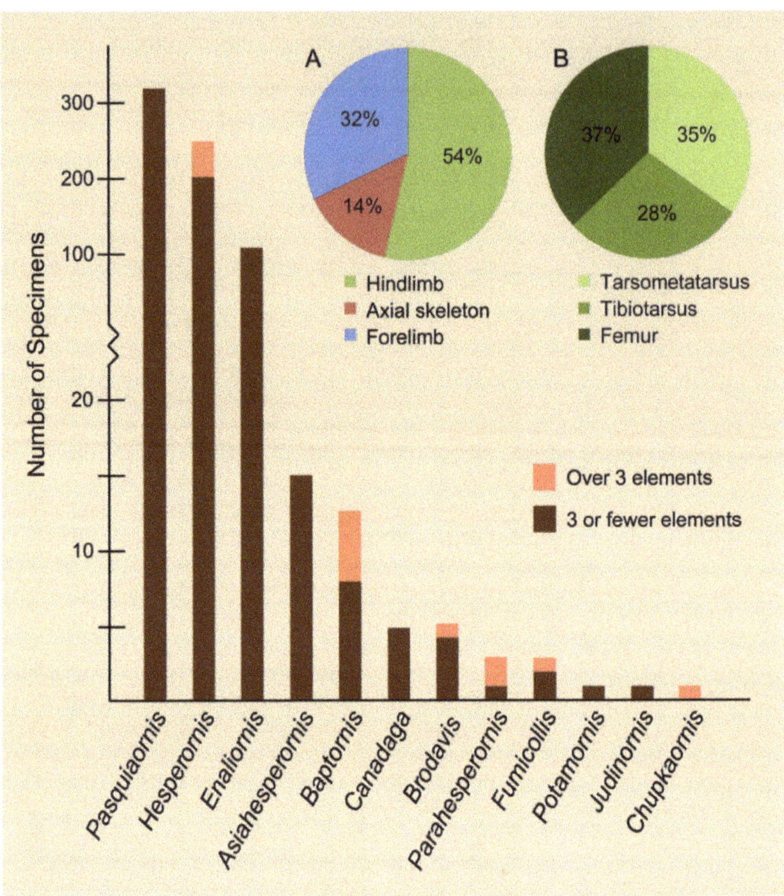

Figure 10. Specimen completeness of hesperornithiform taxa. Insets show percentages of specimens that preserve: (**A**) elements belonging to different regions of the body; and (**B**) main elements of the hindlimb. After Bell and Chiappe [20].

The examples of fraught taxonomy of hesperornithiforms described here illustrate opportunities for future research, as significant synthesis and revision may be achieved from broad-scale studies of currently known material, enhanced and improved with the addition of more recently discovered material that has not been previously published as well as future discoveries [65].

4. Phylogeny of the Hesperornithiformes

Early approaches to hesperornithiform phylogenetics relied on a limited number of taxa (usually under five) and characters (usually under ten) identified as synapomorphies *a priori* and then used as justification for a particular tree topology. One example of this approach is Cracraft's [19] tree of diving birds that showed hesperornithiforms within modern birds, basal to a clade containing loons and grebes and derived from penguins, pelicans, and other seabirds (Procellariiformes). Another example is Martin's [63] work in which he developed two trees showing: (1) Hesperornis and *Parahesperornis* as a monophyletic clade progressively more derived than the Baptornithidae and the Enaliornithidae; and (2) hesperornithiforms as basal to a monophyletic clade of ichthyornithiforms and modern birds but more derived than "Sauriurae". A more modern approach to a phylogenetic analysis was taken by Elzanowski and Galton [37], who developed a matrix of 17 characters for seven taxonomic units, but did not conduct an analysis of this matrix and did not offer any phylogenetic hypotheses from these data.

The explosion of Mesozoic bird fossils in the 1990s from China, South America, and Europe initiated a new wave of research into early bird evolution, including the widespread application of modern phylogenetic methods. In these analyses, hesperornithiforms, usually represented by Hesperornis, consistently placed as fairly derived within Ornithuromorpha, usually as the sister taxon to a clade containing Ichthyornis and modern birds (e.g., [32,66–68]). More recent work involving multiple hesperornithiform taxa has returned the Hesperornithiformes as the sister taxon to Neornithes, or crown clade birds [22,69–71], while other studies retain the placement of the hesperornithiforms as sister to Ichthyornis + Neornithes [24,72,73].

Two things are important to note in comparing the different placement of hesperornithiforms among derived ornithuromorphs. First, the choice of taxa seems to play a role in the placement of hesperornithiforms, with studies that include numerous hesperornithiforms more likely to resolve them as closer to Neornithes (e.g., [24,71]) than studies that include only one or two of the typically more derived hesperornithiforms (i.e., Hesperornis) (e.g., [32,72,73] but see [24]). Second, many matrices have not updated specimen codings in response to the ever-evolving taxonomic changes and in light of new evidence identifying mistakes in previous descriptions of taxa. For example, some studies maintain specimens coded as *Baptornis* that have since been removed from that genus (e.g., [73]) or use characters that have been identified as erroneous such as features of the quadrate of *Baptornis*, which was incorrectly described by Martin and Tate [13] and is, in fact, unknown for that taxon (as described in Bell [65] and Bell and Chiappe [25] (e.g., [24,73]).

There are few studies that have examined phylogenetic relationships among the hesperornithiforms, but these studies generally agree in overall topography (Figure 11). The most derived clade consists of a monophyletic Hesperornis, with *Parahesperornis* as sister taxa, followed by the progressively more basal *Fumicollis*, *Brodavis*, and *Baptornis* [22,24]. There is disagreement at the base of the tree, with Bell and Chiappe [22] resolving *Pasquiaornis* as more basal than *Enaliornis* and Tanaka et al. [24] resolving the reverse relationship and a similar switch between *Baptornis* and *Brodavis* (Figure 12). This disagreement likely results from the incredibly fragmentary nature of the material for both these taxa and stems from coding discrepancies of features easily obscured by weathering. The details of this are reviewed in Bell and Chiappe [25] but are rooted in the very poor preservation of *Enaliornis* as part of a reworked deposit that has resulted in the smoothing of the bones, thus making the observation of many details difficult, if not impossible.

5. Evolutionary Trends

As the oldest known lineage of diving birds, hesperornithiforms allow us to study a remarkable evolutionary transition—a group of birds that gave up the ability to fly in favor of foot-propelled diving. This transition is evident in a unique suite of skeletal adaptations as well as in the size of these birds and the range of environments they occupied.

Perhaps one of the most interesting aspects of the evolutionary trends described below is the absence of a fossil record of early stages of these trends. It remains unclear who the predecessors of the Hesperornithformes were. Despite an abundance of Early Cretaceous lagerstätten preserving both freshwater and estuarine environments in China and Spain, there are no clear foot-propelled diving adaptations in the diverse avifauna. The oldest hesperornithiforms, the species of *Enaliornis*, appear in the earliest Late Cretaceous already equipped with numerous adaptations that support interpretations of a foot-propelled diving lifestyle, including an expanded lateral condyle on the femur and angled articular surface with greatly expanded cnemial expansion on the tibiotarsus, and stacked or shingled metatarsals in the tarsometatarsus. By the middle of the Late Cretaceous (late Coniacian to early Campanian) some of the most abundant deposits of hesperornithiform birds are known from the Smoky Hill Chalk (Kansas, USA) of the Western Interior Seaway, with many roughly coeval taxa ranging from the small *Baptornis* to the large, highly derived *Hesperornis regalis*. Furthermore, all of the hesperornithiforms from the early Late Cretaceous (pre-Campanian) are known from entirely marine deposits, while fewer hesperornithiforms are known from the Maastrichtian, these are primarily known from marginal marine to terrestrial deposits (with a small number of exceptions in the Campanian) (Table 2). It is tempting to interpret this as demonstrating an evolutionary diversification from entirely marine taxa to taxa constituting different species adapted to the marine realm, shallow waters of estuaries, and even freshwater. It should be noted that there may be an element of taphonomic bias at play, as the depositional environment of an animal is not necessarily that in which it lived, and the fact that climate and tectonic-driven transgressions of the late Early to early Late Cretaceous led to a disproportionate abundance of marine deposits for this time period [74].

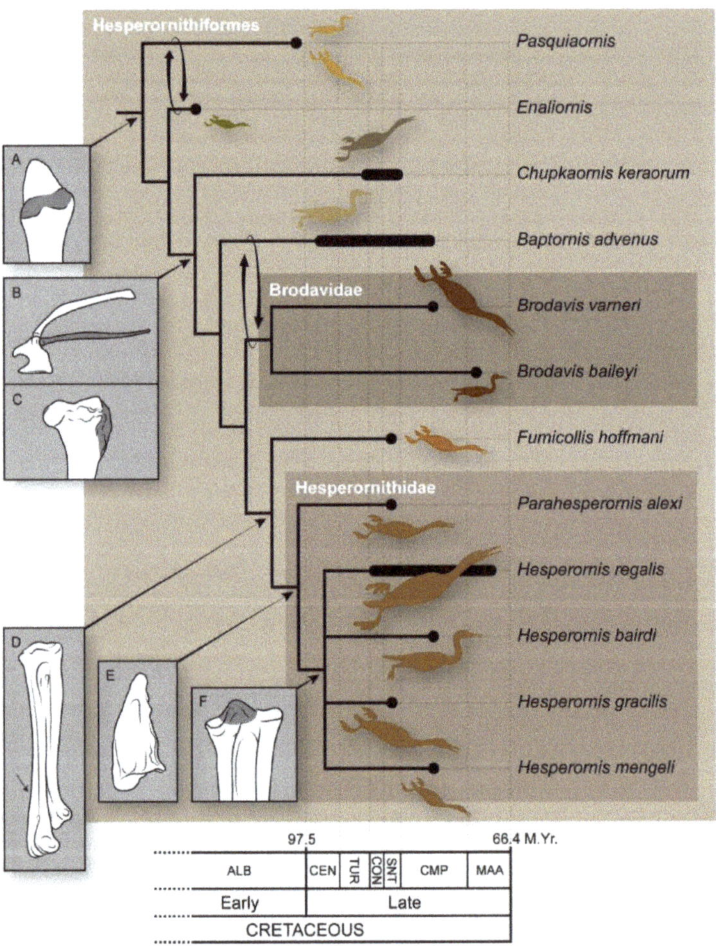

Figure 11. Phylogenetic tree of hesperornithiform birds, after Bell and Chiappe [22], with the addition of *Chupkaornis* from Tanaka et al. [24]. Arrows indicate alternative positions for *Enaliornis* and *Pasquiaornis* and *Baptornis* and *Brodavis* in Tanaka et al. [24]. Diagrams indicate anatomical adaptations proposed to correlate with foot-propelled diving: (**A**) angled articular surface and cnemial expansion on the tibiotarsus; (**B**) reduced humerus; (**C**) expanded femoral trochanter; (**D**) fourth trochlea of the tarsometatarsus forms a sharp ridge; (**E**) enlarged, triangular patella; (**F**) expanded intercotylar eminence on the tarsometatarsus.

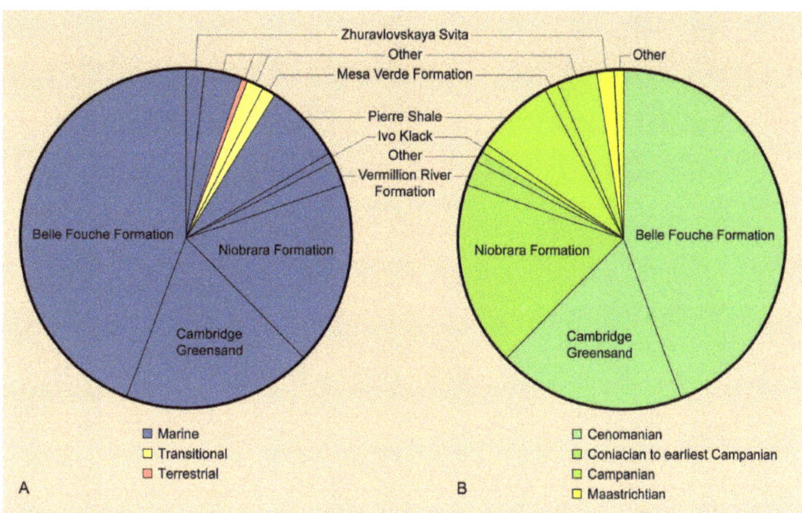

Figure 12. Distribution of hesperornithiform specimens in geologic units and: (**A**) depositional environments; and (**B**) time periods within the Late Cretaceous. All formations from which fewer than five specimens are known are included in "other" for each depositional environment and time period. After Bell et al. [22].

Table 2. Geologic units, divided by depositional environment, with published hesperornithiform taxa. After Bell et al. [22].

Continental		
Mesa Verde Formation (Teapot Sandstone) [57]	*Hesperornis regalis*	Campanian
	Hesperornis sp.	
Nemegt Formation [75]	*Brodavis mongoliensis*	Maastrichtian
	Judinornis nogontsavensis	
	Hesperornithidae indet.	
Lance Formation [76]	*Potamornis skutchi*	Late Maastrichtian
Frenchman Formation [77]	*Brodavis americanus*	Late Maastrichtian
Transitional		
Foremost Formation [78]	Hesperornis sp.	Campanian
Judith River Formation [79]	*Baptornis* sp.	Campanian
	Hesperornis altus	
Dinosaur Provincial Park Formation [80]	*Baptornis* sp.	Late Campanian
Hell Creek Formation [81]	*Brodavis baileyi*	Maastrichtian
	Hesperornis sp.	
Marine		
Cambridge Greensand Member (West Melbury Chalk Formation) [44]	*Enaliornis barretti*	Early Cenomanian
	Enaliornis seeleyi	
	Enaliornis sedgewicki	
Belle Fouche Formation (formerly Ashville Formation) [14]	*Pasquiaornis hardiei*	Late Cenomanian
	Pasquiaornis tankei	

Table 2. *Cont.*

Greenhorn Formation [48]	*Baptornis* sp.	Cenomanian
Kashima Formation [24]	*Chupkaornis keraorum*	Coniacian to Santonian
Vermillion River Formation [82]	*Hesperornis regalis*	Coniacian to Santonian
	Hesperornis sp.	
Ignek Formation [83]	*Hesperornis* sp.	Late Coniacian to Campanian
Smoky Hill Chalk, Niobrara Formation [84]	*Baptornis advenus*	Late Coniacian to early Campanian
	Hesperornis crassipes	
	Hesperornis gracilis	
	Hesperornis regalis	
	Hesperornis sp.	
	Fumicollis hoffmani	
	Parahesperornis alexi	
	Parahesperornis sp.	
Smoking Hills Formation [85]	*Hesperornis regalis*	Middle Santonian to early late Campanian
Eginsaiskaya [49]	*Baptornis advenus*	Latest Santonian to Early Campanian
	Asiahesperornis bazhanovi	
Rybushka Formation [86]	*Hesperornis rossicus*	Early Campanian
	Hesperornis sp.	
Kristianstad Basin (unreported formation) [61]	*Baptornis* sp.	Latest early Campanian
	Hesperornis rossicus	
	Hesperornis sp.	
Kanguk Formation [87]	*Canadaga arctica*	Early to middle Campanian
	Hesperornis sp.	
Clagget Shale [57]	*Hesperornis* sp.	Campanian
Pierre Shale [88]	*Baptornis advenus*	Campanian
	Brodavis varneri	
	Hesperornis bairdi	
	Hesperornis chowi	
	Hesperornis lumgairi	
	Hesperornis macdonaldi	
	Hesperornis mengeli	
	Hesperornis regalis	
	Hesperornis rossicus	
	Hesperornis sp.	
Chico Formation [89]	*Hesperornis* sp.	Campanian
Ozan Formation [90]	*Hesperornis* sp.	Campanian
Kita-ama Formation [91]	Hesperornithiformes undet.	Early Maastrichtian
Zhuravlovskaya Svita [92]	*Asiahesperornis bazhanovi*	Maastrichtian

5.1. Flightlessness

The most dramatic and obvious evolutionary trend in the Hesperornithiformes is the complete adaptation to a foot-propelled diving lifestyle. For the majority of hesperornithiforms that preserve forelimb elements, there is agreement that these birds were entirely flightless [1,8,13,25,63,65,93]. The only hesperornithiform for which a degree of flight capacity has been proposed as plausible is *Enaliornis*, on the basis of the small body size and extensive pneumatization of the braincase [37]. Indeed, the degree of pneumatization in the braincase of *Enaliornis* is much greater than in either *Parahesperornis* or Hesperornis [25]. However, additional data from the forelimb, which is entirely unknown in *Enaliornis*, is required to better evaluate these claims. As described below, some form of limited flight capabilities have also been tentatively suggested for *Pasquiaornis* [15], but again the dearth of fossil evidence for this taxon makes those claims speculative.

The forelimb and shoulder girdle are indicative of the evolutionary pathway leading to loss of flight in hesperornithiforms, with *Pasquiaornis* less derived than *Baptornis* and hesperornithids. The coracoids of hesperornithiforms are less developed as compared to flying birds, with reduced acrocoracoid processes and the absence of a procoracoid process. This trend culminates in hesperornithids, which also have an increasingly short coracoidal neck compared to *Baptornis* and even more than in *Pasquiaornis*. A similar trend is seen in the scapula and clavicle, with the articular surfaces only faintly developed in hesperornithiforms. Another factor associated with flightlessness is the complete absence of a ventral keel on the sternum. Hesperornis and *Parahesperornis* both preserve nearly complete sterna which show the complete absence of a keel.

The humerus is known for several hesperornithiform taxa and consists of an incredibly gracile bone with little to no development of articular surfaces and the deltopectoral and bicipital crests, in both large birds such as Hesperornis and small birds such as *Baptornis* (Figure 7). *Pasquiaornis* is the only hesperornithiform that preserves forelimb material for which rudimentary flight abilities have been suggested as tentatively possible [15], based largely on the less reduced state of flight-related features such as the development of the distal condyles and deltopectoral and bicipital crests on the humerus [15,47]. The ulna and radius are only known in *Baptornis* and *Pasquiaornis*, where both are reduced with faintly developed articular ends compared to flying birds [13,15,93]. The carpometacarpus is only known in *Pasquiaornis*, which is also consistent with flightlessness in the thickened compact bone and the distal placement of the extensor process [15]. As indicated above, only the discovery of more complete material in this basal-most hesperornithiform can provide a reliable interpretation of its potential aerial capabilities.

5.2. Foot-Propelled Diving

In concert with the reduction of the forelimb described above, a number of hesperornithiform skeletal features point to a highly derived foot-propelled diving lifestyle. These features have been assessed morphometrically [22,94] and discussed in detail by Bell and Chiappe [25]. In particular, the articulation of the leg in hesperornithiforms has the femur splayed laterally from the pelvis and possibly contained entirely within the body, with the lower limb extending linearly from the knee joint parallel with the mainline of the body. This orients the feet, the source of propulsion, directly behind the body. Adaptations associated with this in Hesperornis include a robust femoral trochanter that extends evenly to the femoral head, an exaggerated lateral femoral condyle roughly even with the medial condyle, a sharply angled proximal articular surface on the tibiotarsus, twisted shafts on the tibiotarsus and tarsometatarsus, and a dramatically enlarged third toe with expanded lateral condyle for rotation of the toe. These features are variably present, but often to a lesser degree, in more basally diverging hesperornithiform taxa.

In addition to the dramatic restructuring of the hindlimb for foot-propelled diving, hesperornithiforms have a suite of more subtle features that also indicate a diving lifestyle. The number, shape, and arrangement of teeth in the jaws of Hesperornis have trophic implications, with the increased number of teeth in the dentary having been related to a piscivorous diet [28], an interpretation well aligned with the environment these birds lived in and morphological interpretations discussed above. The distinct hooked cranial terminus of the premaxilla, which may have been emphasized by the shape of the keratinous beak, may have also been useful for the retention or capture of larger fish. The wide variation in tooth loss, reduction, and shape seen across Mesozoic birds highlights the trophic diversity present among these early birds [26] (in some cases supported by gut contents). However, specific correlations between a particular dental trait, such as the dramatic mesio-distal recurvature gradient in hesperornithiforms, and specific dietary specializations, have not been identified to date [30].

Several features of the skull of hesperornithiforms have been used to support interpretations of a diving lifestyle. Elzanowski and Galton identified the large size of the auricular fossae, the reduced dorsal pneumatic recess, and the flattened cerebellar fossa as traits shared with modern diving birds [37]. The latter two of these features was noted as possibly associated with the expansion of the dural sinuses [37], a convergent feature found in a wide range of diving birds and mammals [95–97]. Histological work has identified a thick compact bone wall and comparatively small medullary cavity in the femur of Hesperornis, a feature also found in penguins and interpreted as decreasing buoyancy as a diving adaptation [98].

5.3. Gigantism

One of the first things noted about Hesperornis was its very large body size [2,3], which could approach 1.5 m in length. The discovery of the much smaller *Baptornis* soon after, showed the dramatic size range present in hesperornithiforms [99]. Most interestingly, this range of sizes does not appear to be correlated to any particular evolutionary trend and is unrelated to the degree of diving specialization [22,25]. While large-bodied taxa are missing among the most basally diverging hesperornithiforms, *Enaliornis* and *Pasquiaornis*, there are more derived taxa that are also small. Within the brodavids, for example, the midshaft of the tarsometatarsus of the smallest species, *B. mongoliensis*, is less than half the diameter of that of the largest species, *B. varneri* (Figure 9). Similarly, there are several small species of Hesperornis, and while it is not possible to make direct comparisons between them due to lack of overlap in preserved elements, they can be compared to larger species such as *H. regalis* and *H. rossicus* (Figure 9). The tarsometatarsus of *H. mengeli* is half the size of *H. rossicus*, while the femur of *H. macdonaldi* is less than half the length of that of *H. regalis*. All of these species of Hesperornis show similar development of the features described above associated with diving specializations, thus decoupling the evolution of foot-propelled diving from changes in body size [22,25] which we see varying within lineages of hesperornithiforms, indicating the independent evolution of gigantism [22]. This occurred at least twice, once in the brodavids and at least once in several species of Hesperornis, with miniaturization possible in some of the smallest species of Hesperornis as well.

The topic of body size goes hand-in-hand with that of growth rates and ontogenetic patterns. Very little histological work has been done on hesperornithiforms, so the manner or timing in which gigantism (or lack thereof) was achieved in these birds is largely unknown. The first histological study of hesperornithiforms was conducted to address the discrepancies at the time regarding the treatment of hesperornithiforms as either ratites or neognaths within neornithines (modern birds) and found that the bone microsctructure of the hindlimb of Hesperornis was like that of neognaths [100]. This study did not address growth rates. The next histological study of hesperornithiforms characterized the microstructure of the bone from a femur of Hesperornis, identifying the individual as a subadult from the lack of peripheral lamellar bone [98]. Significantly, this study did not identify lines of arrested growth in Hesperornis, indicating there was no evidence for cyclical growth as seen in more basal Mesozoic birds (e.g., Archaeopteryx, *Sapeornis*, *Confuciusornis*, enantionithines, and many stem ornithuromorphs). Thus, continuous growth and the resulting absence of lines of arrested growth is a derived feature that

Hesperornis shares with modern birds [98]. This has important implications for physiology and life history, as it may indicate a fully endothermic physiology consistent with the interpretation of hesperornithids as venturing far offshore into deep marine waters [98] and growth patterns comparable to those of modern birds (i.e., reaching full-grown size within the first year). The capability for rapid, sustained growth would also contribute to the gigantic body size attained in multiple lineages of these birds.

More recently, hindlimb bones from Hesperornis specimens discovered along a latitudinal gradient from Kansas to the Arctic were examined to investigate the effects of climate and possible migration on bone microstructure [101]. This study found continuous bone deposition and did not identify cyclic growth marks [101], supporting the previous results of Chinsamy et al. [98] and indicating that migratory patterns to different climates is either not recorded in bone microstructure or that these birds achieved skeletal maturity before migrating [101]. The lack of histological work may be in part complicated by taphonomic processes in some of the more prolific hesperornithiform sedimentary units. For example, several members of the Pierre Shale are characterized by calcite crystallization in the preserved bones, destroying histological information.

6. Paleoecology

The obvious and dramatic diving adaptations in hesperornithiforms described above, combined with their wide distribution across the Northern Hemisphere (Figure 2), have led to a large body of work involving the paleoecology of these birds, including the aquatic environments they occupied [22,94,101], modern ecological analogues [22,94], and niche partioning [22,24].

The diversity of hesperornithiform taxa in terms of size, morphological features that are interpreted as diving specializations, and the range of environments in which they are preserved all point to habitat or trophic specializations among hesperornithiforms. Interpretations of habitat preference for hesperornithiforms are limited to the depositional environment in which their fossils were discovered. While these interpretations may not precisely align with the environment in which these birds actually lived, some general conclusions can be drawn. Hesperornithiforms are predominantly known from marine environments, with some specimens known from continental and transitional environments (Figure 12 and Table 2). As mentioned above, only in the latter half of the Late Cretaceous (Campanian-Maastrichtian) do fossils occur in nonmarine sediments, thus suggesting a possible colonization of these environments later in their evolutionary history. Body size does not appear to correlate well with depositional environment, with large and small taxa reported from continental, transitional, and marine environments. There may be an underlying trend of large-bodied birds restricted to marine environments that is obscured by taphonomic processes, particularly in regards to preservation in the reworked marine Cambridge Greensand and Belle Fouche Formation.

Of particular interest is the overlap of multiple taxa in single geologic units such as that seen in the Niobrara Formation and the Pierre Shale of the United States. Both of these units are widely deposited deep-water marine sediments of the Western Interior Seaway, with the older Niobrara Formation grading into the Pierre Shale in some places [102]. While the number of taxa reported may be inflated (see Section 3 above), the range of body sizes preserved in both is striking, with the small *Baptornis* and *Fumicollis*, the large *H. regalis*, and taxa such as *Parahesperornis* of intermediate size, known from the Niobrara Formation, and some of the smallest Hesperornis species, *H. lumgairi* and *H. macdonaldi*, found with large species such as *H. chowi* and *H. regalis* in the Pierre Shale. This juxtaposition should not necessarily be interpreted as direct evidence of niche partioning, as it may result in full or part from taphonomic processes or time averaging. However, it does raise the possibility of ecological specializations to reduce interspecific competition. Ecologic niche segregation is common among modern diving birds in which sympatric species differentiate in either diet (prey type) or foraging range [103].

7. Summary and Future Directions

Hesperornithiforms became the first birds (and dinosaurs) to adapt to a fully aquatic lifestyle. A number of morphological features highlight this evolutionary pathway, resulting in a highly streamlined body optimized for diving through the water, propelled by powerful hindlimbs. Adaptations include an elongate neck that allowed for increased maneuverability of a skull with sharp teeth and an expanded rostrum, ideal of capturing fish and other mobile prey. The pelvis was also elongate, allowing for the attachment of larger muscles for powering the feet. The femur was reduced to varying degrees in the different species of hesperornithiform, but the shortened length and horizontal articulation with the pelvis allowed for the orientation of the feet in line with and directly behind the body, optimizing power production by reducing drag. As evidence of the degree to which these birds optimized foot-propelled diving, the forelimbs of all hesperornithiforms were reduced to the point of flightlessness, with the hesperornithids showing the most extreme reduction.

This suite of morphological adaptations are present in varying degrees among the different specimens assigned to the group, suggesting a progression of diving specializations and even the evolution of niche partitioning among these birds. This is supported by the many geologic units where multiple species showing ranges in size and interpreted diving capabilities have been discovered.

The broad morphological diversity present among hesperornithiform specimens has been interpreted as representing an increasing number of taxa over the years. These taxonomic interpretations are complicated by the highly fragmentary nature of the fossil record of these birds. While the addition of more specimens, particularly of the most basal taxa, through future fieldwork is something to look forward to, there remains much to be done with the existing global collection of specimens. Much of the taxonomic work published to date has been limited to the geographic area of the authors (i.e., [13,47,56]) or relied on photographs and loans of select specimens (i.e., [15,24,50]). Very few studies have incorporated direct observations of unpublished material from multiple continents, but even these studies were not able to access much of the Asian and Canadian material (i.e., [22,25,65]). Digitization and publication of a more complete record of current museum collections as measurements, photographs, written morphological descriptions, and three-dimensional datasets such as those from computed tomography or laser scans, would enable broad-scale studies not limited by geography (or travel funds). Such studies could test taxonomic hypotheses that have remained largely untested over the past 150 years of taxonomic work. The creation of digital specimens might also enable digital reconstructions of the skulls, which are disarticulated and often deformed to some degree. Such reconstructions might provide insight into the shape of the hesperornithiform brain and allow comparisons to Mesozoic and modern birds as well as inferences about the sensory capabilities of these birds.

While much of the modern work on these birds has focused on taxonomy (e.g., [15,24,50,58]) or ecology [22,94,101], very little has been done regarding the ontogeny of these birds [98]. Additionally, histological studies might better inform taxonomic studies involving specimens that range widely in size but have less variability in morphology, such as the species of Hesperornis. There are a surprisingly large number of isolated hesperornithiform bones, primarily from the hindlimb, in museum collections across North America, and additional histological work to characterize growth patterns and rates as well as life history as a whole seem to be a potentially fruitful line of inquiry, despite complications from poor preservation.

In conclusion, taxonomic, phylogenetic, and paleoecologic studies on the Hesperornithiformes for the past 150 years have led us to an understanding of these birds as a fascinating chapter in adaptive evolution. Hesperornithiforms are the first group of marine diving birds to evolve, and while the origins of this group remain elusive, a large body of work documents their spread across Laurasia and their expansion from marine to estuarine and even to freshwater environments by the Maastrichtian. While we have a large body of previous research on these birds, there is much to be done in the future.

Author Contributions: Conceptualization, literature review, and initial draft of the manuscript and figures were drafted by A.B. Conceptualization, contribution to the manuscript, and review was provided by L.M.C. All authors have read and agreed to the published version of the manuscript.

Funding: This research was funded in part by a generous donation from the Augustyn Family.

Acknowledgments: The authors would like to thank Stephanie Abramowicz for assistance with figures as well as the collections management staff at all of the museums for their assistance with accessing specimens.

Conflicts of Interest: The authors declare no conflict of interest.

Abbreviations

AMNH, American Museum of Natural History, New York, NY, USA; IZASK, Institute of Zoology, Ministry of Science, Almaty, Kazakhstan; KUVP, University of Kansas Museum of Natural History, Lawrence, KS, USA; LACM, Natural History Museum of Los Angeles County, Los Angeles, CA, USA; MCM, Mikasa City Museum, Mikasa, Japan; RSM, Royal Saskatchewan Museum, Regina, SK, Canada; SDSM, Museum of Geology, South Dakota School of Mines and Technology, Rapid City, SD, USA; UNSM, University of Nebraska State Museum, Lincoln, NE, USA; YPM, Yale Peabody Museum, New Haven, CT, USA; YPM PU, Princeton University (collections now housed in the Yale Peabody Museum).

References

1. Marsh, O.C. *Odontornithes: A Monograph on the Extinct Toothed Birds of North America*; United States Government Printing Office: Washington, DC, USA, 1880; 312p.
2. Marsh, O.C. Discovery of a remarkable fossil bird. *Am. J. Sci. Ser.* **1872**, *3*, 56–57. [CrossRef]
3. Marsh, O.C. ART. XLVIII.—Preliminary Description of *Hesperornis regalis*, with Notices of Four Other New Species of Cretaceous Birds. *Am. J. Sci. Arts* **1872**, *3*, 360. [CrossRef]
4. Marsh, O.C. Fossil birds from the Cretaceous of North America. *Am. J. Sci. Arts* **1872**, *5*, 229–231.
5. Marsh, O.C. On a new subclass of fossil birds. *Am. J. Sci. Arts* **1873**, *5*, 161.
6. Marsh, O.C. On the Odontornithes, or birds with teeth. *Am. J. Sci. Arts* **1873**, *10*, 403.
7. Marsh, O.C. Notice of new Odontornithes. *Am. J. Sci. Arts* **1876**, *11*, 509–511. [CrossRef]
8. Seeley, H.G. On the British Fossil Cretaceous Birds. *Q. J. Geol. Soc.* **1876**, *32*, 496–512. [CrossRef]
9. Furbinger, M. *Untersuchungen Zur Morphologie und Systematik der Vögel, Zugleich ein Beitrag Zur ANATOMIE der Stütz- und Bewegungsorgan*; T. van Holkema: Amsterdam, The Netherlands, 1888.
10. Lydekker, R. *Catalogue of the Fossil Birds in the British Museum (Natural History) United Kingdom, Order of the Trustees*; Longmans & Co.: London, UK, 1891.
11. Wetmore, A. The fossil birds of the AOU Check-list. *Condor* **1930**, *32*, 12–14. [CrossRef]
12. Storer, R.W. Evolution in the Diving Birds. In *Proceedings of the XII International Ornithological Congress*; Tilgmannin Kirjapaino: Helsinki, Finland, 1958; Volume 2, pp. 694–707.
13. Martin, L.D.; Tate, J. The Skeleton of *Baptornis advenus* (Aves: Hesperornithiformes). *Smithson. Contrib. Paleobiol.* **1976**, *27*, 35–66.
14. Cumbaa, S.L.; Schröder-Adams, C.; Day, R.G.; Phillips, A.J. Cenomanian Bonebed Faunas from the Northeastern Margin, Western Interior Seaway, Canada. In *Late Cretaceous Vertebrates from the Western Interior. New Mexico Museum of Natural History and Science Bulletin 35*; Lucas, S., Sullivan, R., Eds.; Kansas Academy of Science: Kansas, MO, USA, 2006; pp. 139–155.
15. Sanchez, J. Late Cretaceous (Cenomanian) Hesperornithiformes from the Pasquia Hills, Saskatchewan, Canada. Master's Thesis, Carleton University, Ottawa, ON, Canada, 2010.
16. Tokaryk, T.T.; Harington, C.R. *Baptornis* sp. (Aves: Hesperornithiformes) from the Judith River Formation (Campanian) of Saskatchewan, Canada. *J. Paleontol.* **1992**, *66*, 1010–1012. [CrossRef]
17. Lambrecht, K. *Handbuch der Palaeomithologie*; Gebruder Borntraeger: Berlin, Germany, 1933; 1022p.
18. Brodkorb, P. *Origin and Evolution of Birds*; Farner, D.S., King, J.R., Eds.; Avian Biology: New York, NY, USA; Academic Press: London, UK, 1971; Volume 1.
19. Cracraft, J. Phylogenetic Relationships and Monophyly of Loons, Grebes, and Hesperornithiform Birds, with Comments on the Early History of Birds. *Syst. Zool.* **1982**, *31*, 22. [CrossRef]
20. Hackett, S.J.; Kimball, R.T.; Reddy, S.; Bowie, R.C.; Braun, E.L.; Braun, M.J.; Chojnowski, J.L.; Cox, W.A.; Han, K.; Harshman, J.; et al. A phylogenomic study of birds reveals their evolutionary history. *Science* **2008**, *320*, 1763–1768. [CrossRef]

21. Jetz, W.; Thomas, G.H.; Joy, J.B.; Hartmann, K.; Mooers, A.O. The Global Diversity of Birds in Space and Time. *Nature* **2012**, *491*, 444–448. [CrossRef]
22. Bell, A.; Wu, Y.-H.; Chiappe, L.M. Morphometric Comparison of the Hesperornithiformes and Modern Diving Birds. *Palaeogeogr. Palaeoclimatol. Palaeoecol.* **2019**, *513*, 196–207. [CrossRef]
23. Bell, A.; Chiappe, L.M. A Species-Level Phylogeny of the Cretaceous Hesperornithiformes (Aves: Ornithuromorpha): Implications for Body Size Evolution amongst the Earliest Diving Birds. *J. Syst. Palaeontol.* **2016**, *14*, 239–251. [CrossRef]
24. Tanaka, T.; Kobayashi, Y.; Kurihara, K.; Fiorillo, A.R.; Kano, M. The Oldest Asian Hesperornithiform from the Upper Cretaceous of Japan and the Phylogenetic Reassessment of Hesperornithiformes. *J. Syst. Palaeontol.* **2017**, *16*, 689–709. [CrossRef]
25. Bell, A.; Chiappe, L. Anatomy of *Parahesperornis*: Evolutionary Mosaicism in the Cretaceous Hesperornithiformes (Aves). *Life* **2020**, *10*, 62. [CrossRef] [PubMed]
26. O'Connor, J.K. The trophic habits of early birds. *Palaeogeo. Palaeoclim. Palaeoeco.* **2019**, *513*, 178–195. [CrossRef]
27. O'Connor, J.K.; Wang, M.; Hu, H. A new ornithuromorph (Aves) with an elongate rostrum from the Jehol Biota, and the early evolution of rostralization in birds. *J. Syst. Palaeontol.* **2016**, *14*, 939–948. [CrossRef]
28. Wu, Y.H.; Chiappe, L.M.; Bottjer, D.J.; Nava, W.; Martinelli, A.G. Dental replacement in Mesozoic birds: Evidence from newly discovered Brazilian enantiornithines. *Sci. Rep.* **2021**, *11*, 1–12. [CrossRef] [PubMed]
29. Scotese, C.R. *Atlas of Earth History*; PALEOMAP Project: Arlington, TX, USA, 2001; Volume 1.
30. Dumont, M.; Tafforeau, P.; Bertin, T.; Bhullar, B.-A.; Field, D.; Schulp, A.; Strilisky, B.; Thivichon-Prince, B.; Viriot, L.; Louchart, A. Synchrotron Imaging of Dentition Provides Insights into the Biology of Hesperornis and Ichthyornis, the "Last" Toothed Birds. *BMC Evol. Biol.* **2016**, *16*, 178. [CrossRef] [PubMed]
31. Howgate, M.E. The teeth of Archaeopteryx and a reinterpretation of the Eichstätt specimen. *Zoo. J. Linn. Soc.* **1984**, *82*, 159–175. [CrossRef]
32. Clarke, J.A. Morphology, Phylogenetic Taxonomy, and Systematics of Ichthyornis and *Apatornis* (Avialae: Ornithurae). *Bull. Am. Mus. Nat. Hist.* **2004**, *286*, 1–179. [CrossRef]
33. Clarke, J.A.; Zhou, Z.; Zhang, F. Insight into the evolution of avian flight from a new clade of Early Cretaceous ornithurines from China and the morphology of Yixianornis Grabaui. *J. Anat.* **2006**, *208*, 287–308. [CrossRef]
34. O'Connor, J.K.; Chiappe, L.M. A revision of enantiornithine (Aves: Ornithothoraces) skull morphology. *J. Syst. Palaeo.* **2011**, *9*, 135–157. [CrossRef]
35. Sander, P.M. The microstructure of reptilian tooth enamel: Terminology, function, and phylogeny. *F. Pfeil* **1999**, *38*, 1–102.
36. Martin, L.D.; Stewart, J.D. Implantation and replacement of bird teeth. *Smiths. Cont. Paleobio.* **1999**, *89*, 295–300.
37. Elzanowski, A.; Galton, P. Braincase of *Enaliornis*, an Early Cretaceous Bird from England. *J. Vertebr. Paleontol.* **1991**, *11*, 90–107. [CrossRef]
38. Ribak, G.; Weihs, D.; Arad, Z. Consequences of Buoyancy to the Maneuvering Capabilities of a Foot-Propelled Aquatic Predator, the Great Cormorant (*Phalacrocorax carbo sinensis*). *J. Exp. Biol.* **2008**, *211*, 3009–3019. [CrossRef]
39. Lovvorn, J.R. Mechanics of Underwater Swimming in Foot-Propelled Diving Birds. *Proc. Int. Ornithol. Congr.* **1991**, *20*, 1868–1874.
40. Wang, M.; O'Connor, J.K.; Pan, Y.; Zhou, Z. A bizarre Early Cretaceous enantiornithine bird with unique crural feathers and an ornithuromorph plough-shaped pygostyle. *Nat. Commun.* **2017**, *8*, 14141. [CrossRef] [PubMed]
41. Machalski, M. The Cenomanian ammonite *Schloenbachia varians* (J. Sowerby, 1817) from the Cambridge Greensand of eastern England: Possible sedimentological and taphonomic implications. *Cret. Res.* **2018**, *87*, 120–125. [CrossRef]
42. Lyell, C. *Manual of Elementary Geology, Supplement to the 5th Edn*, 3rd ed.; Murray: London, UK, 1859.
43. Storer, R.W. P. Brodkorb, Catalogue of Fossil Birds, Part 1. *Auk* **1965**, *82*, 657–658. [CrossRef]
44. Galton, P.M.; Martin, L.D. Postcranial Anatomy and Systematics of *Enaliornis* SEELEY, 1876, a Foot-Propelled Diving Bird (Aves: Ornithurae: Hesperornithiformes) from the Early Cretaceous of England. *Rev. Paleobiol.* **2002**, *21*, 489–538.
45. Brodkorb, P. Catalogue of Fossil Birds. *Bull. Fla. St. Mus. Bio. Sci.* **1963**, *7*, 179–293.
46. Kurochkin, E.N. Mesozoic Birds of Mongolia and the Former USSR. In *The Age of Dinosaurs in Russia and Mongolia*; Benton, M., Shishkin, M., Unwin, D., Kurochkin, E., Eds.; Cambridge University Press: Cambridge, UK, 2000; pp. 544–559.
47. Tokaryk, T.T.; Cumbaa, S.L.; Storer, J.E. Early Late Cretaceous Birds from Saskatchewan, Canada: The Oldest Diverse Avifauna Known from North America. *J. Vertebr. Paleontol.* **1997**, *17*, 172–176. [CrossRef]
48. Everhart, M.J.; Bell, A. A Hesperornithiform Limb Bone from the Basal Greenhorn Formation (Late Cretaceous; Middle Cenomanian) of North Central Kansas. *J. Vertebr. Paleontol.* **2009**, *29*, 952–956. [CrossRef]
49. Zelenkov, N.V.; Panteleyev, A.V.; Yarkov, A.A. New Finds of Hesperornithids in the European Russia, with Comments on the Systematics of Eurasian Hesperornithidae. *Paleontol. J.* **2017**, *51*, 547–555. [CrossRef]
50. Martin, L.D.; Kurochkin, E.N.; Tokaryk, T.T. A New Evolutionary Lineage of Diving Birds from the Late Cretaceous of North America and Asia. *Palaeoworld* **2012**, *21*, 59–63. [CrossRef]
51. Martin, J.; Cordes-Person, A. A New Species of the Diving Bird *Baptornis* (Ornithurae: Hesperornithiformes) from the Lower Pierre Shale Group (Upper Cretaceous) of Southwestern South Dakota. *Geol. Soc. Am. Spec. Pap.* **2007**, *427*, 227–237.
52. Nessov, L.A.; Prizemlin, B. A Large Advanced Flightless Marine Bird of the Order Hesperornithiformes of the Late Senonian of Turgai Strait-the First Finding of the Group in the USSR. *Tr. Zool. Inst. SSSR* **1991**, *239*, 85–107.
53. Hou, L.-I. New Hesperornithid (Aves) from the Canadian Arctic. *Gu Ji Zhui Dong Wu Xue Bao* **1999**, *37*, 231–237.
54. Nessov, L.A.; Yarkov., A.A. Hesperornis in Russia, Russk. *Ornitol. Zh* **1993**, *2*, 37–54.
55. Martin, L.D.; Lim, J.-D. New Information on the Hesperornithiform Radiation. In Proceedings of the 5th Symposium of the Society of Avian Paleontology and Evolution, Beijing, China, 1–4 June 2000; Science Press: Beijing, China, 2002.

56. Aotsuka, K.; Sato, T. Hesperornithiformes (Aves: Ornithurae) from the Upper Cretaceous Pierre Shale, Southern Manitoba, Canada. *Cretac. Res.* **2016**, *63*, 154–169. [CrossRef]
57. Gill, J.; Cobban, W. Regional unconformity in late cretaceous, Wyoming. *Am. Geol. Surv. Prof. Pap.* **1966**, *550-B*, B20–B27.
58. Bell, A.; Chiappe, L.M. Identification of a New Hesperornithiform from the Cretaceous Niobrara Chalk and Implications for Ecologic Diversity among Early Diving Birds. *PLoS ONE* **2015**, *10*, e0141690. [CrossRef]
59. Lucas, F.A. Notes on the Osteology and Relationship of the Fossil Birds of the Genera Hesperornis, *Hargeria, Baptornis,* and *Diatryma. Proc. U. S. Natl. Mus.* **1903**, *26*, 545–556. [CrossRef]
60. Elzanowski, A.; Paul, G.S.; Stidham, T.A. An Avian Quadrate from the Late Cretaceous Lance Formation of Wyoming. *J. Vertebr. Paleontol.* **2000**, *20*, 712–719. [CrossRef]
61. Rees, J.; Lindgren, J. Aquatic Birds from the Upper Cretaceous (Lower Campanian) of Sweden and the Biology and Distribution of Hesperornithiforms: Cretaceous Aquatic Birds. *Palaeontology* **2005**, *48*, 1321–1329. [CrossRef]
62. Shufeldt, R.W. The Fossil Remains of a Species of Hesperornis Found in Montana. *Auk* **1915**, *32*, 290–294. [CrossRef]
63. Martin, L.D. A New Hesperornithid and the Relationships of the Mesozoic Birds. *Trans. Kans. Acad. Sci.* **1984**, *87*, 141–150. [CrossRef]
64. Olson, S.L. *Neogaeornis wetzeli* Lambrecht, a cretaceous loon from Chile (Aves: Gaviidae). *J. Vertebr. Paleontol.* **1992**, *12*, 122–124. [CrossRef]
65. Bell, A.K. Evolution and Ecology of Mesozoic Birds: A Case Study of the Derived Hesperornithiformes and the Use of Morphometric Data in Quantifying Avian Paleoecology. Master's Thesis, The University of Southern California, Los Angeles, CA, USA, 2013.
66. Chiappe, L.M. The first 85 million years of avian evolution. *Nature* **1995**, *378*, 349–355. [CrossRef]
67. Chiappe, L.M.; Jorge, O.C. *Neuquenornis volans*, a new Late Cretaceous bird (Enantiornithes: Avisauridae) from Patagonia, Argentina. *J. Vert. Paleont.* **1994**, *14*, 230–246. [CrossRef]
68. Padian, K.; Chiappe, L.M. The origin and early evolution of birds. *Bio. Rev.* **1998**, *73*, 1–42. [CrossRef]
69. Liu, D.; Chiappe, L.M.; Zhang, Y.; Bell, A.; Meng, Q.; Ji, Q.; Wang, X. An advanced, new long-legged bird from the Early Cretaceous of the Jehol Group (northeastern China): Insights into the temporal divergence of modern birds. *Zootaxa* **2014**, *3884*, 253–266. [CrossRef] [PubMed]
70. O'Connor, J.K.; Zhou, Z. A redescription of *Chaoyangia beishanensis* (Aves) and a comprehensive phylogeny of Mesozoic birds. *J. Syst. Palaeont.* **2013**, *11*, 889–906. [CrossRef]
71. 71. O'Connor, J. K.; Chiappe, L. M.; Bell, A. Pre-modern birds: avian divergences in the Mesozoic. In *Living Dinosaurs: The Evolution-Ary History of Modern Birds*; John Wiley & Sons: Hoboken, NJ, USA, 2011; pp. 39–114.
72. Pittman, M.; O'Connor, J.; Tse, E.; Makovicky, P.; Field, D.J.; Ma, W.; Turner, A.H.; Norell, M.A.; Pei, R.; Xu, X. The Fossil Record of Mesozoic and Paleocene Pennaraptorans. *Bull. Am. Mus. Nat. Hist.* **2020**, *440*, 37–95.
73. Wang, M.; O'Connor, J.K.; Zhou, Z. A new robust enantiornithine bird from the Lower Cretaceous of China with scansorial adaptations. *J. Vert. Paleont.* **2014**, *34*, 657–671. [CrossRef]
74. Fluteau, F.; Ramstein, G.; Besse, J.; Giraud, R.; Masse, J.P. Impacts of palaeogeography and sea level change on Mid-Cretaceous climate. *Palaeogeog. Palaeoclim. Palaeoeco.* **2007**, *247*, 357–381. [CrossRef]
75. Gradzinski, R.; Kielan-Jaworowska, Z.; Maryanska, T. Upper Cretaceous Djadokhta, Barun Goyot and Nemegt formations of Mongolia, including remarks on previous subdivisions. *Acta Geol. Pol.* **1977**, *27*, 281–326.
76. Breithaupt, B. Paleontology and paleoecology of the Lance Formation (Maastrichtian), east flank of Rock Springs Uplift, Sweetwater County, Wyoming. *Rocky Mt. Geol.* **1982**, *21*, 123–151.
77. Kupsch, W. Frenchman formation of eastern Cypress Hills, Saskatchewan, Canada. *GSA Bull.* **1957**, *68*, 413–420. [CrossRef]
78. Ogunyomi, O.; Hills, L. Depositional environments, foremost formation (Late Cretaceous), Milk river area, southern Alberta. *Bull. Can. Petrol. Geol.* **1977**, *25*, 929–968.
79. Wood, J.; Thomas, R.; Vasser, J. Fluvial processes and vertebrate taphonomy: The upper cretaceous Judith River formation, south-central dinosaur Provincial Park, Alberta, Canada. *Palaeogeogr. Palaeoclimatol. Palaeoecol.* **1988**, *66*, 127–143. [CrossRef]
80. Eberth, D.A. Stratigraphy and sedimentology of vertebrate microfossil sites in the uppermost Judith River formation (Campanian), Dinosaur Provincial Park, Alberta, Canada. *Palaeogeogr. Palaeoclimatol. Palaeoecol.* **1990**, *78*, 1–36. [CrossRef]
81. Johnson, K.R.; Nichols, D.J.; Hartman, J.H. Hell Creek formation: A 2001 synthesis. In *The Hell Creek Formation and the Cretaceous-Tertiary Boundary in the Northern Great Plains: An Integrated Continental Record of the End of the Cretaceous*; Geological Society of America Special Paper, 361; Hartman, J.H., Johnson, K.R., Nichols, D.J., Eds.; Geological Society of America: Boulder, CO, USA, 2002; pp. 503–510.
82. Bardack, D. Fossil vertebrates from the marine Cretaceous of Manitoba. *Can. J. Earth Sci.* **1968**, *5*, 145–153. [CrossRef]
83. Reifenstuhl, R.R. Gilead sandstone, northeastern Brooks Range, Alaska: An Albian to Cenomanian marine clastic succession. In *Short Notes on Alaskan Geology*; Professional Report; Reger, R.D., Ed.; Alaska Division of Geological & Geophysical Surveys: Fairbanks, AK, USA, 1991; Volume 111, pp. 69–76.
84. Barlow, L.; Kauffman, E. Depositional cycles in the Niobrara formation, Colorado front range. In *The Society of Economic Paleontologists and Mineralogists (SEPM) Fine-Grained Deposits and Biofacies of the Cretaceous Western Interior Seaway: Evidence of Cyclic Sedimentary Processes (FG4)*; AAPG: Tulsa, OK, USA, 1985.
85. Russell, D. Cretaceous vertebrates from the Anderson River, N.W.T. *Can. J. Earth Sci.* **1967**, *4*, 21–43. [CrossRef]
86. Panteleyev, A.V.; Popov, E.V.; Averianov, A.O. New record of *Hesperornis rossicus* (Aves, Hesperornithiformes) in the campanian of Saratov Province, Russia. *Paleontol. Res.* **2004**, *8*, 115–122. [CrossRef]

87. Hills, L.; Strong, W. Multivariate analysis of late cretaceous Kanguk formation (Arctic Canada) Palynomorph assemblages to identify nearshore to distal marine groupings. *Bull. Can. Petrol. Geol.* **2007**, *55*, 160–172. [CrossRef]
88. Shultz, L.G.; Tourtelot, H.A.; Gill, J.R.; Boerngen, J.G. Composition and properties of the Pierre Shale and equivalent rocks, northern Great Plains region. *Geol. Surv. Prof. Pap.* **1980**, *1064-B*, 123.
89. Hilton, R. *Dinosaurs and Other Mesozoic Reptiles of California*; University of California Press: Berkeley, CA, USA, 2003.
90. Bell, A.; Irwin, K.J.; Davis, L.C. Hesperornithiform birds from the Late Cretaceous (Campanian) of Arkansas, USA. *Trans. Kans. Acad. Sci.* **2015**, *118*, 219–229. [CrossRef]
91. Tanaka, T.; Kobayashi, Y.; Ikuno, K.; Ikeda, T.; Saegusa, H. A marine hesperornithiform (Avialae: Ornithuromorpha) from the Maastrichtian of Japan: Implications for the paleoecological diversity of the earliest diving birds in the end of the Cretaceous. *Cretac. Res.* **2020**, *113*, 104492. [CrossRef]
92. Dyke, G.; Malakhov, D.; Chiappe, L.M. A re-analysis of the marine bird *Asiahesperornis* from northern Kazakhstan. *Cretac. Res.* **2006**, *27*, 947–953. [CrossRef]
93. Marsh, O.C. A New Cretaceous Bird Allied to Hesperornis. *Am. J. Sci.* **1893**, *45*, 81–82. [CrossRef]
94. Hinić-Frlog, S.; Motani, R. Relationship between osteology and aquatic locomotion in birds: Determining modes of locomotion in extinct Ornithurae. *J. Evol. Bio.* **2010**, *23*, 372–385. [CrossRef]
95. Galantsev, V.P. Adaptational changes in the venous system of diving mammals. *Can. J. Zool.* **1991**, *69*, 414–419. [CrossRef]
96. Girgis, S. Anatomical and functional adaptations in the venous system of a diving reptile, Trionyx triunquis. *Proc. Zool. Soc. Lond.* **1962**, *138*, 355–377. [CrossRef]
97. Jessen, C. Selective brain cooling in mammals and birds. *Jpn. J. Physiol.* **2001**, *51*, 291–301. [CrossRef] [PubMed]
98. Chinsamy, A.; Martin, L.D.; Dobson, P. Bone microstructure of the diving Hesperornis and the voltant Ichthyornis from the Niobrara Chalk of western Kansas. *Cretac. Res.* **1998**, *19*, 225–235. [CrossRef]
99. Marsh, O.C. ART. XII. Characters of the Odontornithes, with notice of a new allied genus. *Am. J. Sci. Arts* **1877**, *14*, 79–84.
100. Houde, P. Histological evidence for the systematic position of Hesperornis (Odontornithes: Hesperornithiformes). *Auk* **1987**, *104*, 125–129. [CrossRef]
101. Wilson, L.E.; Chin, K.; Cumbaa, S.; Dyke, G. A High Latitude Hesperornithiform (Aves) from Devon Island: Palaeobiogeography and Size Distribution of North American Hesperornithiforms. *J. Syst. Palaeontol.* **2011**, *9*, 9–23. [CrossRef]
102. Carpenter, K. Vertebrate biostratigraphy of the Smoky Hill Chalk (Niobrara Formation) and the Sharon Springs Member (Pierre Shale). In *High-Resolution Approaches in Stratigraphic Paleontology*; Springer: Dordrecht, The Netherlands, 2008; pp. 421–437.
103. Petalas, C.; Lazarus, T.; Layoie, R.A.; Elliott, K.H.; Guigueno, M.F. Foraging niche partitioning in sympatric seabird populations. *Sci. Res.* **2021**, *11*, 2493. [CrossRef] [PubMed]

Article

Reappraisal on the Phylogenetic Relationships of the Enigmatic Flightless Bird (*Brontornis burmeisteri*) Moreno and Mercerat, 1891

Federico L. Agnolin [1,2]

1   Laboratorio de Anatomía Comparada y Evolución de los Vertebrados, Museo Argentino de Ciencias Naturales "Bernardino Rivadavia"-CONICET, Av. Ángel Gallardo 470, Buenos Aires C1405DJR, Argentina; fedeagnolin@yahoo.com.ar

2   Fundación de Historia Natural "Félix de Azara", Departamento de Ciencias Naturales y Antropología, CEBBAD–Universidad Maimónides, Hidalgo 775 piso 7, Buenos Aires C1405BDB, Argentina

Abstract: The fossil record of birds in South America is still very patchy. One of the most remarkable birds found in Miocene deposits from Patagonia is *Brontornis burmeisteri* Moreno and Mercerat, 1891. This giant flightless bird is known by multiple incomplete specimens that represent a few portions of the skeleton, mainly hindlimb bones. Since the XIX century, *Brontornis* was considered as belonging to or closely related to phorusrhacoid birds. In contrast to previous work, by the end of 2000 decade it was proposed that *Brontornis* belongs to Galloanserae. This proposal was recently contested based on a large dataset including both phorusrhacoids and galloanserine birds, that concluded *Brontornis* was nested among cariamiform birds, and probably belonged to phorusrhacoids. The aim of the present contribution is to re-evaluate the phylogenetic affinities of *Brontornis*. Based on modified previous datasets, it is concluded that *Brontornis* does belong to Galloanserae, and that it represents a member of a largely unknown radiation of giant graviportal birds from South America.

Keywords: *Brontornis*; Phorusrhacoidea; Galloanserae; South America; Neogene

Citation: Agnolin, F.L. Reappraisal on the Phylogenetic Relationships of the Enigmatic Flightless Bird (*Brontornis burmeisteri*) Moreno and Mercerat, 1891. *Diversity* **2021**, *13*, 90. https://doi.org/10.3390/d13020090

Academic Editor: Eric Buffetaut

Received: 19 January 2021
Accepted: 17 February 2021
Published: 20 February 2021

Publisher's Note: MDPI stays neutral with regard to jurisdictional claims in published maps and institutional affiliations.

1. Introduction

The genus *Brontornis* was originally described by Moreno and Mercerat (1891) based on several specimens coming from Lower-Middle Miocene localities at Santa Cruz province, Patagonia, Argentina [1]. This genus contains a single species: *B. burmeisteri* Moreno and Mercerat [2,3]. *Brontornis* was a giant flightless bird of about 2.8 m tall that may have weighed about 350 to 400 kg [2]. Its limb proportions and shape of elements indicate that *Brontornis* was a graviportal bird [4–7], probably a carrion eater [7], or even herbivorous [4,6,8].

On its original description, Moreno and Mercerat [3] include *Brontornis* on its own family Brontornithidae in the Order Stereornithes (this later included several genera now known as phorusrhacoids). In their concept, the Stereornithes were carinate birds with a shared combination of characters between anseriformes, coconiiforms (Herodiones therein), and accipitriforms, probably "intermediate" between Anatidae and Cathartidae. Moreno and Mercerat also noted the persistence of "reptilian" (i.e., plesiomorphic) characters in phorusrhacoids. Ameghino [9] made a revision of fossil Patagonian birds and partially resolved the confusion created by Moreno and Mercerat's [3] work. Ameghino considered the Stereornithes as belonging to Ratitae, and included *Brontornis* among phorusrhacids, a criterion was followed by most authors until Dolgopol de Sáez [10]. She revalidated the Brontornithidae (as Brontorniidae) and based on morphological grounds coined the Order Brontornithes to separate them from remaining phorusrhacoids (encompassed by her in the Order Stereornithes). Despite that, Dolgopol de Sáez was not able to recognize the suprageneric relationships of *Brontornis* and kin and considered that *Gastornis* may be closely related to it. Kraglievich [6,11] followed Dolgopol de Sáez and retained *Brontornis* on its own order Brontornithes (Brontornitiformes for Kraglievich, [6]). Subsequent authors followed Moreno and Mercerat and Ameghino views and considered *Brontornis* and kin as belonging to a different family or subfamily of phorusrhacoid birds [1,12–14], without regard of the distinctive anatomical features cited by Dolgopol de Sáez and Kraglievich.

Posteriorly, Agnolin [8,15,16] proposed that *Brontornis* may not be closely related to phorusrhacoids, but may be included among Galloanseres as a basal member of Anseriformes, a criterion followed by several authors [17–22]. However, Alvarenga et al. [23] returned to previous ideas and sustained that *Brontornis* belongs to Phorusrhacoidea. The arguments exposed by Alvarenga et al. [23] were contested by Agnolin [16], who supported the anseriform affinities for Brontornis again.

More recently, Worthy et al. [24] made a comprehensive phylogenetic analysis of Galloanseres, with special emphasis on extinct and flightless fowls. In their impressive analysis, Worthy et al. concluded that *Brontornis* is closely related to phorusrhacoids and considered it as part of Cariamiformes, far from Galloanseres. They argued that the strong differences observed in the postcranial anatomy of *Brontornis* and other cariamiforms are the result of the gigantism and graviportal locomotion of the former.

The aim of the present contribution is to describe and re-describe some materials that has referred to *Brontornis*, as well as to review Worthy et al.'s [24] analysis and re-consider the phylogenetic affinities of *Brontornis*.

2. Materials and Methods

2.1. Nomenclature

I follow the taxonomic nomenclature employed by Agnolin [16]. In that contribution I regard as valid the genus *Tolmodus* Ameghino, 1891 instead of *Patagornis* Moreno and Mercerat, 1891, following Patterson and Kraglievich ([14]; contra [2]). Following Agnolin (2006), the genus *Onactornis* is restricted to the species *O. depressus* Cabrera 1939, and probably *O. pozzi* Kraglievich, 1931, and the genus *Devincenzia* is considered as distinct from *Onactornis* and represented by its type species *D. gallinali* Kraglievich, 1932 [2,16,25].

The terms Phorusrhacoidea Ameghino, 1889, and *Phorusrhacos* Ameghino, 1887 are used instead of Phororhacoidea Patterson, 1941 and *Phororhacos* Ameghino, 1889 following Brodkorb [12] and Buffetaut [26].

I follow the anatomical nomenclature employed by Baumel and Witmer [27], with details on muscular attachments and syndesmology taken from Zinoviev [28].

2.2. Phylogenetic Analysis

With the aim to test the phylogenetic relationships of *Brontornis* proposed by Worthy and collaborators [24], I followed the character definition and numbers of Worthy et al. [24] (see Appendix A). The resulting data matrix was composed by 290 characters and 48 taxa.

The matrix was analyzed using TNT 1.5 [29], with all characters weighted equally. The dataset was analyzed under equally weighted parsimony. A total of 1,800,000 trees was set to be retained in memory. A first search using the algorithms Sectorial Searches, Ratchet (perturbation phase stopped after 20 substitutions), and Tree Fusing (5 rounds) was conducted, performing 1000 replications in order to find all tree islands (each replication starts from a new Wagner tree). The best tree or trees obtained at the end of the replicates were subjected to a final round of TBR (tree-branch-swapping) algorithm.

Two different phylogenetic analyses were performed (Figure 1). The first one follows strictly that of Worthy et al.'s [24] unconstrained analysis. This resulted in the recovery of 13 Most Parsimonious Trees (MPTs), of 1567 steps, with a consistency index of 0.26, and a retention index of 0.65, which were summarized using a strict consensus tree (see Discussion).

As a branch support measure, Bremer support was calculated, and as a measure of branch stability, a bootstrap resampling analysis was conducted, performing 10,000 pseudoreplicates. Bremer support was calculated after searching for suboptimal trees and not with the script that accompanies the program. Both absolute and GC bootstrap frequencies are also reported (Figure 1).

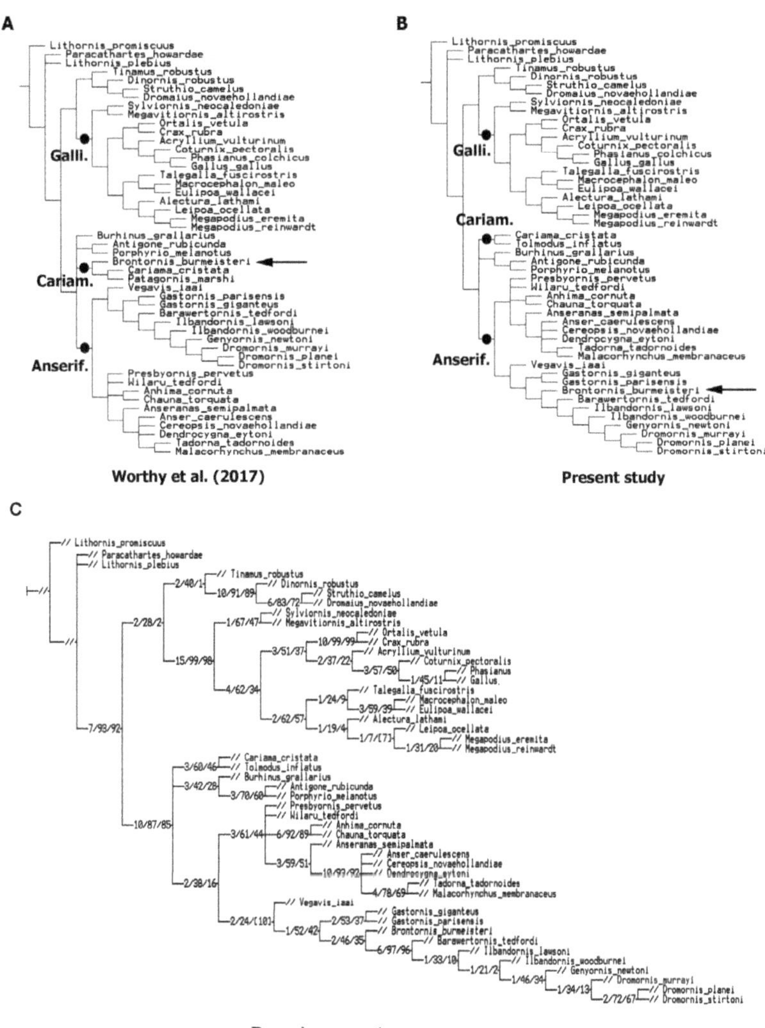

Figure 1. Phylogenetic analysis depicting the position of *Brontornis burmeisteri*. (**A**) hypothesis proposed by Worthy et al. [24]; (**B**) hypothesis proposed in the present study; (**C**) consensus tree showing branch support measures. From left to right: Bremer support, absolute bootstrap frequency, and GC bootstrap frequency. The arrow indicates the position of *Brontornis*. Abbreviations. Anserif., Anseriformes; Cariam., Cariamiformes; Galli., Galliformes.

The second analysis was carried out with the modifications in the scorings of *Brontornis* and *Gastornis* remarked in the "Discussion" section. This resulted in the recovery of four most parsimonious trees (MPTs) of 1564 steps, with a consistency index of 0.26, and a retention index of 0.65, which were summarized using a strict consensus tree (see Discussion).

2.3. Institutional Abbreviations

FM-P, Field Museum of Natural History, Vertebrate Palentology Collection; MACN A, Colección Nacional Ameghino, Museo Argentino de Ciencias Naturales "Bernardino Rivadavia", Buenos Aires, Argentina; MACN Pv, Museo Argentino de Ciencias Naturales "Bernardino Rivadavia", Buenos Aires, Argentina; MLP, Museo de La Plata, Buenos Aires,

Argentina; NHMUK, Natural History Museum of the United Kingdom, London, United Kingdom.

SYSTEMATIC PALEONTOLOGY

Neornithes Gadow, 1893

Galloanseres Sibley and Ahlquist, 1990

Brontornithes Dolgopol de Sáez, 1927

Brontornithidae Moreno and Mercerat, 1891

Brontornis burmeisteri Moreno and Mercerat, 1891

Synonymy. *Rostrornis floweri* Moreno and Mercerat, 1891; *Brontornis platyonyx* Ameghino, 1895; *Liornis floweri* Ameghino, 1895; *Callornis giganteus* Ameghino, 1895 *in part*; *Eucallornis giganteus* (Ameghino, 1895) Ameghino, 1901 *in part* [1,2,12,18,19].

Lectotype. MLP-88-91, left femur, tibiotarsus, fibula, and tarsometatarsus belonging to the same individual [12,30].

Diagnosis. Giant bird with graviportal proportions (tibiotarsus/tarsometatarsus ratio: 1.88) and the following unique combination of derived characters: distal end of tibiotarsus strongly anteroposteriorly compressed and with lateral margin forming an acute ridge of bone; distal end of tibiotarsus lacking supratendinal bridge [19], extensor groove shallow, poorly defined and medially tilted, retinacular tubercles feebly developed, prominent pyramidal-shaped prominence (central tubercle for attaching the *lig. meniscotibiale intertarsi*; [19,31]); tarsometatarsus having hypotarsus situated distal to the articular level of proximal cotylae [2], absence of posterior opening of the distal vascular foramen due to the unbifurcated condition of the *canalis interosseous distalis* [10], absence of fossa or scar for the first metatarsal [19], and proximodorsal margin of metatarsal trochlea III strongly projected [32].

Remarks. To date, the only certain member of Brontornithes and Brontornithidae is *Brontornis burmeisteri* [15]. However, recent finding of an incomplete distal tibiotarsus from the Oligocene of Bolivia [33] suggests that *Brontornis*-like taxa were probably more geographically and temporally widespread than thought.

Referred material. MLP 20-110, distal half of a left tibiotarsus with abraded distal condyles (Figure 2); MLP 20-581, distal end of left tibiotarsus without distal condyles (Figure 3).

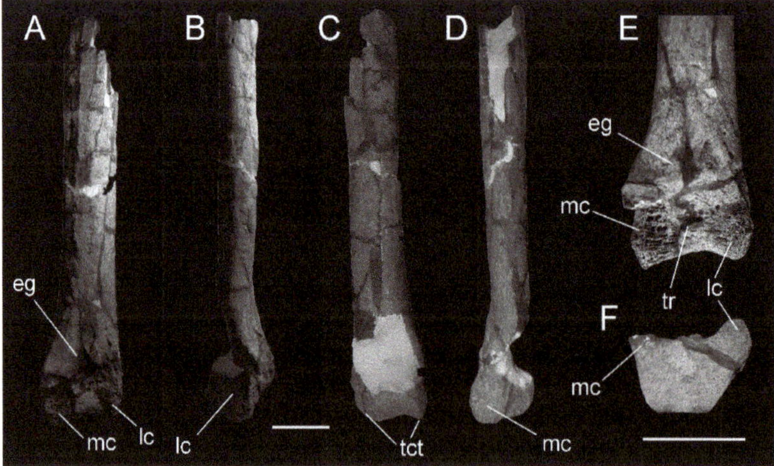

Figure 2. *Brontornis burmeisteri* (MLP 20-110) distal half of left tibiotarsus in (**A**) anterior; (**B**) lateral; (**C**) posterior; (**D**) medial views; (**E**) detail of its distal end in anterior view; and (**F**) distal view. **Abbreviations.** eg, extensor groove; lc, lateral condyle; mc, medial condyle; tct, *trochlea cartilaginis tibialis*; tr, transverse ridge. Scale bar: 5 cm.

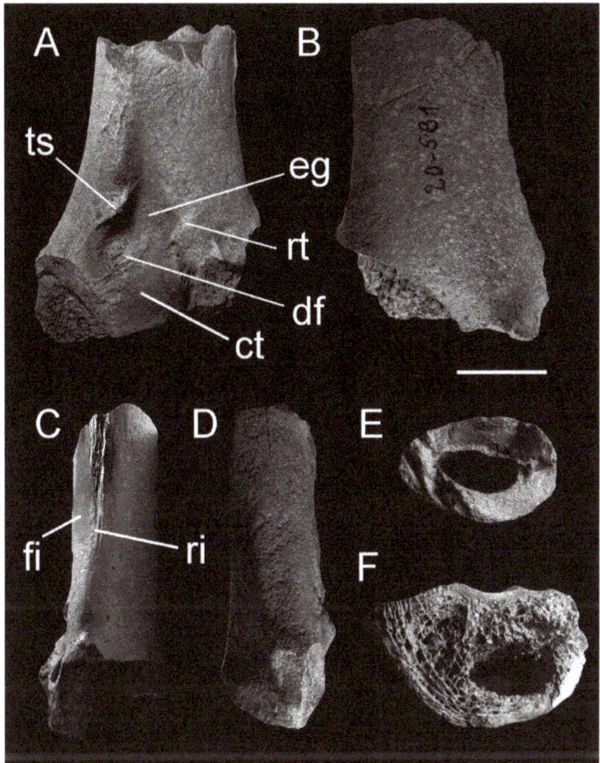

Figure 3. *Brontornis burmeisteri* (MLP 20-581) distal end of left tibiotarsus in (**A**) anterior; (**B**) posterior; (**C**) lateral; (**D**) medial; and (**F**) distal views; and (**E**) cross section of the shaft. **Abbreviations**. ct, central tubercle for the *lig. meniscotibiale intertarsi*; df, distal fossa; eg, extensor groove; fi, surface for fibula; ri, proximodistally extended lateral ridge; rt, possible lateral retinacular tubercle; ts, ridge representing medial retinacular tubercle. Scale bar: 5 cm.

Locality and horizon. The specimens come from old collections at the La Plata Museum, and thus, collecting data are scarce. MLP 20-110, originally referred by Moreno and Mercerat [3] to *Rostrornis floweri* (a junior synonym of *B. burmeisteri*; [1,9,12]), comes from the Santa Cruz Formation (Middle Miocene), at Santa Cruz province; more details on provenance are not available [3,30]. MLP 20-581 only figures in the catalogue as "?*Liornis* sp." without any additional data. However, it is possible to infer that it corresponds to the distal end of tibiotarsus mentioned, and was briefly described by Dolgopol de Sáez [10]. If this is the case, MLP 20-581 was collected by Federico Berry in the Santa Cruz Formation (Middle Miocene) in Santa Cruz province.

3. Description

MLP 20-110 and MLP 20-581 represent the incomplete distal end of tibiotarsi lacking distal condyles. Because both materials are similar in all features, the description is based on the most complete individual (MLP 20-110) and is complemented in some cases by MLP 20-581.

The tibiotarsus shows a nearly straight shaft that is proximally ellipsoidal in cross-section, with convex anterior and posterior surfaces. Distally, the anterior surface of the bone becomes transversely flat. Although poorly preserved, the distal intercondylar fossa is transversely expanded and weakly undercuts the proximal margin of the distal condyles, forming a shallow transverse ridge of bone. Although distal condyles are abraded, they appear to be not strongly posteriorly extended. The posterior *trochlea cartilaginis tibialis* is poorly-defined, and is dorsoventrally low, with shallow delimiting crests. The extensor groove (*linea extensoria* in Buffetaut [33]) is aligned with the medial condyle, it is transversely wide and is poorly delimited by very shallow ridges of bone. Although there is no bony bridge on the extensor groove, there is a well-developed ridge of bone on the medial surface of the groove that indicates the insertion of a tendinal sling, which represents a low retinacular tubercle. Limiting the lateral surface of the distal end of the extensor groove there is a pyramidal bump (tubercule central of Ameghino [9]; central tubercle of Buffetaut [19,33]; attachment of the *lig. meniscotibiale intertarsi*; Zinoviev [28]), that is indistinguishable in size and shape from the ascending process of the astragalus [34] fused to the tibia and present in some ratite birds (e.g., *Rhea, Aepyornis*; see [33]) and basal ornithurines [35]. The distal crest for the attachment of the transverse ligament appears to be absent. In medial view, the shaft is smoothly convex, whereas in lateral view it shows a prominent proximodistally extended, sharp and acute bony crest.

4. Discussion

4.1. Comments on the Genus Liornis Ameghino, 1895

The genus *Liornis* was erected by Ameghino with the aim to include the single species *L. floweri* [2]. The material on which Ameghino based his species was the incomplete distal end of tibiotarsus, tarsometatarsus, and pedal phalanges of a single individual (Figure 4). Ameghino [9,36] assigned it to the Phorusrhacoidea, and distinguished *Liornis* from other terror birds by having the tibiotarsus with anteroposteriorly compressed and transversely expanded shaft, flat anterior surface of the distal shaft without deep muscular ridges and scars, poorly defined extensor groove, and absence of supratendinal bridge. The tarsometatarsus was characterized by its wide and anteroposteriorly compressed shaft and the absence of impression for the hallux. Due to these unique features, Dolgopol de Sáez [10], in his overview of phorusrhacoid birds, considered it as a valid genus, probably related to the genus *Brontornis* within the Brontornithidae (considered by that author as the Order Brontornithes). Kraglievich [6,11] retained *Liornis* as a valid taxon, and considered that due to its hindlimb proportions, it must be distinguished from *Brontornis* at the subfamily level at least, and thus, established the ad hoc subfamily Liorninae within the Brontornithidae [6]. More recently, Brodkorb, in his renowned *"Catalogue of fossil birds"* [12] synonymized *Liornis floweri* to *Brontornis burmeisteri* without discussing this in detail, a point of view followed by Tonni [1] among other authors. Later, Alvarenga [37] and Alvarenga and Hofling [2] considered that *Liornis* was a synonym of *Phorusrhacos longissimus*, but they did not discuss this point of view in detail; a criterion was followed by Bertelli et al. [38] and Alvarenga et al. [23]. More recently, Buffetaut [18,19,33] analyzed the materials of *Liornis floweri* and included it as a junior synonym of *Brontornis burmeisteri*. Ameghino [9] noted that the tarsometatarsus of *Liornis* differs from *Brontornis* in lacking any sign of scar for the hallux and because its tibiotarsus lacks a supratendinal bridge. However, both differences appear to be misinterpretations, probably due to the paucity of available specimens at that time. The presence of a hallux scar proposed by Ameghino (and followed by Agnolin [8]) was most probably a mistake based on artifact bone preservation. As noted by Buffetaut [19], there is no evidence of such scar in any of the available *Brontornis* and *"Liornis"* specimens. Further, the presence of a supratendinal bridge in all available specimens cannot be corroborated; instead, a pyramidal-shaped tubercle delimiting the extensor groove is present [19]. In this way, the differences reported by Ameghino between *Brontornis* and *Liornis* are not valid, and thus, *Liornis* should be considered its junior synonym, following previous authors.

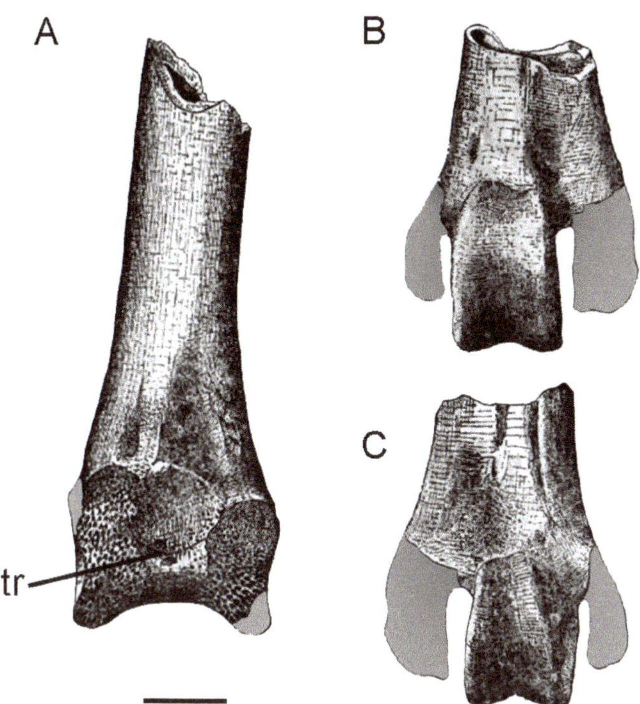

Figure 4. *Brontornis burmeisteri*; specimens on which Ameghino [9] based the species *Liornis floweri* Ameghino, 1895 (NHMUK PV A9058 and NHMUK PV A580). (**A**) right distal half of tibiotarsus in anterior view; (**B**,**C**) distal end of left tarsometatarsus in (**A**) anterior; and (**B**) posterior views. (**A–C**) modified from Ameghino (1895). **Abbreviations**. tr, transverse ridge. Scale bar: 5 cm.

In sum, the tibiotarsus of *Brontornis* (including *Liornis*) clearly departs from that of phorusrhacids (and most birds; Figure 5) in a unique combination of characters, including distal end strongly anteroposteriorly compressed with its lateral margin forming an acute ridge, strongly medially oriented medial condyle, small and rounded distal condyles that are joined by a transversely oriented ridge, low and poorly defined *trochlea cartilaginis tibialis*, absence of a supratendinal bridge, poorly excavated extensor groove that is medially tilted, feebly developed retinacular tubercles, and the presence of a prominent pyramidal-shaped prominence for attaching the *lig. meniscotibiale intertarsi* [3,10,19,33]. With this new evidence at hand, I re-scored the tibiotarsus of *Brontornis* in the Worthy et al. [24] data matrix (see below).

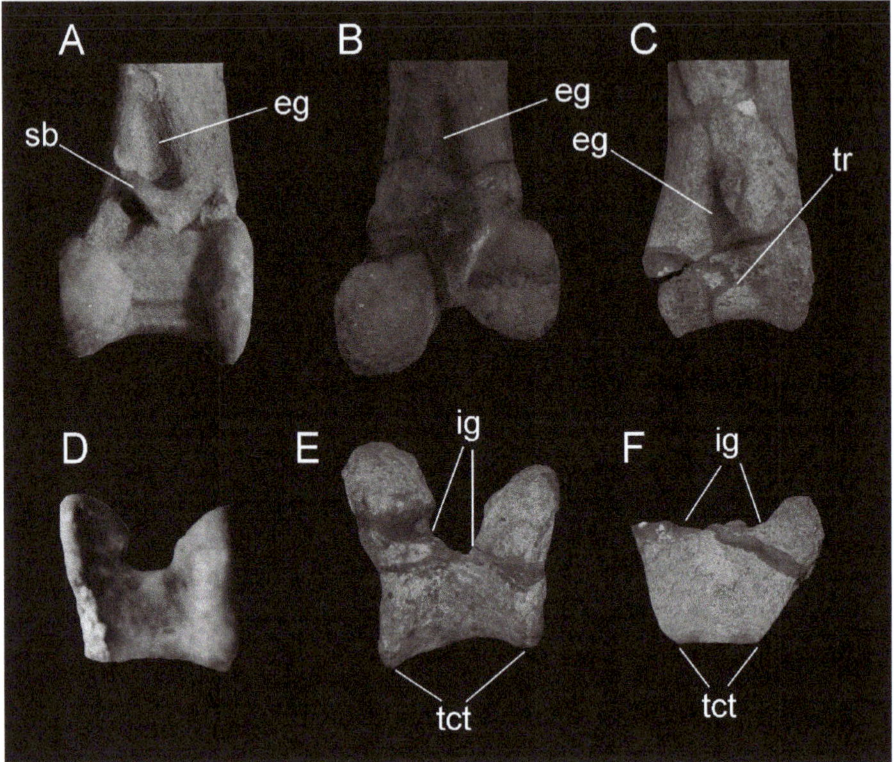

Figure 5. Distal end of left tibiotarsus of (**A,D**) *Onactornis pozzii* (MACN Pv-6554); (**B,E**) *Brontornis burmeisteri* (MLP 20-92/93); (**C,F**) *Brontornis burmeisteri* (MLP 20-110) in (**A–C**) anterior; and (**D–F**) distal views. **Abbreviations.** eg, extensor groove; ig, anterior intercondylar groove; sb, supratendinal bridge; tct, *trochlea cartilaginis tibialis*; tr, transverse ridge. Not to scale.

4.2. The Quadrate Bone Referred to Brontornis burmeisteri Moreno and Mercerat, 1891

Skull material referred to as *Brontornis* is very scarce and consists of isolated and incomplete jaws, as well as a single and incompletely preserved right quadrate bone [2]. The latter was ambiguously associated with *Brontornis* remains [3], and as such, it was not included in their data matrix by Worthy et al. [24], a criterion with which I concur.

In any way, this quadrate shows several features that are worth analyzing. This element was interpreted by Agnolin [8] as having only two condyles, constituting an important piece of evidence for galloanserine affinities of *Brontornis*. However, Agnolin misinterpreted the quadrate bone anatomy of *Brontornis* as demonstrated by Worthy et al. [24]. The later authors compared the quadrate with that of the phorusrhacid *Tolmodus* and found some similarities, including the presence of three quadrate condyles. In the view of these authors, the quadrate indicates that *Brontornis* belongs to Neoaves and not to Galloanseres.

However, the *Brontornis* quadrate (MLP 20-111; Figure 6) is different from the homologue of any known bird, especially with those of phorusrhacoids such as *Tolmodus* and *Patagorhacos* [39,40].

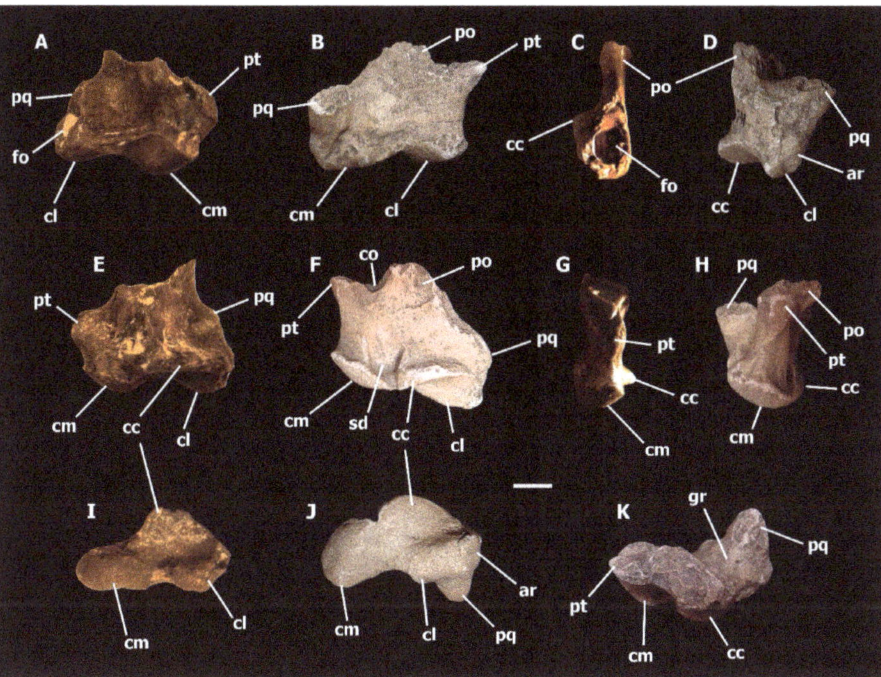

Figure 6. Comparisons between the quadrate referred to *Brontornis burmeisteri* (**B,D,F,H,J,K**; plaster copy of MLP 20-111) and the phorusrhacid *Patagorhacos terrificus* (**A,C,E,G,I**) in (**A,B**) anterior; (**C,D**) lateral; (**E,F**) posterior; (**G,H**) medial; (**I,J**) distal; and (**K**) proximal views. **Abbreviations**. ar, extended articular surface; cc, caudal condyle; cl, lateral condyle; cm, medial condyle; co, concave surface separating the pterygoid and orbital processes; fo, quadratojugal fossa or fovea; gr, wide groove separating the quadratojugal process and the orbital process; po, orbital process; pq, quadratojugal process; pt, pterygoid process; sd, supracondylar depression or fossa. Scale bar: 2 cm for B,D,F,H,J,K; 1 cm for A,C,E,G,I.

The distal end of the quadrate shows two well-defined condyles that are relatively elongate and differently oriented from that of phorusrhacids (Figure 6I,J). Worthy et al. [24] recognized the existence of a caudal condyle. However, I am not certain about the homology of this structure. At first, in contrast with phorusrhacids and other birds, this "caudal condyle" is represented by a shelf-like prominence that is dorsally positioned with respect to the distal condyles and shows a flattened to slightly concave distal "articular" surface. This condition is very different from that known in most other birds, such as in *Patagorhacos*, in which this condyle is at level with the medial and lateral condyles and is notably convex (Figure 6I). A bony flange somewhat similar to that present in *Brontornis* is exhibited by dromornithid anseriforms [41]. In *Brontornis* the "caudal condyle" is medially separated from the medial condyle by an oval-shaped and well-defined supracondylar depression that is unique to this taxon.

The pterygoid condyle is represented by an acute and prominent process that differs from that of most birds, including phorusrhacids, in which it is represented by a rounded articular surface. In pseudodontornithids and some anseriforms such as *Anseranas* and *Dendrocygna*, this condyle is also represented by a prominent and relatively acute process [31,42]. It is separated from the orbital process by a well-defined concave surface that is only represented by its base.

A particular trait of *Brontornis* is its unique and massive pyramidal-shaped quadratojugal process that is very different from the condition reported for most birds. Further, there is no evidence of a quadratojugal fossa or fovea, contrasting with the condition of most birds. Remarkably, the presence of a robust quadratojugal process and the absence of a quadratojugal fovea are features only known in conjunction in dromornithid anseriforms [41,43] and some ratites [43]. Further, *Brontornis* quadrate lacks any sign of pneumaticity, resembling also in this aspect dromornithids and ratite birds [43].

In sum, the quadrate of *Brontornis* is very apomorphic and is not matched by any known bird. The existence of a third condyle, the "caudal condyle" is somewhat dubious. The morphology of this condyle clearly departs from that of other birds, and because of its position it is possible that it does not contact the mandible; it is very similar to a bony flange present in dromornithid anseriforms. Further, as remarked by Worthy et al. [24] the association of this bone with those unambiguously belonging to *Brontornis* is not clear.

Characters modified from Worthy et al. (2017)

As indicated above, and based on the detailed review of new specimens, several postcranial features of *Brontornis* should be reinterpreted, and this has impact on the codifications carried out by Worthy et al. [24]. As follows, we discuss the changes made on *Brontornis* scorings.

Femur

ch#213. *Brontornis burmeisteri* re-scored from 1 to 0. As observed in the femur of *Brontornis* the patellar groove of the distal end of femur is notably transversely wide (see pl. III Figure 1 in Moreno and Mercerat [3]), being much wider than the lateral condyle. In this way, I re-score *Brontornis* as 0.

Tibiotarsus

ch#240. *Brontornis burmeisteri* re-scored from 1 to 0. As indicated in the description above, there exists a pyramidal-shaped prominence at the lateral surface of the extensor groove that represents the attachment for the lig. meniscotibiale intertarsi of Zinoviev [28]. The presence of such prominence is uncommon among birds (it is present in some flightless ratites as *Emeus*, among others) and may be considered an autapomorphic feature of *Brontornis* (see [19]; Figure 3).

ch#246. *Brontornis burmeisteri* re-scored from ? to 0. In the distal end of tibiotarsus MLP 20-581 the groove for the *m. fibularis* is anteriorly extended, as shown by the concave impression located at the lateral surface of the extensor groove (Figures 2 and 3).

ch#247. *Brontornis burmeisteri* re-scored from ? to 0. In the distal end of tibiotarsus MLP 20-581 a ridge located adjacent to the extensor groove, represents in all probability the lateral retinacular tubercle (Figure 3).

ch#248. *Brontornis burmeisteri* re-scored from 1 to 0. This character is somewhat difficult to score, especially because of the absence of a supratendinal bridge in *Brontornis*. However, as can be extrapolated from the distal end of tibiotarsus MLP 20-581, the distal aperture of the extensor groove shows a subvertically oriented main axis, and thus, it is here scored as such (Figure 3).

Tarsometatarsus

ch#253. *Brontornis burmeisteri* re-scored from 1 to 0. In a completely preserved *Brontornis* tarsometatarsus (FM-P13259) the lateroplantar margin of the cotyle is notably dorsally projected, and consequently, it is re-scored as 0 (Figure 7). This character was previously considered by Bourdon [44] as a synapomorphy of the clade Anseriformes + Pelagornithidae.

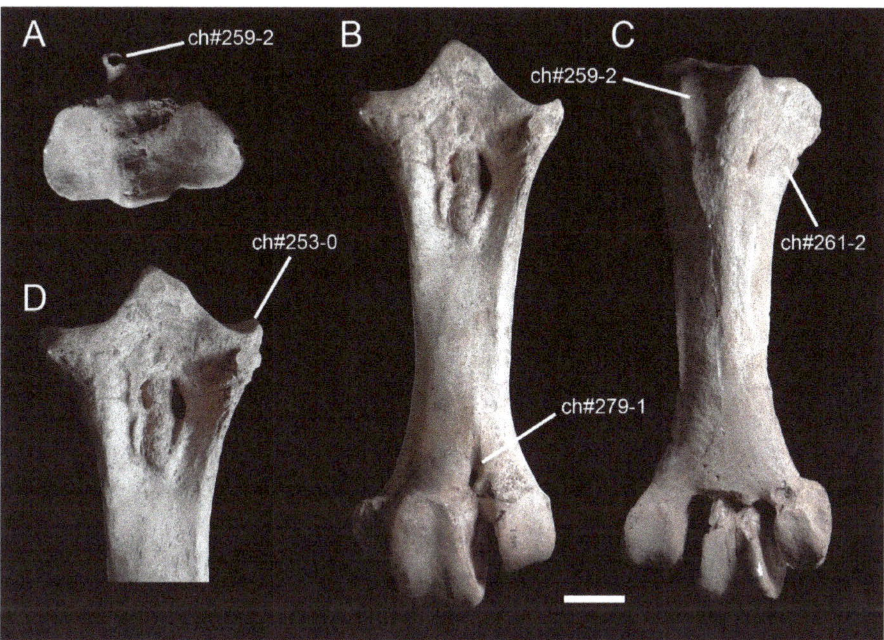

Figure 7. Left tarsometatarsus of *Brontornis burmeisteri* (plaster copy of FM-P13259) in (**A**) proximal, (**B**) anterior, and (**C**) posterior views. (**D**) detail of the proximal region of the tarsometatarsus in anterior view. The notation includes the characters that were modified from Worthy et al. [24]. ch#253-0, plantar-lateral side of *cotyla medialis* elevated proximally; ch#259-2, hypotarsus with two ridges; ch#261-2, surface from medial calcaneal ridge to anterior margin of medial shaft concave (shallow fossa parahypotarsalis medialis); and ch#279-1, *foramen vasculare distale* small and distinct. Scale bar for A to C: 4 cm; D, 3.5 cm.

ch#254. *Gastornis parisiensis* re-scored from 1 to 0. A complete tarsometatarsus of *G. parisiensis* described and illustrated by Martin [45], Buffetaut and Angst [46], and Mourer Chauviré and Bourdon [47] clearly showed that the intercotylar eminence of the tarsometatarsus in this taxon was prominent and proximally extended. In addition, *G. geiselensis* (a species closely related or even a synonym of *G. parisiensis*) shows prominent intercotylar eminence [48].

Gastornis giganteus was re-scored from 1 to ?. The tarsometatarsus of *G. giganteus* is known by fragmentary material with eroded intercotylar prominence [49,50]. Because of that, the morphology of this eminence in *G. giganteus* is considered as unknown.

ch#259. *Brontornis burmeisteri* re-scored from 4 to 2. Worthy et al. [24] consider the block-like hypotarsus as a derived trait shared between *Brontornis* and phorusrhacoids. In the same line of thought, Alvarenga and Hofling [2] include as diagnostic of phorusrhacoids a block-like hypotarsus that is subquadrangular in proximal view and subtriangular in posterior view, lacking crests and grooves. However, as recognized by Worthy et al. [24] the hypotarsus of *Brontornis* is distinctive and very different from the condition exhibited by phorusrhacids, (e.g., *Phorusrhacos, Tolmodus* [9,24,49]). In *Brontornis* the hypotarsus in proximal view is subtriangular in contour, showing a prominent and thick medial crest, and a slightly pronounced lateral edge, both separated by a longitudinal tendinal groove (Figure 7). This morphology is indistinguishable from that of *Gastornis* [8,45,49], and thus, is codified as such (state 2).

ch#261. *Brontornis burmeisteri* re-scored from 3 to 2. *Brontornis* was scored as having a flat or convex surface between the medial calcaneal ridge and the medial margin of the shaft. However, in *Brontornis* (FM-P13259) there exists a notable concave surface medial to the medial calcaneal ridge (Figure 7), and thus is scored as "2".

ch#271. *Gastornis giganteus* re-scored from 0 to ?. The incomplete nature of the distal tarsometatarsus of *Gastornis giganteus* precludes the clear recognition of a surface for articulation with digit I. In this way, this character is coded as "?".

ch#279. *Brontornis burmeisteri* re-scored from 3 to 1. The tarsometatarsus of *Brontornis* was scored as lacking a distal vascular foramen by Worthy et al. [24]. However, such foramen is present in available specimens [9,19] (Figure 7).

Finally, I included in the data matrix the codifications of characters from 280 to 283. These refer to the shape of pedal phalanges and were scored by Worthy et al. [24] as "?". Probably, Worthy et al. [24] did not include these scorings because there was no direct evidence indicating the phalanges previously referred to *Brontornis* unambiguously belong to this taxon. However, two phalanges are preserved in the single associated specimen on which *Liornis floweri* (a junior synonym of *Brontornis burmeisteri*) is based (NHMUK PV A580) [9,19]. These phalanges are massive, transversely wide, and ventrally flat, a combination of features that are exhibited by phalanges previously referred to *Brontornis* [2,3,9]. In this way, the specimen NHMUK PV A580 confirms previous referral of pedal phalanges to *Brontornis*, and thus, are coded in the data matrix as such.

4.3. Phylogenetic Results

With the aim to test Worthy et al.'s [24] analysis, only hindlimb material was included in the data matrix. Worthy et al. [24] did not include in their work several bones that have doubtful association with material unambiguously belonging to *Brontornis*. These materials include vertebrae [3,9], quadrate [3], and mandible [2,3,9,18]. These elements, particularly the mandible, show several features reminiscent to giant galloanseres such as dromornithids and *Gastornis* [8,18], and their inclusion in the data matrix may give additional support to the galloanserine affinities of *Brontornis*. In any case, it is preferred to exclude the codification of these elements in the data matrix following Worthy et al. [24].

The phylogenetic analysis here performed resulted in the nesting of *Brontornis* among Anseriforms in a clade formed by gastornithids and dromornithids, in a position similar to that proposed by Agnolin [8] (Figure 1). It is worthy of mentioning that forcing the position of *Brontornis* as a cariamiform results in a tree 1569 in length, having five additional steps.

The clade grouping dromornithids and gastornithids was named by Worthy et al. [24] as Gastornithiformes, to which, based on present analysis, *Brontornis* may belong. In any case, this clade formed by giant graviportal fowls is sustained almost by hindlimb features (characters 202, 211, 215) and it is not improbable that this group may be the result of convergent features related to graviportality (see discussion in [24]). In their work, Worthy et al. [24] concluded that *Brontornis* resolved as sister to Cariamiformes, but with very low support. They recognized that *Brontornis* was very different from other birds, and indicated in several parts of the text that the position of *Brontornis* in the phylogenetic tree is unstable. Because I agree with Worthy et al. [24] in that *Brontornis* is still incompletely known, it is possible that its inclusion within Gastornithiformes is not strongly warranted.

Worthy et al. [24] listed some similarities shared by the hindlimb of *Brontornis* and phorusrhacids, including a lateral excavation at the medial surface of the lateral condyle of femur, and a block-like hypotarsus. The first condition is known to occur in *Gastornis* and dromornithids [50,51], suggesting that it is not only exclusive of phorusrhacids, but is also widespread among giant anseriforms. On the other side, as indicated above and as recognized by Worthy et al. [24], the morphology of the hypotarsus of *Brontornis* is very different from that of phorusrhacids, being very similar to the condition exhibited by *Gastornis* and dromornithids [19,41] (see above, analysis of character 259). Both in *Gastornis* and *Brontornis*, the hypotarsus is subtriangular-shaped in proximal view, with a prominent medial crest and a reduced lateral edge. Further, Worthy et al. [24] recognize that mandibular and hindlimb shape structure of *Brontornis* differ substantially from phorusrhacids. The same seems to be true for the quadrate bone, as indicated in the descriptive section of the present contribution.

5. Conclusions

A review of the character codifications for *Brontornis burmeisteri* carried out in the comprehensive work of Worthy et al. [24] resulted in a change in the phylogenetic position of this taxon. This change argues against the sentence of Worthy et al. [24] that declares that it was conclusively shown that *Brontornis* is not a galloanserine bird. After few changes in the data matrix, *Brontornis* results as part of a clade composed by the giant anseriforms designated by Worthy et al. [24] as Gastornithiformes. This result is in agreement with recent proposals that excluded *Brontornis* from phorusrhacoid cariamiforms (where it was traditionally nested) and included it among Anseriformes [8,16,18].

Graviportal anseriforms of the clade Gastornithiformes (*sensu* [24]) are represented by Eurasian and North American Paleogene *Gastornis* and kin [17] and by Paleogene and Neogene Australasian members of the Dromornithidae [41,52]. To these, it should now be added the Paleogene-Neogene brontornithes from South America [17,33]. If this phylogenetic grouping is correct, a widespread radiation of giant anseriforms occurred along several landmasses during the Paleogene. The paucity of the fossil record of these giant birds still precludes a detailed framework to understand the palaeobiogeographic history of these birds.

Finally, the nesting of *Brontornis* among herbivorous giant anseriforms [5,17,24,46], together with several aspects of its mandibular morphology [8,18] reinforces previous thoughts that *Brontornis* was herbivorous in habits (Figure 8).

Figure 8. Life reconstruction of *Brontornis burmeisteri*. Artwork by Agustín Agnolín.

Funding: This research received no external funding.

Institutional Review Board Statement: Not applicable.

Informed Consent Statement: Not applicable.

Data Availability Statement: Not applicable.

Acknowledgments: Special thanks to M.D. Ezcurra and A.G. Martinelli (MACN), and M. Reguero and M. de los Reyes (MLP) for allowing to study material under their care, and for their help during the revision of paleontological collections. E. Buffetaut shared comments regarding the status and affinities of *Liornis*. I thank T. Worthy for discussions about phylogeny and anatomy of basal birds. Special thanks to A. Agnolin for the life reconstruction of *Brontornis* in Figure 1. Special acknowledge to E. Buffetaut and D. Angst for inviting me to participate on the Special Issue "Evolution and Palaeobiology of Flightless Birds". I thank two anonymous reviewers whose comments improved the quality of the present work. This contribution was financed by PICT 2018-01390 (ANPCyT).

Conflicts of Interest: The authors declare no conflict of interest.

Appendix A

Data matrix modified from Worthy et al. [24]. The number of characters and taxa (290 and 48, respectively) as well as character description and states follow Worthy et al. [24]. Scoring modifications are almost restricted to *Brontornis* and *Gastornis* species (see above).

Tinamus_robustus
20001000100220000000000001000012000000000- 330000000?001111100001001001000000
00222211002111110012003111001100100100021-110212012110012121111010110101130001
01000210000010100100000220021200001000000000101101000111000010002200000001001
01100011001100010111000110012200311021010010110101100020012000

Vegavis_iaai
?????????????????????????0?????????????????1001????????????????1????0??????011???0?????????
0010001001000010100?10?????0000-1000000001220-000?000???????????01102011210101?010
1021?0????20?????001000000100120011004011010001120011010100200?????????00100??1????
???12200????????????????????????????111

Chauna_torquata
20001111000110110211010110011011110111111100101011110100000001000300010220001010
0001001011000021001000000010010110111-1000000000100000-00000[30]0001100011011111
01200200- 0100000110110000000011000000100010010001010000001110000100011000011010
000010121021000020011000122021100101000010010200000010?101

Anhima_cornuta
20001110000110100211010100011001010101111100101011110100010110003000102200100100
01000210100001101100000000100110111-1000000000?00020-001000000110001101111011200
100-011000110001100110011000000010001100100101001001011101000010100001101001
1010121021000020011001120022100101000000101000000102101

Wilaru_tedfordi
??
011000100130011011000001101001 0-0000000111020-000001110110003001100020101002011 01
0002????????????????110100001100100100001011011000100 01?????????000000101110210000
200010010?20312001010010100102 00??????2??1

Presbyornis_pervetus
00111?110112?11010110?01??0221??1000????1???11001110110111011103100 11?1??????0???1
0000001100010013000101 10000??1?10010-0000000101010-0000010001 110030011100211 0210
101001011201020??1001100000100001100100101101001011000110001100010000012000001 110 1110
21000020001001200011 1001001 10020010200????3?2101

Anseranas_semipalmata
0110121100011?10101 1010010022102110111111022110110101000000010113100112010111 11
0011000010000001100200000011000111000010000000000000000-00000000011000000011 10010
101101000000000011100110111000011000000001001000010000111000000001000001 10100001
0011102100002000000000011110000100011000020000002021-0

Dendrocygna_eytoni
211112111002021010110100100221021101111110221101111010001001011310011101001011
111000011110002100110000111201011001011101010000100000-00010110011000001100011
2020010000000110111101201100011100000000010010000101101100100001110000110100000
001210210000200100002101311001011110100112000000302??0
Cereopsis_novaehollandiae
211112111012211110110100002200211011111100211011101100010110113110112000010101
111000210000110211001101100101101100100000000100000-00000110011100000111011020
101200101001200110102011001011000010011001101020010100000100111100011100100000
121021000020000000120121101101111011011100000030 2100
Anser_caerulescens
011112111002101010110101100221021101111110221101111010001001111311011201001 01011
011010101000210211000101100001101101100001000010000 0-0000111001100020011001 1120
101101102000200120002021000011000001000100111002001011001010011100000100000 0 001
012102100000000100012200311001012010210111000000302100
Malacorhynchus_membranaceus
021112111102221010111101100221021101111110221101111010001011113100111010010 0111012
10001110110021000300101111010110010111100100001112 20-0001112001110021011001 11202
101110020012101200120211001110000010001101000020110100000000111100011010000 0 00
121021000010000001200031100001111021011200000?0302100
Tadorna_tadornoides
211112111002111010111100002210211011111110221111111010001011113110111010110 1012
100012101100210001000101100101100101110001001011011 0-00011120011100201110011 12
010011010101020112001202100001100000100010011100200101000101000111000111100000
101210210000100000002200311000011110210112000000302100
Leipoa_ocellata
210010001000111001010101000120110000?010100200001101010111010000200110000001 0102
322110220111112020011110000010000202 1-0102120121110011002100100100012000000 1200
210100010120000000121111011001100100010101101000000010000001000001010011110001 0
11112001011010001111110110001101100020000021000210 22??0
Megapodius_reinwardt
210010001000101000010101000120110000000101012000011010101110100020011000001010 2
322110220111112020011110000011000202 1-0102120121110011002100100101012000000120 0
21010001012011000012011001100110010000010110000000010000001000001010010100001
01112001011011001111110110001100100020000020000210 22??0
Eulipoa_wallacei
010010002000101000010101000120110000000101012000011010101110100200?100000101 ?
23221102201101120210111000000110112021-0102120121110111002100100101012000000 11
102102000010120100000220001011001101100010001001000001010010000100101011010100
00101011001011011001111010111001100100020000021000110 ?????0
Megapodius_eremita
220010001000101000010101000120110000000101012000011010?011101000020?110000010
11232211022011101202001111000000110002021-0102120121110011002100100 1010120000
001201210000010120100000120010011001100100000010110000000100100010000010 10
0100000010111200101101100111110101100011001000200000 2000021022110
Alectura_lathami
22001000100010100101010100012011000000101012000011010101110100020011000 0010
1023211102201111120200111100000100002021 0102110121110111002100100 1010120000
001200210100010120100000121111011001100100000010110000000010000010000010 10
0111000010101200101101100011111011000010110002100002000021022100
Talegalla_fuscirostris
220010011000101000010100000120110000????100200001101010111010002001100000 10
1023211102201111120210111000000100002021-0202110121110211002100100 1010120000

001200210000010110100000121110011001100001000101101000000010010001000101010
011010001011200101101100011112011000100110002000001100001022??0

Macrocephalon_maleo
120010012000101001010100100120110000001010120000110101111101000200?10000010
102322110220110112021011000000110012021-01021101211102110021001001000120000
001100210100010120100000121101?011?01100000000010010000010100100010001000010
010000001010120010110110001111101111000100021000021000010?2??0

Gallus_gallus
200010000101121001010101010120110000001010120000110101011111000300210000
10102[26]221102201110120201111101000110002021-1102121121110221112100011
101012000010120021010001012110000012000101000110000001000110100001011010
000110000101001100[9]0001010110010110100011011011110000110101100002100000
022100

Phasianus_colchicus
210010000101011001010101000120110000????1012000021010001111100030001000011
11020221102201110120200111100000100002031-110212112111002111210011110101200
10101201211200010121100001120101010001100100010001201000011110000011200001
010011110011010110010110100011011011211001110100100002000000022??0

Coturnix_pectoralis
210010000001111001010101010120110000???0101200001101010111110003002100000
101?22221102200110120200311101100100002031-131212112111001111210010010101
200101012002102000101210000002201010100010000000100011010000110000000011
000000100110100110101200101101000111111011210000110110100002000000022??0

Acryllium_vulturinum
210010000101001010010101000120110000????101200001101010111110003000100001
101123221102201110120200311101100100002021-1202121101100111112100000101012
0000101200210000010110000000120100010001100000010011101000000010010001000
000010011010001010120010110[30]100111011011110000011001100002000000022??0

Megavitiornis_altirostris
?22010?21000?0?0000?0????0012011??00????1????0001101000100[30]1000200010?000
1?1?2321?01???011012020—1—10-1-0?????–0—1—1–0–1-00100————010—2002?
-000-1—0-01000??01??0?1??1000000001101101001000010101011200000011011110000
10120010110100001111101111101011000201000100100??????0

Crax_rubra
21001000000201100101010000012011000000110120001110111011111000300010000
0[30]01123021102201111120200311100000110002021-12021201211101111121000101
00002200000120121000101012100000012000101110100000001000100100000001010
000110000001001022-0010101200101001100111112011010000110002000002100001
22??0

Ortalis_vetula
20001000000201100101010001012011000000110120011110101011111000300210000
0101123021102201111120200111101000110002031-12021201211102211121000001000
0200100012012100010101210000001200010101011000000100011010000001011000
1000000000010110011010120010100110011101101101?000110002000002100000122
100

Sylviornis_neocaledoniae
?22010022000001000010100100120110001????100210001101?001010100?200?10000
1101?2??2??????11111202————1-0?????1–10–2—1–0—02100————010—20021
1001-10110102?0???????211??10001000111010010000000100000011000001101101
00000001200201101010011111211111010000000001010220001???2??0

Dromaius_novaehollandiae
000100001102201000000001001000002000000000-331010001000010100000100100
301011100222101102–1-0————1—1————————0—00————1————?————

```
————1012001202000001100010100110000010400100100110110000111110111012010 1-
101211-2110011102303220010121101101012110200200-
         Dinornis_robustus
         220000001002201000000010110000110000000000-3310000010?001?1000000001003
0100?100222100021-1————————1————————————————————?————1
01200120201100010010000110000101132011001111000000010001110 11020000-001211
02110111112203120 2-00001001-1-130101?0120??
         Struthio_camelus
         0001000001021010000000000100000200000000 0-3310100000200101000001 0110030
0000000221101011-1————————1-0-1————————————————?————101
01002020000011000101000000000104001021001100000101021021111120001-1-021-211
011010230221111012110-0-1-1210200?200-
         Genyornis_newtoni
         ???21?????????????????????????10????????????? ?0021010?0??00100??????0?00?????
02121000020-10-2————100-1————————0————0————111??
0??0011100010001101100 01?01103201120011101011100111101010010002001121011001 0
101021021202110001010010111112??2???-
         Dromornis_stirtoni
         022110?2?00?001?021?????100??010??????? ?1102110021001000000100210?0?0??0010
??02221000020-10-2————000-1————————0————————1112
00?20211100010000111100012011032011200011010110001111110100110020010210110
0101010230312001100010101101111 12???1??-
         Dromornis_planei
         02211022?00?00110211?100100110100??????? 100211002101?1000??10021000?0??0??????
?????????-1?-2————-???-1————————0————————????????????
???0010000111100 01?01103201120001101011101?1???1010011002001021011001010102
303120?110001010110111112???1??-
         Dromornis_murrayi
         0?2110?2?00?00110211?100000???10????????????? 11002001?0000??100210?0?0?20??????2
121000020-10-2————000-1————————0————————????????????
???001000010110000201103201120011201 01?1???????1010010002001021011001010102 3
031202?100010001101111?2???1??-
         Ilbandornis_woodburnei
         0?2??????00?00110211?1001001?010???????? ?100211002101??0?1??1002?0?0?0??0??????
2211000020-?0-2————?—-1————————0————————011?0
0??0211100010000101100 00?0110320112001110101110111110101001000200102101200 10
101023031203111001010010121112???0??-
         Ilbandornis_lawsoni
         ??????????????????0211??????0?????????????? 1002?1002101??0?1??1??????????0?????????
?????0-?0-2————?—-1————————0————————????????????
???00100001011010 0?011032011210111010111011112110100110020010210120010101 02
3031202101101010[30]10011??????0??-
         Barawertornis_tedfordi
         0?2???????????1?0???????100??010???????? ?002??????????????1000100000??0?????????
???????-??-?————?—-?————————?————————??????????
????001000010?1010 0?011030011200111010111001112010100?100200102?012001010 102
30?110311?001000?10111??????0??-
         Lithornis_promiscuus
         000000000102200?000?0?????00001?00000000 0033?0100010?00101000101001 00?200101
10011100122011010103030010010001010002 1-0000000000100120-0220010102100030 10
0112002?1100010000101000?200212011?11100000000 01100000111000010002210000011
01011011111011110000 0-00001111223011200001000010000200?????????0
         Lithornis_plebius
?????????????000000?010000000 01?????????00331?000000001101 0?????????????2?010?10?
```

????????0110101030300110101210???0?1-1000000000100120-0211010102100030000112 0
02?2100010100101?00??00??200??111000000000110000011101001000221000001001111 01
01110111100000-00011111223011200101000010000200?????????0
Paracathartes_howardae
00000????????000000???000?????????????033?000?000?0110100??????1?????0??0????? ????????1010102030010010021 0???0???1000000000?00120-022101000?100000000111202
121010101001??????????????111000001000110002011101001 0??22100000100111100011
10111100000-10000111222011200101000010000200000??????0
Burhinus_grallarius
0001100001120210011001 10001–012011000111100001010002010110000000000
00100111012211001200010001000300111102210111020010000000000001110-00200111
02100130001002002020020100001110002001200110010000000011101002000000010100
1011100010000 2202-00010011101011102000100122103210010111101111021100000?2100
Porphyrio_melanotus
010110010002201001000100001–002011000111100001011012010010000000100
0100111101221101121001000100000000010001000030210020200000010 00-0021
11200000000011111201102001000000012001000001010001000000000010 12000001
1110011000110000000002101000100012002100002001100010100110010101 0111-0
01000000002110
Antigone_rubicunda
10011000102101001000100011–00201100011110000101100201001000000000 0020?
10100000010000200100012010000000001 0000-031-1000000000000000-10200021021
00000011010111011011000000110010001010111010000000100010000001001001100 0
11001100000210100000101100110010200110012211221001011110200002000000002110
Brontornis_burmeisteri
???
??
????????????????????????0????1001001?010?3101120010100001100?1110100001?000-1112
0000000000112202100010101101001-111102???????
Gastornis_parisensis
?22110???101100?0001?01002100000??????1?????
??????1-10-0———??0??1?????——————0———————————????????
?????001000?0??0?100?011?3?011?0?101?10?0000101100000011012102 10?010000000?02?
022100?0?001010110 11??????????-
Gastornis_giganteus
022110120001?010010101001?0020100101????100211101100?10011?100210000000011011
0?????????0-10-2————100-1?????——————?———0-011——-?-1-1——21
002??12101110000000011100000201103001100010120000101?0111000001101210200000 00
010?011220212001010 0?000110220102??????-
Tolmodus_inflatus
02201010010210110001010010001010 0111????1?1?10001110?100010000000100002?00010
0001100002111002010–00-00100010?????????????????????????????10001?1000000?1111010
2000000100101011010001 01011??1010000111110000111010111001110000010111 1101001
0000110010100111001001124021102101101100101 10??1???????
Cariama_cristata
0000100101021210010001 00101–002011000111100100011000000010000000002000200
001000211001210110021 1[25]13001011001 10000021-10-0000010001210-02111010010
1110001111012201001010100100001[9]00101011101000000011111 00001110101110011
100010110011201000120011001010002000100122121202101211010010110000 0002??0

References

1. Tonni, E.P. The present state of knowledge of the Cenozoic birds of Argentina. *Contrib. Sci.* **1980**, *330*, 105–114.
2. Alvarenga, H.M.; Höfling, E. Systematic revision of the Phorusrhacidae (Aves: Ralliformes). *Pap. Avulsos Zool.* **2003**, *43*, 55–91. [CrossRef]
3. Moreno, F.P.; Mercerat, A. Paleontología Argentina I. *Catálogo Pájaros Rep. Argent. Conserv. Mus. Plata* **1891**, *1*, 8–71.

4. Angst, D.; Chinsamy, A. Ecological implications of the revised locomotory habits of the giant extinct South American birds (Phorusrhacidae and Brontornithidae). *Contrib. Mus. Argent. Cienc. Nat.* **2017**, *7*, 17–38.
5. Angst, D.; Buffetaut, E.; Lecuyer, C.; Amiot, R. A new method for estimating locomotion type in large ground birds. *Palaeontology* **2015**, *59*, 217–223. [CrossRef]
6. Kraglievich, L. Una gigantesca ave fósil del Uruguay, *Devincenzia gallinali* n. gén. n. sp., tipo de una nueva familia Devincenziidae del orden Stereornithes. *Anal. Mus. Hist. Nat. Montev.* **1932**, *2*, 323–353.
7. Tonni, E. El rol ecológico de algunas aves fororracoideas. *Ameghiniana* **1977**, *14*, 316–317.
8. Agnolín, F.L. *Brontornis burmeisteri* Moreno & Mercerat, un Anseriformes (Aves) gigante del Mioceno medio de Patagonia, Argentina. *Rev. Mus. Argent. Cienc. Nat.* **2007**, *9*, 15–25.
9. Ameghino, F. Sur les oiseaux fossiles de Patagonie. *Bol. Inst. Geogr. Argent.* **1895**, *15*, 501–602.
10. de Sáez, M.D. Las aves corredoras fósiles del Santacrucense. *Anal. Soc. Cient. Argent.* **1927**, *103*, 145–160.
11. Kraglievich, L. Contribución al conocimiento de las aves fósiles de la época arauco-entrerriana. *Physis* **1931**, *10*, 304–315.
12. Brodkorb, P. Catalogue of fossil birds. Part III (Gruiformes, Ichthyornithiformes, Charadriiformes). *Bull. Flo. State Mus.* **1967**, *11*, 99–220.
13. Lambrecht, K. *Handbuch der Palaeornithologie*; Gebrüder Borntraeger: Berlin, Germany, 1933; 988p.
14. Patterson, B.; Kraglievich, J.L. Sistemática y nomenclatura de las Aves Fororracoideas del Plioceno Argentino. *Publ. Mus. Cienc. Nat. Tradic. Mar. Plata* **1960**, *1*, 1–52.
15. Agnolin, F.L. *Sistemática y Filogenia de las aves Fororracoideas (Gruiformes, Cariamae)*; de Azara, F., Mazzini, V., Eds.; Fundación de Historia Natural: Buenos Aires, Argentina, 2009; p. 82. Available online: https://www.fundacionazara.org.ar/img/libros/sistematica-y-filogenia-de-aves-fororracoideas-ok.pdf (accessed on 19 January 2021).
16. Agnolin, F. La posición sistemática de *Hermosiornis* (Aves, Phororhacoidea) y sus implicancias filogenéticas. *Rev. Mus. Argent. Cienc. Nat.* **2013**, *15*, 39–60. [CrossRef]
17. Angst, D.; Buffetaut, É. *Palaeobiology of Giant Flightless Birds*; Elsevier: Amsterdam, The Netherlands; ISTE Press: Great Britain, UK, 2017; 282p.
18. Buffetaut, E. Tertiary ground birds from Patagonia (Argentina) in the Tournouër collection of the Muséum National d'Histoire Naturelle, Paris. *Bull. Soc. Géol. Franc.* **2014**, *185*, 207–214. [CrossRef]
19. Buffetaut, E. A reassessment of the giant birds *Liornis floweri* Ameghino, 1895 and *Callornis giganteus* Ameghino, 1895, from the Santacrucian (late Early Miocene) of Argentina. *Palaeovertebrata* **2016**, *40*, 3. [CrossRef]
20. Mayr, G. *Paleogene Fossil Birds*; Springer International Publishing: Berlin/Heidelberg, Germany, 2009; 262p.
21. Tambussi, C.P. Palaeoenvironmental and faunal inferences based on the avian fossil record of Patagonia and Pampa: What works and what does not. *Biol. J. Linn. Soc.* **2011**, *103*, 458–474. [CrossRef]
22. Tambussi, C.P.; Degrange, F. South American and Antarctic Continental Cenozoic Birds. In *Paleobiogeographical Affinities and Disparities*; Springer: Dordrecht, Switzerlamd, 2013; 114p.
23. Alvarenga, H.M.F.; Chiappe, L.; Bertelli, S. Phorusrhacids: The terror birds. In *Living Dinosaurs. The Evolutionary History of Modern Birds*; Dyke, G., Kaiser, G., Eds.; John Wiley & Sons, Ltd.: London, UK, 2011; pp. 187–208
24. Worthy, T.H.; Degrange, F.J.; Handley, W.D.; Lee, M.S. The evolution of giant flightless birds and novel phylogenetic relationships for extinct fowl (Aves, Galloanseres). *R. Soc. Open Sci.* **2017**, *4*, 170975. [CrossRef]
25. Agnolín, F.L. Posición sistemática de algunas aves fororracoideas (Gruiformes; Cariamae) Argentinas. *Rev. Mus. Argent. Cienc. Nat.* **2006**, *8*, 27–33.
26. Buffetaut, E. Phororhacoidea or Phorusrhacoidea? A note on the nomenclature of the "terror birds". *Ann. Paléontol.* **2013**, *99*, 157–161. [CrossRef]
27. Baumel, J.J.; Witmer, L.M. Osteology. In *Handbook of Avian Anatomy: Nomina Anatomica Avium*; Baumel, J.J., King, A.S., Brazile, J.E., Evans, H.E., Van den Berge, J.C., Eds.; Nuttal Ornithological Club: Cambridge, MA, USA, 1993; pp. 45–132.
28. Zinoviev, A.V. Notes on pelvic and hindlimb myology and syndesmology of Emeus crassus and Dinornis robustus (Aves: Dinornithiformes). In *Proceedings of the 8th International Meeting of the Society of Avian Paleontology and Evolution, Vienna, Austria, 11–16 June 2012*; Verlag Naturhistorisches Museum Wien: Wien, Austria, 2013; Volume 2013, pp. 253–278.
29. Goloboff, P.A.; Catalano, S.A. TNT version 1.5, including a full implementation of phylogenetic morphometrics. *Cladistics* **2016**, *32*, 221–238. [CrossRef]
30. Hospitaleche, C.A.; Tambussi, C.P.; Reguero, M. Catálogo de los Tipos de aves fósiles del Museo de La Plata. *Serie Técnica y Didáctica del Museo de La Plata* **2001**, *41*, 1–28.
31. Ono, K. A bony-toothed bird from the Middle Miocene, Chichibu Basin, Japan. *Bull. Nat. Sci. Mus.* **1989**, *15*, 33–38.
32. Alvarenga, H.M. Paraphysornis novo gênero para *Physornis brasiliensis* Alvarenga, 1982 (Aves: Phorusrhacidae). *An. Acad. Bras. Ciênc.* **1993**, *65*, 403–406.
33. Buffetaut, E. A brontornithid from the Deseadan (Oligocene) of Bolivia. *Contrib. Mus. Argent. Cienc. Nat.* **2017**, *7*, 39–47.
34. McGowan, C. Tarsal development in birds: Evidence for homology with the theropod condition. *J. Zool.* **1985**, *206*, 53–67. [CrossRef]
35. Agnolín, F.L.; Martinelli, A.G. Fossil birds from the Late Cretaceous Los Alamitos Formation, Río Negro Province, Argentina. *J. South Am. Earth Sci.* **2009**, *27*, 42–49. [CrossRef]
36. Ameghino, F. L'âge des formations sédimentaires de Patagonie. *Anal. Soc. Cient. Argent.* **1901**, *51–52*, 65–91.

37. Alvarenga, H.M. Revisao Sistemática das aves Phorusrhacidae. Ph.D. Thesis, Universidade de Sao Paulo, Sao Paulo, Brazil, 1999; 95p.
38. Bertelli, S.; Chiappe, L.M.; Tambussi, C. A new phorusrhacid (Aves: Cariamae) from the middle Miocene of Patagonia, Argentina. *J. Vertebr. Paléontol.* **2007**, *27*, 409–419. [CrossRef]
39. Agnolín, F.L.; Chafrat, P. New fossil bird remains from the Chichinales Formation (Early Miocene) of northern Patagonia, Argentina. *Ann. Paléontol.* **2015**, *101*, 87–94. [CrossRef]
40. Andrews, C.W. On the Extinct Birds of Patagonia-I. The Skull and Skeleton of *Phororhacos inflatus* Ameghino. *Trans. Zool. Soc. Lond.* **2010**, *15*, 55–86. [CrossRef]
41. Worthy, T.H.; Handley, W.D.; Archer, M.; Hand, S.J. The extinct flightless mihirungs (Aves, Dromornithidae): Cranial anatomy, a new species, and assessment of Oligo-Miocene lineage diversity. *J. Vertebr. Paléontol.* **2016**, *36*, e1031345. [CrossRef]
42. Elzanowski, A.; Stidham, T.A. Morphology of the quadrate in the Eocene anseriform *Presbyornis* and extant galloanserine birds. *J. Morphol.* **2010**, *271*, 305–323. [PubMed]
43. Murray, P.F.; Megirian, D. The skull of dromornithid birds: Anatomical evidence for their relationship to Anseriformes. *Rec. South Aust. Mus.* **1998**, *31*, 51–97.
44. Bourdon, E. The Pseudo-toothed Birds (Aves, Odontopterygiformes) and their bearing on the early evolution of modern birds. In *Living Dinosaurs. The Evolutionary History of Modern Birds*; Dyke, G., Kaiser, G., Eds.; John Wiley & Sons, Ltd.: London, UK, 2011; pp. 209–234.
45. Martin, L.D. *The Status of the Late Paleocene Birds Gastornis and Remiornis*; Science Series; Natural History Museum of Los Angeles County: Los Angeles, CA, USA, 1992; Volume 36, pp. 97–108.
46. Buffetaut, E.; Angst, D. Stratigraphic distribution of large flightless birds in the Palaeogene of Europe and its palaeobiological and palaeogeographical implications. *Earth Sci. Rev.* **2014**, *138*, 394–408. [CrossRef]
47. Mourer-Chauviré, C.; Bourdon, E. The *Gastornis* (Aves, Gastornithidae) from the Late Paleocene of Louvois (Marne, France). *Swiss J. Palaeontol.* **2015**, *135*, 327–341. [CrossRef]
48. Hellmund, M. Reappraisal of the bone inventory of *Gastornis geiselensis* (Fischer, 1978) from the Eocene "Geiseltal Fossillagerstätte" (Saxony-Anhalt, Germany). *N. Jahrb. Geol. Paläontol. Abh.* **2013**, *269*, 203–220. [CrossRef]
49. Andors, A. Reappraisal of the Eocene groundbird *Diatryma* (Aves: Anserimorphae). In *Papers in Avian Paleontology Honoring Pierce Brodkorb*; Science Series; Campbell, K.E., Ed.; Natural History Museum of Los Angeles County: Los Angeles, CA, USA, 1992; Volume 36, pp. 109–125.
50. Matthew, W.D.; Granger, W. The skeleton of *Diatryma*, a gigantic bird from the Lower Eocene of Wyoming. *Bull. Am. Mus. Nat. Hist.* **1917**, *37*, 307–326.
51. Rich, P.V. The Australian Dromornithidae: A group of large extinct ratites. *Los Angel. Cty. Nat. Hist. Mus Contrib. Sci.* **1980**, *330*, 93–103.
52. Park, T.; Fitzgerald, E.M.G. A late Miocene–early Pliocene Mihirung bird (Aves: Dromornithidae) from Victoria, southeast Australia. *Alcheringa Aust. J. Palaeontol.* **2012**, *36*, 419–422. [CrossRef]

Review

The Evolution and Fossil Record of Palaeognathous Birds (Neornithes: Palaeognathae)

Klara Widrig [1,*] and Daniel J. Field [1,2,*]

1   Department of Earth Sciences, University of Cambridge, Cambridge CB2 3EQ, UK
2   Museum of Zoology, University of Cambridge, Cambridge CB2 3EJ, UK
*   Correspondence: kew66@cam.ac.uk (K.W.); djf70@cam.ac.uk (D.J.F.); Tel.: +44-(0)1223-768329 (D.J.F.)

Citation: Widrig, K.; Field, D.J. The Evolution and Fossil Record of Palaeognathous Birds (Neornithes: Palaeognathae). *Diversity* **2022**, *14*, 105. https://doi.org/10.3390/d14020105

Academic Editors: Michael Wink, Eric Buffetaut and Delphine Angst

Received: 31 December 2021
Accepted: 27 January 2022
Published: 1 February 2022

Publisher's Note: MDPI stays neutral with regard to jurisdictional claims in published maps and institutional affiliations.

Abstract: The extant diversity of the avian clade Palaeognathae is composed of the iconic flightless ratites (ostriches, rheas, kiwi, emus, and cassowaries), and the volant tinamous of Central and South America. Palaeognaths were once considered a classic illustration of diversification driven by Gondwanan vicariance, but this paradigm has been rejected in light of molecular phylogenetic and divergence time results from the last two decades that indicate that palaeognaths underwent multiple relatively recent transitions to flightlessness and large body size, reinvigorating research into their evolutionary origins and historical biogeography. This revised perspective on palaeognath macroevolution has highlighted lingering gaps in our understanding of how, when, and where extant palaeognath diversity arose. Towards resolving those questions, we aim to comprehensively review the known fossil record of palaeognath skeletal remains, and to summarize the current state of knowledge of their evolutionary history. Total clade palaeognaths appear to be one of a small handful of crown bird lineages that crossed the Cretaceous-Paleogene (K-Pg) boundary, but gaps in their Paleogene fossil record and a lack of Cretaceous fossils preclude a detailed understanding of their multiple transitions to flightlessness and large body size, and recognizable members of extant subclades generally do not appear until the Neogene. Despite these knowledge gaps, we combine what is known from the fossil record of palaeognaths with plausible divergence time estimates, suggesting a relatively rapid pace of diversification and phenotypic evolution in the early Cenozoic. In line with some recent authors, we surmise that the most recent common ancestor of palaeognaths was likely a relatively small-bodied, ground-feeding bird, features that may have facilitated total-clade palaeognath survivorship through the K-Pg mass extinction, and which may bear on the ecological habits of the ancestral crown bird.

Keywords: Palaeognathae; ostrich; tinamou; ratite; emu; kiwi; moa; elephant bird; rhea; Lithornithidae

1. Introduction

Crown birds (Neornithes) comprise roughly 11,000 extant species [1]. They are divided into the reciprocally monophyletic Palaeognathae and Neognathae, with the latter including the hyperdiverse clade Neoaves [1]. At no point in time do total group palaeognaths appear to have been particularly diverse, especially in comparison with contemporaneous neognath diversity. Despite their relatively sparse taxonomic diversity, however, the position of palaeognaths as the sister group to all other neornithines makes them critical to efforts to understand the early evolutionary history of crown birds. Palaeognathae is diagnosed by several traits including a unique palatal structure characterized by enlarged basipterygoid processes and fused pterygoids and palatines (Figure 1), a grooved rhamphotheca, a single articular facet for the otic capitulum of the quadrate, and open ilioischiadic foramina (Figure 2) [2–6]. The palatal structure of palaeognaths was traditionally considered plesiomorphic for Neornithes [7], though recent evidence regarding the palatal structure of the near-crown Ichthyornithes may indicate that the palaeognathous palate is in fact a synapomorphy of Palaeognathae [8,9].

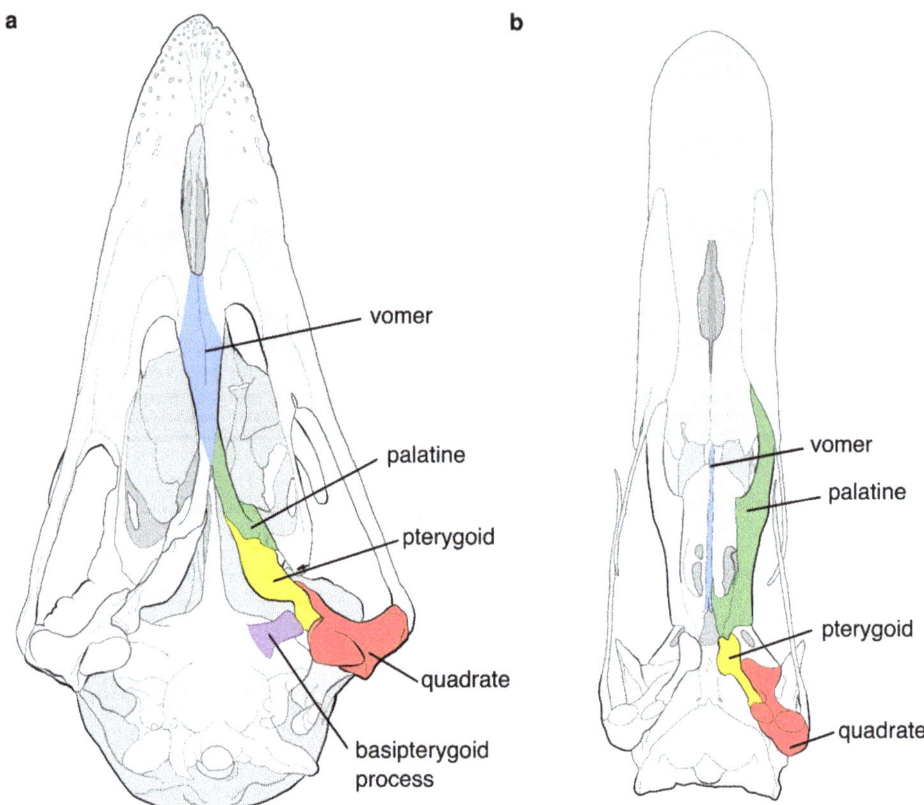

Figure 1. Comparison of the palate of a palaeognathous and a neognathous bird. (**a**) Palate of the palaeognathous Emu *Dromaius novaehollandiae*. The basipterygoid process is elongate, and the pterygoid and palatine are fused (demarcation between them is approximate). (**b**) Palate of the neognathous Mute Swan *Cygnus olor*. The pterygoid and palatine are connected by an intrapterygoid joint, and the short basipterygoid processes are mostly obscured by the pterygoids.

Extant palaeognaths are represented by 46 species of tinamou (Tinamidae) and two species of rhea (Rheidae) in Central and South America, two species of ostrich (Struthionidae) in Africa, the monotypic emu and three species of cassowaries (Casuariidae) in Australia and New Guinea, and approximately five species of kiwi in New Zealand (Apterygidae) [10]. Nine species of moa (Dinornithiformes) [11] and four species of elephant bird (Aepyornithidae) [12] survived into the Holocene in New Zealand and Madagascar respectively, before their extinction which may have been related to human activity that had a disproportionate impact on insular flightless birds [13].

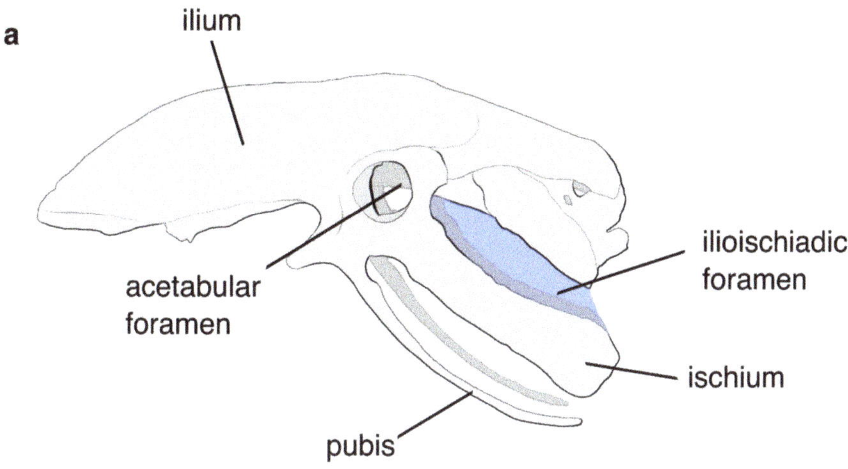

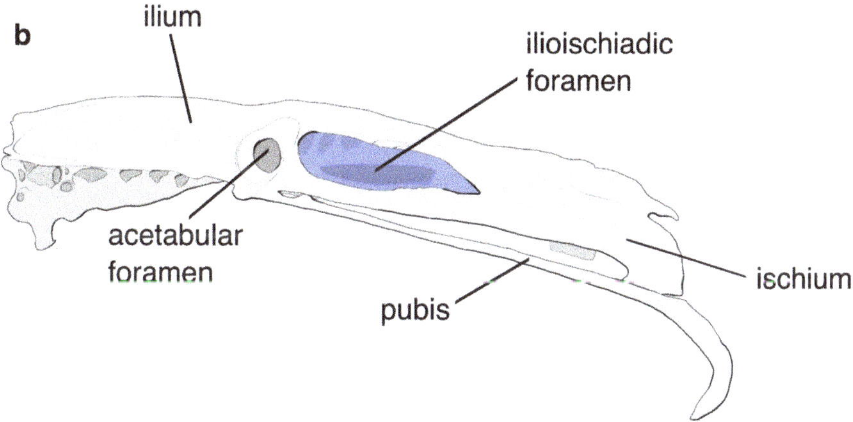

Figure 2. Comparison of the pelvis of a palaeognathous and a neognathous bird. The ilioischiadic foramen is highlighted in blue. (**a**) Pelvis of the Little Spotted Kiwi *Apteryx owenii*. The ilium and ischium are unfused throughout their lengths, leaving the ilioischiadic foramen open. (**b**) Pelvis of the Mute Swan *Cygnus olor*. The ilioischiadic foramen is closed due to the fusion of the posterior ilium and ischium.

Despite being relatively species-poor, extant and recently extinct palaeognaths encompass an impressive range of body sizes and ecologies. The group contains both cursorial open habitat specialists (e.g., emu) and graviportal forest dwellers (e.g., cassowaries), and feeding strategies ranging from cryptic nocturnal invertivores (e.g., kiwi) to megaherbivorous browsers (e.g., moa). Out of all extant palaeognaths, only tinamous (Tinamidae) are capable of flight [14]. This clade comprises small to medium-sized birds, ranging from 43 g in the smallest species (the Dwarf Tinamou *Taoniscus nanus*) [15], to 2080 g in the largest females of the Gray Tinamou (*Tinamus tao*) [16]. By contrast, flightless palaeognaths, from here on referred to collectively as "ratites" (acknowledging the paraphyletic nature of the group), are renowned for their gigantism. The Common Ostrich *Struthio camelus* is the world's largest extant bird in both height and weight, with large males reaching sizes up to

2.8 m and 156 kg [17]. Recently extinct ratites were even larger: A body mass of 860 kg was estimated from femur measurements of an exceptionally large individual of the elephant bird *Vorombe titan*, making this species the heaviest-known bird ever discovered [12]. Females of the moa *Dinornis robustus* were less massive but appear to have constituted the tallest birds yet discovered, attaining heights of 3.6 m [18,19].

Several early authors argued that 'ratites' represented a non-monophyletic assemblage of large-bodied, flightless birds, and debate regarding the potential non-monophyly of ratites persisted through much of the 20th Century [4,20–24]. Opinion shifted with the widespread acceptance of continental drift theory in the latter half of the 20th century, as a monophyletic "Ratitae" became enshrined as a classic example of Gondwanan vicariance biogeography, a hypothesis stipulating that stem group ratites became flightless prior to the breakup of Gondwana, and that Gondwanan fragmentation drove the divergence of the extant ratite lineages as populations became geographically isolated from one-another [25–27]. This hypothesis of a monophyletic "Ratitae", sister to Tinamidae, was supported by a number of phenotypic features such as the absence of a triosseal canal and sternal keel, and the presence of a fused scapulocoracoid (Figure 3) [5]. Indeed, the term "ratite" refers to the flat, raft-like sterna of taxa lacking a sternal keel (Figure 4) [28]. This consensus opinion was upheld for several decades by most phylogenetic analyses of morphological characters [29–31], though analyses of cranial characters recovered alternative relationships [32–34]. However, over the past twenty years, molecular phylogenetic analyses have forced a wholescale revision of the Gondwanan vicariance paradigm of palaeognath evolution and historical biogeography. Evidence from analyses of both nuclear [35–43] and mitochondrial DNA [41,42,44–46], as well as large-scale phylogenomic analyses [47–50], demonstrate that tinamids are in fact phylogenetically nested within ratites, rendering "Ratitae" paraphyletic, once again reviving the early hypothesis of ratite non-monophyly [4,20–24] (Figure 5).

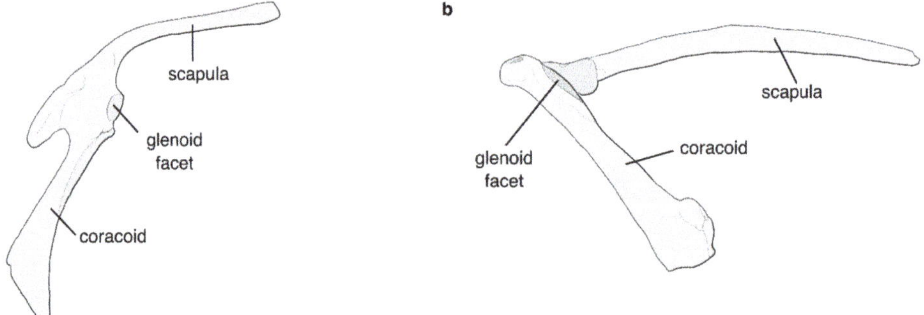

Figure 3. Comparison of the shoulder girdle of a flightless palaeognath displaying the fused 'ratite' condition, and that of a volant palaeognath in left lateral view. (**a**) Fused scapulocoracoid of the flightless Greater Rhea *Rhea americana*. (**b**) Unfused scapula and coracoid of the volant Andean Tinamou *Nothoprocta pentlandii*.

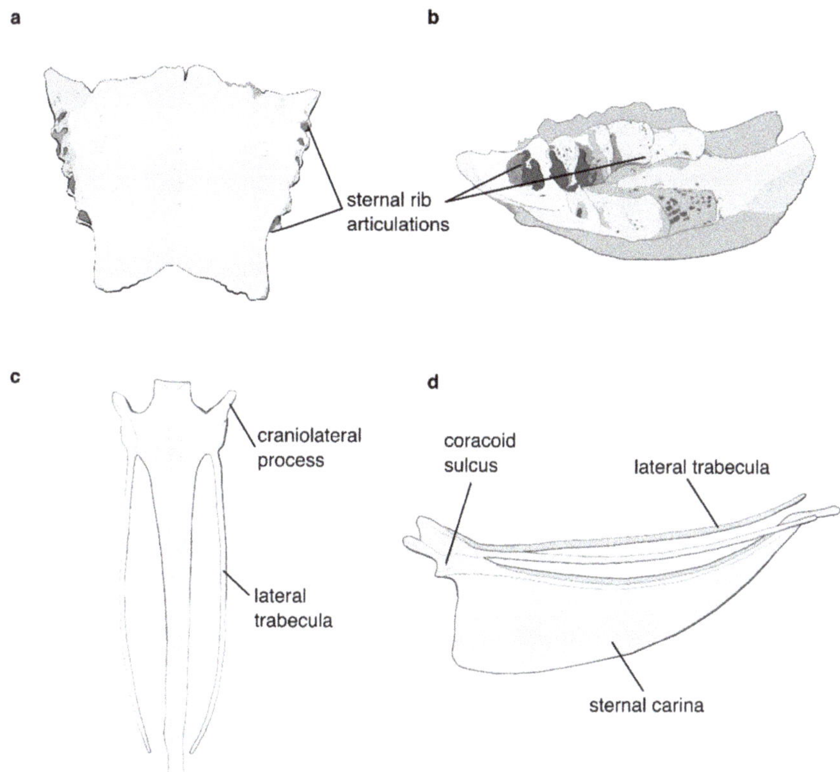

Figure 4. Comparison of the sterna of a flightless palaeognath, the Common Ostrich *Struthio camelus* and a volant palaeognath, the Andean Tinamou *Nothoprocta pentlandii*. (**a**) Sternum of *S. camelus* in dorsal view. (**b**) Sternum of *S. camelus* in left lateral view. A sternal keel is absent. (**c**) Sternum of *N. pentlandii* in dorsal view. (**d**) Sternum of *N. pentlandii* in left lateral view. A deep sternal keel provides an attachment area for the pectoralis and supracoracoideus muscles.

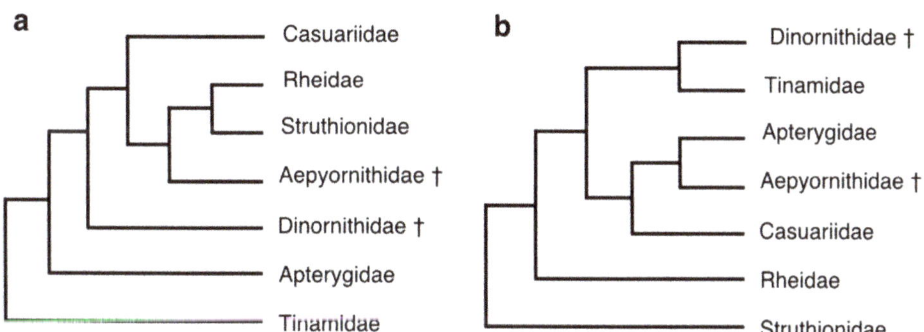

Figure 5. Old and new hypotheses of palaeognath interrelationships. Extinct clades are indicated by †. (**a**) Ratite monophyly based on the morphological study of Livezey and Zusi [30]. (**b**) Molecular phylogeny suggesting ratite paraphyly recovered by Mitchell, et al. [45], Grealy, et al. [41], Yonezawa, et al. [49], Urantówka, et al. [46], and Almeida, et al. [42].

The most parsimonious interpretation of this revised tree topology would be that the most recent common ancestor of crown Palaeognathae was flightless, with a reacquisition of flight arising along the tinamou stem lineage. This interpretation is indeed favoured by maximum likelihood analyses [44] and cannot be definitively rejected; however, this hypothesis would seem to be unlikely from first principles (after all, strong evidence exists for only four independent acquisitions of powered flight throughout the entire evolutionary history of animals [51]). By contrast, multiple independent transitions to flightlessness within the same crown bird subclade are not uncommon. For example, flightlessness has arisen dozens of times in Rallidae among island-dwelling taxa [52,53]. According to some recent molecular topologies, transitions to flightlessness arose a minimum of six times in palaeognaths, and transitions to gigantism a minimum of five [41,45].

The recent revival of a phylogenetic hypothesis stipulating that ratites repeatedly and independently lost the capacity to fly has largely been driven by molecular phylogenetic analyses [36–46,48–50,54–58], but has accrued supporting evidence from independent datasets. For instance, embryological studies have demonstrated important differences in patterns of wing growth among ostriches and emu, suggesting that alternative heterochronic mechanisms may underlie the acquisition of flightlessness in disparate ratite taxa and potentially supporting independent evolutionary transitions to flightlessness among ratites [59]. Furthermore, misexpression of the cardiac transcription factor *Nkx2.5* is associated with reduced wing growth in chicken embryos, and this transcription factor is expressed in the wings of emu embryos but not ostriches—again indicating the potential non-homology of flightlessness in emu and ostriches [60]. Sackton, et al. [50] found that many similarities in ratite forelimb morphology may be the result of convergence in gene regulatory networks, rather than the product of homologous changes to protein coding genes. Overall, the existing body of evidence is congruent with the hypothesis that 'ratites' are indeed paraphyletic, and have repeatedly converged on a suite of remarkably similar morphologies that were long interpreted as synapomorphies for the group. Much remains to be understood about the underlying drivers of these independent transitions to large size and flightlessness, as well as the developmental underpinnings of convergent ratite morphologies.

The recognition of ratite paraphyly, coupled with phylogenomic time trees that indicate an origin of crown palaeognaths long after the breakup of Gondwana commenced (e.g., [41,42,45,48,49,55]), makes the classic vicariance hypothesis untenable. Instead, present-day palaeognath biogeography must be the product of dispersal of volant ancestral palaeognaths to multiple landmasses preceding independent origins of flightlessness (Figure 6). However, this interpretation raises many questions regarding the nature of the volant last common ancestor of crown palaeognaths. Tinamous are the only extant volant palaeognaths available for reference, but they are primarily ground-dwelling and are only capable of flight over relatively short distances to flee predators or roost in trees [14,61]. It is difficult to imagine a burst-flying tinamou-like bird undertaking the transoceanic journeys needed to explain the distribution of extant palaeognaths (Figure 6), thus they would appear to be a poor analogue for hypothetical dispersive ancestral palaeognaths. Fossil evidence further suggests that the specialized burst flying of extant tinamous was not plesiomorphic for palaeognaths. The extinct lithornithids (Lithornithidae), known from the Paleocene and Eocene of Europe and North America, were apparently volant and appear to represent the oldest and most stemward known total-clade palaeognaths [49,62–65]. Importantly, they also appear to have been more capable long-distance fliers than extant tinamids are [62,65], and, as the earliest known palaeognaths in the fossil record, they may provide the best models for informing reconstructions of the dispersive ancestral palaeognaths that gave rise to extant palaeognath diversity.

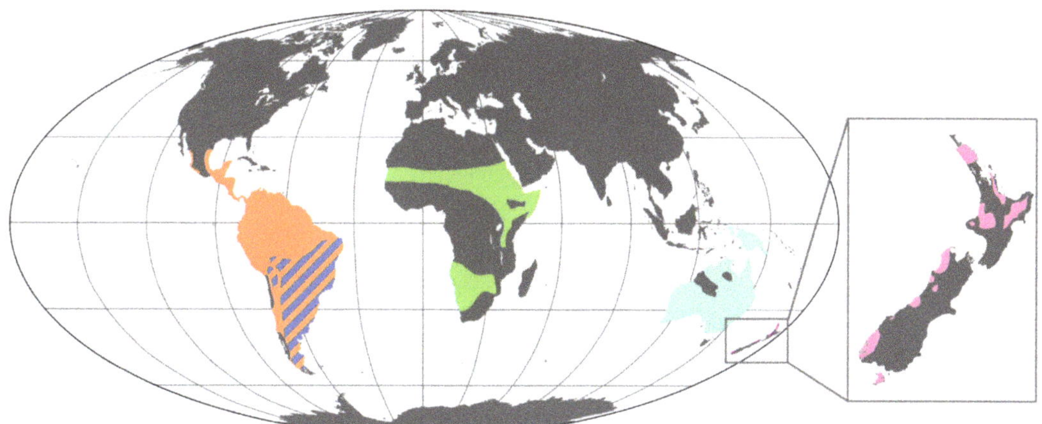

Figure 6. Present-day geographic ranges of extant palaeognath subclades. Range of Rheidae in dark blue, Tinamidae in orange, Struthionidae in green, Casuariidae in aqua, and Apterygidae in pink [10,14,66–68].

In order to probe deeper into the origin and early evolution of total group Palaeognathae, an in-depth understanding of the palaeognath fossil record is necessary. Early fossil palaeognaths are rare, and the phylogenetic interrelationships among them are poorly understood. For example, the monophyly and phylogenetic position of lithornithids are debated, and thus their relevance for clarifying the pattern and timing of the extant palaeognath radiation remains unclear. Due to the phylogenetic position of palaeognaths as the extant sister taxon of all other Neornithes, stem palaeognaths, which may include lithornithids, should provide key insight into the nature of the ancestral crown bird. Recent time-scaled phylogenies suggest that total-group palaeognaths were one of just a small number of extant neornithine lineages that passed through the Cretaceous-Paleogene (K-Pg) mass extinction event (e.g., [48,69–72]). A better understanding of the ecology and biology of early stem palaeognaths could therefore help clarify the biological attributes of avian survivors of the end-Cretaceous mass extinction, which appears to have eliminated all non-neornithine avialans [73]. Early palaeognath fossils from around the world will also be critical for illustrating how the remarkable convergent evolution of flightlessness and gigantism arose among crown palaeognaths, as well as providing insight into the biogeographic origins of extant palaeognath subclades and their responses to Cenozoic shifts in climate and environment [74,75].

Here, we summarize the current state of knowledge regarding the palaeognath fossil record. Useful reviews on palaeognath fossils and the evolutionary history of this group have previously been published, e.g., [76–78], and we refer interested readers to these excellent summaries, but the present review is the first attempt to systematically address the fossil record of palaeognaths in its entirety. We present the most specific locality data reported in the literature for each fossil occurrence, necessarily limited by the differential specificity available for certain records. We outline key lingering gaps in the known palaeognath fossil record, and suggest potential ways forward in hopes of narrowing those gaps. In addition, we provide an overview of strong inferences about palaeognath macroevolution that can be made on the basis of current molecular phylogenies and estimated divergence times, and summarise what can be reasonably inferred about the most recent common ancestor of crown group palaeognaths. We hope that this review provides both a solid base of information for those interested in the evolution and fossil record of palaeognaths, and helps inspire further work clarifying the evolutionary history of these remarkable birds.

Institutional abbreviations are as follows: AM—Australian Museum, Darlinghurst, Australia; AIM—Auckland Institute and Museum, Auckland, New Zealand; AMNH—American Museum of Natural History, New York, New York, USA; AU—Auckland University, Auckland, New Zealand; AUG—Aristotle University School of Geology, Thessaloniki, Greece; BGR—Bundesanstalt für Geowissenschaften Und Rohstoffe, Hanover, Germany; CICYTTP—Centro de Investigación Científica y de Transferencia Tecnológica a la Producción, Diamante, Argentina; CPC—Commonwealth Palaeontological Collections, Canberra, Australia; DK—Danekrae collections, Geological Museum, University of Copenhagen, Copenhagen, Denmark; FMNH—Field Museum of Natural History, Chicago, Illinois, USA; GHUNLP—Universidad Nacional de La Pampa, Santa Rosa, Argentina; GMB—Geological Museum of Budapest, Budapest, Hungary; GMH—Geiseltalmuseum, Martin Luther University, Halle, Germany; HLMD—Hessisches Landesmuseum, Darmstadt, Germany; IGM—Institute of Geology, Mongolian Academy of Sciences, Ulaan Baatar, Mongolia; IRSNB—Institut royal des Sciences naturelles de Belgique, Brussels, Belgium; IVPP—Institute of Vertebrate Paleontology and Paleoanthropology, Beijing, People's Republic of China; KNM—Kenya National Museum, Nairobi, Kenya; MACN—Museo Argentino de Ciencias Naturales "Bernardino Rivadavia", Buenos Aires, Argentina; MASP—Colección del Museo de Ciencias Naturales y Antropológicas, Paraná, Argentina; MFN—Museum für Naturkunde, Berlin, Germany; MGL—Geological Museum of Lausanne, Lausanne, Switzerland; MGUH—palaeontology type collection, Geological Museum, University of Copenhagen, Copenhagen, Denmark; MHNT—Muséum de Toulouse, Toulouse, France; MLP—Museo de La Plata, La Plata, Argentina; MNHN—Muséum National d'Histoire Naturelle, Paris, France; MPCN—Museo Patagónico de Ciencias Naturales, General Roca, Argentina; MPM—Museo Regional Provincal Padre Manuel Jesús Molina, Río Gallegos, Argentina; MUFYCA—Museo Florentino y Carlos Ameghino (Instituto de Fisiografía y Geología), Rosario, Argentina; MV—Musée Vivenel, Compiègne, France; NHMUK—Natural History Museum, London, UK; NJSM—New Jersey State Museum, Trenton, New Jersey, USA; NMNHS—National Museum of Natural History, Bulgarian Academy of Sciences, Sofia, Bulgaria; NMNZ—Museum of New Zealand Te Papa Tongarewa, Wellington, New Zealand; NNPM—National Museum of Natural History of the National Academy of Sciences, Kyiv, Ukraine; ONU—Odes'kiy Natsional'niy Universitet, Odessa, Ukraine; PIN—Palaeontological Institute, Russian Academy of Sciences, Moscow, Russian Federation; PU—Princeton University Collection (now at Yale Peabody Museum), Princeton, New Jersey, USA; QM—Queensland Museum, Brisbane, Australia; RAM—Raymond Alf Museum, Claremont, California, USA; ROM—Royal Ontario Museum, Toronto, Ontario, Canada; SAM—South Australian Museum, Adelaide, Australia; SGPIMH—Geologisch-Paläontologisches Institut und Museum der Universität Hamburg, Hamburg, Germany; UCMP—University of California Museum of Paleontology, Berkeley, California, USA; UCR—University of California Riverside, Riverside, California, USA; UM—Museum of Paleontology, University of Michigan, Ann Arbor, Michigan, USA; UNSW—University of New South Wales; Sydney, Australia; USNM—Smithsonian Museum of Natural History, Washington, DC, USA; WN—Michael C.S. Daniels collection, Essex, UK; YPM—Yale Peabody Museum, New Haven, Connecticut, USA; ZIUU—Zoologiska Museum, Uppsala Universitet, Sweden.

2. Overview of the Palaeognath Fossil Record

2.1. Lithornithidae

Lithornithids were small bodied, presumably volant birds that were first recognized as palaeognaths by Houde and Olson [79], and described in detail as a clade by Houde [62]. Thus far, they are only known from Europe and North America, contrasting with the Gondwanan distribution of extant palaeognaths. At first glance, they appear remarkably similar to tinamous, particularly in the shape of the skull. Fossil eggshells attributed to lithornithids are also very reminiscent of those of tinamous, and it has been hypothesized that lithornithids shared the same polygynandrous breeding behaviour of many extant palaeognaths [62]. However, numerous characters distinguish tinamous and lithornithids, which are detailed by Houde [62]. On the basis of a more distally positioned deltopectoral crest, longer and more curved humeral shaft, and a less distally elongated sternum in lithornithids compared with tinamous, Houde [62] also speculated that lithornithids were much more capable long-distance fliers than extant tinamous are. This idea received further support from a reconstruction of the wing of a specimen of the Eocene lithornithid *Calciavis grandei* with preserved carbonized feather traces, which indicated that this species may have been capable of long-distance flapping flight [65].

Since their fossils are most often recovered from nearshore lacustrine or marine environments, it was suggested that lithornithids may have exhibited a shorebird-like ecology [62], though this may be coincidental as these depositional settings are most likely to produce fossils in general. The lithornithid jaw apparatus appears well suited to distal rhynchokinesis, which allows a bird to capture food items in the ground without having to fully open the jaws [62]. This suggests they could have used their bills for probing the substrate for food items, in a manner more similar to kiwi than tinamous [62]. Additional evidence for this type of foraging behaviour comes from the recognition of mechanoreceptors known as Herbst corpuscles in the rostrum of lithornithids [80], which form a tactile bill-tip organ that picks up mechanical vibrations to detect buried prey.

A major unresolved question is whether Lithornithidae predate the K-Pg mass extinction. The cranial end of a right scapula with a distinctive pointed acromion was recovered from the latest Maastrichtian or earliest Danian Hornerstown Formation in New Jersey, USA [63]. If this material indeed belongs to a lithornithid, it would provide compelling evidence that the clade survived across the boundary. However, it should be noted that several Mesozoic stem ornithurines also have a hooked acromion that approaches the condition seen in Lithornithidae [64,81,82]. Thus, the identity of this fossil remains uncertain, and more material needs to be recovered from both this formation and other contemporaneous localities to clarify which groups of total-clade palaeognaths persisted across the K-Pg boundary.

2.1.1. North American Lithornithids

Definitive lithornithid fossils are known from North America from the middle Paleocene to the early Eocene (Figure 7, Table 1) [62,83–88]. The earliest uncontroversial record on this continent is *Lithornis celetius*, from the middle Paleocene (early to middle Selandian) Fort Union Formation of Montana and the Polecat Bench Formation of Wyoming [62]. The entire skeleton of this species is known from a composite series of individuals [62]. Slightly younger than *L. celetius* is a proximal end of a humerus from the middle Paleocene (Tiffanian) Goler Formation in southern California. Despite being fragmentary, its large, dorsally positioned humeral head and subcircular opening to the pneumotricipital fossa diagnose it as a probable lithornithid, and it was assigned to the genus *Lithornis* [88]. As nearly all North American lithornithids derive from the Rocky Mountain region, this fossil extends their known range significantly further west.

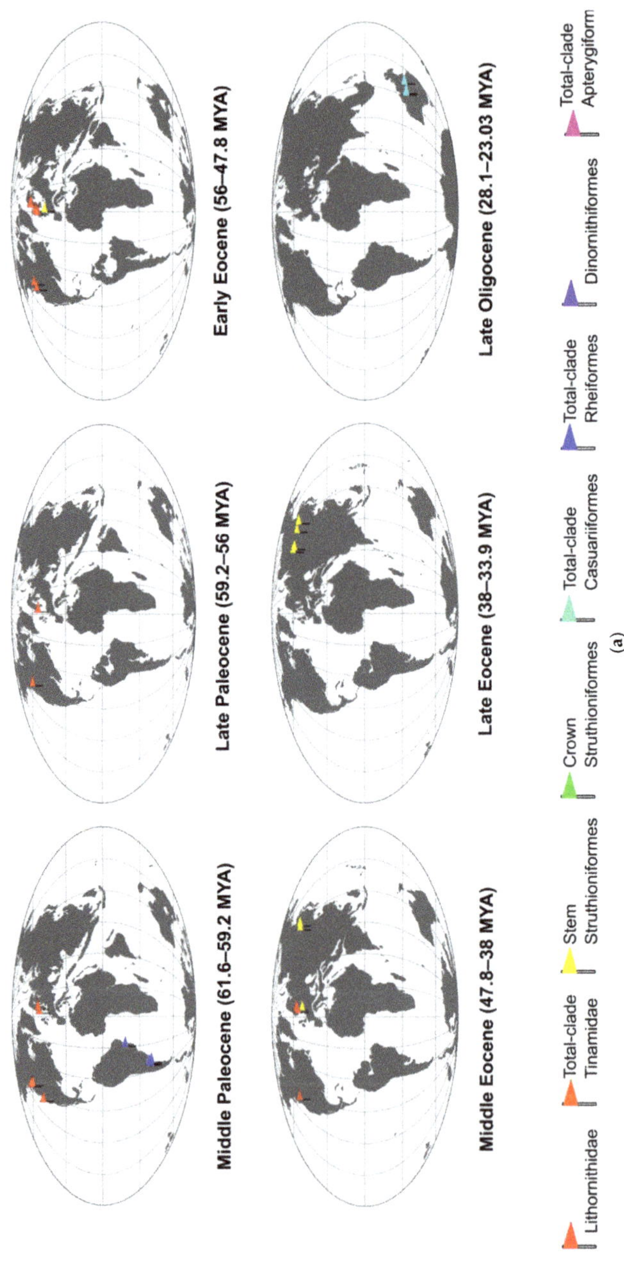

Figure 7. *Cont.*

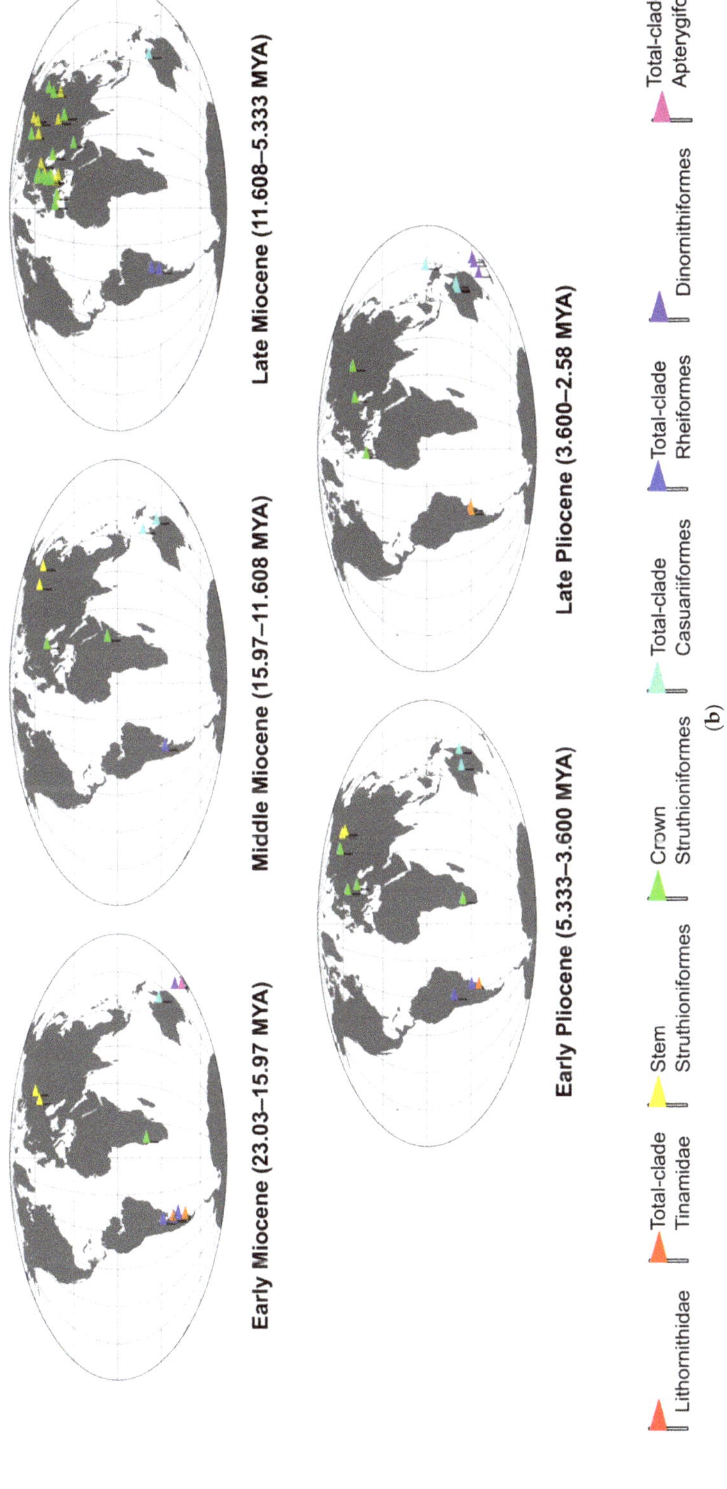

Figure 7. Geographic distribution of palaeognath fossils illustrated on palaeogeographic globes. (**a**) Middle Paleocene, late Paleocene, early Eocene, middle Eocene, late Eocene, and late Oligocene. (**b**) Early Miocene, middle Miocene, late Miocene, early Pliocene, and late Pliocene. Palaeomaps modified from GPlates (www.gplates.org) [83,84].

Table 1. Lithornithid fossil record.

Continent	Geological Unit	Location	Epoch	Stage	Age Reference	Taxa	Institutions	Reference
North America	Hornerstown Formation	New Jersey, USA	Late Cretaceous–early Paleocene	Maastrichtian-Danian	Olson and Parris [85]; Staron, et al. [86]	?Palaeognathae	NJSM	Parris and Hope [63]
	Fort Union Formation	Park County, Montana, USA	middle Paleocene	Selandian	Lofgren, et al. [87]; Stidham, et al. [88]	*Lithornis celetius*	USNM, PU	Houde [62]
	Polecat Bench Formation	Wyoming, USA	middle Paleocene	Selandian	Lofgren, et al. [87]; Stidham, et al. [88]	*Lithornis celetius*	PU, UM	Houde [62]
	Goler Formation	Kern County, California, USA	middle Paleocene	Selandian	Lofgren, et al. [89]; Albright, et al. [90]; Lofgren, et al. [91]	*Lithornis sp.*	RAM	Stidham, et al. [88]
	Willwood Formation, Sand Coulee beds	Park County, Wyoming, USA	late Paleocene	Thanetian	Lofgren, et al. [87]	*Lithornis promiscuus, Lithornis plebius*	USNM, UM, AMNH	Houde [62]
	Willwood Formation	Basin, Wyoming, USA	early Eocene	Ypresian	Lofgren, et al. [87]	*Lithornis nasi* (provisional), *Paracathartes howardae*	UM, ROM, USNM	Houde [62]
	Green River Formation, Fossil Butte member	Lincoln County, Wyoming, USA	early Eocene	Ypresian	Smith, et al. [92]	*Calciavis grandei, Pseudocrypturus cercanaxius*	AMNH, USNM	Houde [62]; Nesbitt and Clarke [64]
	Bridger Formation	Bridger Basin, Wyoming, USA	middle Eocene	Ypresian-Lutetian	Murphey and Evanoff [93]	*incertae sedis*	YPM	Houde [62]
Europe	Heers Formation, Orp Sand member	Maret, Belgium	middle Paleocene	Selandian	Smith and Smith [94], De Bast, et al. [95]	cf. Lithornithidae	IRSNB	Mayr and Smith [96]
	Fissure filling of Walbeck	Helmstedt, Germany	middle Paleocene	Selandian	Aguilar, et al. [97]	*Fissuravis weigelti*	GMH	Mayr [98]
	Tuffeau de Saint-Omer	Templeuve, France	late Paleocene	Thanetian	Steurbaut [99]; Moreau and Mathis [100]; Smith and Smith [94]	Lithornithidae gen. et sp. indet.	IRSNB	Mayr and Smith [96]
	Ølst Formation	Limfjord region, Denmark	early Eocene	Ypresian	Heilmann-Clausen and Schmitz [101]	*Lithornis nasi, Lithornis vulturinus*	MGUH	Bourdon and Lindow [102]
	Fur Formation	Denmark	early Eocene	Ypresian	Chambers, et al. [103]	*Lithornis vulturinus*	DK, MGUH	Leonard, et al. [104]; Bourdon and Lindow [102]

Table 1. *Cont.*

Continent	Geological Unit	Location	Epoch	Stage	Age Reference	Taxa	Institutions	Reference
	London Clay Formation	Kent, Essex, Sussex, England	early Eocene	Ypresian	King [105]; Ellison, et al. [106]; Friedman, et al. [107]	*Lithornis vulturinus, Lithornis nasi, ?Lithornis hookeri, Pseudocrypturus cercanaxius* (provisional)	NHMUK, WN, PU	Houde [62]
	Messel Formation	Messel, Germany	middle Eocene	Ypresian-Lutetian	Franzen and Haubold [108]; Schaal and Ziegler [109]; Lenz, et al. [110]	*Lithornis sp.*	SGPIMH, IRSNB	Mayr [111]; Mayr [112]

Two sympatric species are known from the late Paleocene (late Thanetian) Sand Coulee Beds of the Willwood Formation in Wyoming. *Lithornis promiscuus* was the larger of the two, and is the largest species in its genus [62]. Like *L. celetius*, virtually all bones of the skeleton are known from a composite series [62]. The holotype, USNM 336535, preserves the entire forelimb skeleton. The smaller *Lithornis plebius* is known from all major appendicular elements [62]. Houde [62] acknowledged the possibility that *L. promiscuus* and *L. plebius* may belong to a single sexually dimorphic species, but erred on the side of a more conservative species diagnosis and retained them as separate taxa. Houde [62] tentatively referred specimen NHMUK A 5303 from the London Clay on the Isle of Sheppey, UK to the latter species. Owing to both the homogeneity of the global hothouse climate and the shorter distance across the North Atlantic at the time, North American and European avifaunas were remarkably similar during the late Paleocene and early Eocene (e.g., [76,113,114]). Finding the same species on both sides of the Atlantic should therefore not come as a surprise, and if NHMUK A 5303 is indeed an example of *L. plebius* it would hint towards the dispersal capabilities of these birds.

The remaining North American lithornithids are Eocene in age. *Paracathartes howardae* [115] was found in early Eocene strata of the Willwood Formation [62]. With the exception of the sternum and pelvis, all bones of this species are again known from a composite series [62]. The lacustrine Green River Formation deposited by the Gosiute, Uinta, and Fossil palae-olakes in what is now Utah, Wyoming, and Colorado has yielded an enormous wealth of fossils, most often preserved as slabs [116]. The Fossil Butte member of the formation, deposited by the short-lived early Eocene Fossil Lake [116], has produced the greatest number of lithornithid specimens thus far [64], as well as a great wealth of other bird fossils (e.g., [117–128]). A minimum of two lithornithid species have been found in this Lagerstätte [64]. The holotype of *Pseudocrypturus cercanaxius* [62] is a complete skull and mandible, with nine cervical vertebrae in articulation [62]. A spectacular crushed articulated specimen missing only the pelvis and caudal vertebrae is owned privately by Siber and Siber, and a cast of this specimen is in the collections of the USNM. Two skeletons collected from the London Clay in England were provisionally referred to this species [62], making it another lithornithid with a possible transatlantic distribution. The recently named *Calciavis grandei* [64] was described from a complete, mediolaterally compressed skeleton with preserved soft tissue including feathers, pedal scales, and claw sheaths. A referred specimen includes most of the postcranial skeleton minus the femora and pelvic region, and a disarticulated skull [64].

2.1.2. European Lithornithids

The fossil record of lithornithids in Europe also begins in the middle Paleocene, and stretches to the middle Eocene (Figure 7, Table 1) [96,111,112]. The Orp Sand member (early to middle Selandian) of the Heers Formation in Maret, Belgium yielded a distal humerus fragment and a partial carpometacarpus that were assigned to Lithornithidae, but the fossils are too incomplete to be assigned at a generic level [96]. The next oldest European lithornithid, *Fissuravis weigelti*, is also known from fragmentary remains, in this case the omal end of an isolated coracoid from the late middle Paleocene (Selandian) of the fissure filling of Walbeck, Germany [98]. A lack of clear diagnostic features has cast some level of doubt to this assignment. The coracoid lacks any lithornithid character other than similarity in size, and seems to be missing the small foramina on the posteroventral surface of the hooked acrocoracoid process that is an apomorphy of this clade [64]. Regardless of the true affinities of *Fissuravis weigelti*, the Maret fossils demonstrate that Lithornithidae stretch at least as far back in time in Europe as they do in North America.

As noted by Houde, one of the first fossil birds known to science was *Lithornis vulturinus* [62,129], the holotype specimen of which was purchased by the Royal College of Surgeons in 1798. The holotype was sadly destroyed in the Second World War, though detailed woodcut drawings of the holotype [130] allowed for the identification of a neotype by Houde [62]. The neotype, from the early Eocene (Ypresian) London Clay, was originally identified as an early relative of turacos and named *Promusophaga magnifica* by Harrison and Walker [131]. It consists of a right humerus, radius, ulna, and carpometacarpus, all missing the distal ends, a right scapula, partial sternum, distal left radius and ulna, proximal left femur, proximal right tibiotarsus, a vertebral series, and ribs within a clay nodule [62]. A large amount of fragmentary material from the London Clay, mainly hindlimb elements, has been referred to this species [102]. A slightly younger specimen from the early Eocene Fur Formation of Denmark preserves a three-dimensional skull in articulation with a nearly complete postcranial skeleton and has been described in great detail [102,104]. Another Danish fossil, a distal left humerus from the latest Paleocene-earliest Eocene Olst Formation, was also referred to this taxon [102].

Lithornis nasi [132], also from the early Eocene London Clay Formation, was considered a junior synonym of *L. vulturinus* by Bourdon and Lindow [102]. As the material comes from the type locality of *L. vulturinus*, these authors interpreted the differences between *L. nasi* and *L. vulturinus* as intraspecific variation. The holotype consists of proximal fragments of a left humerus and right ulna, distal fragments of a right femur and a right tibiotarsus, and two thoracic vertebrae [62]. Houde [62] tentatively assigned two specimens from Early Eocene Willwood Formation to *L. nasi*. Another bird from the London Clay, *?Lithornis hookeri* [132], was tentatively referred to the genus by Houde [62]. The holotype, a distal end of a tibiotarsus, suggests it was smaller than all currently known lithornithids [62]. The Messel lithornithid from the middle Eocene of Germany (47–48 MYA) is the youngest lithornithid material yet discovered [111,112]. Known from a partial postcranial skeleton and a skull that appear to represent the same species, it was assigned to the genus *Lithornis* but not to a species-level taxon [112].

2.1.3. Systematics of Lithornithidae

While it is generally accepted that lithornithids are indeed total-clade palaeognaths, important questions regarding their systematics remain: Do lithornithids represent a monophyletic radiation of volant stem or crown palaeognaths? Do they represent a paraphyletic grade of stem palaeognaths? Or, are they polyphyletic, with some taxa more closely related to certain extant palaeognath lineages than others (Figure 8)? All three scenarios would seem to be possible considering that the earliest members of several extant palaeognath subclades would most likely have been relatively small and volant. Houde [62] argued that lithornithids are not monophyletic and placed *Paracathartes* closer to other ratites on the basis of similar histological growth patterns, and the reduced, rounded postorbital process of its frontals. More recent authors have speculated that this histological similarity exists because *Paracathartes* is larger than other lithornithids, reaching approximately the size of a turkey [76].

The phylogenetic analyses of both Nesbitt and Clarke [64] and Yonezawa, et al. [49] recovered lithornithids as a monophyletic group. The character matrix used by Nesbitt and Clarke [64] contained 182 characters combined from the morphological datasets of Cracraft [5], Bledsoe [133], Lee, et al. [29], Mayr and Clarke [134], Clarke [81], Clarke, et al. [135], and new observations gathered by the authors for 38 terminal taxa. In their unconstrained analyses, Lithornithidae was recovered as the sister taxon to Tinamidae at the base of Palaeognathae, congruent with previous morphological phylogenetic hypotheses. This is unsurprising, given that lithornithids and tinamids share numerous skeletal similarities that often optimize as synapomorphies of a lithornithid + tinamou clade. When *Paracathartes* was constrained as sister to ratites, the resultant nonmonophyly of Lithornithidae added a significant number of steps to the analysis. The only character that supported this relationship was the reduction of the postorbital process of the frontal, which the authors considered to be convergent. When relationships of living palaeognaths were constrained to match those recovered by molecular phylogenies, lithornithids were recovered as a clade of stem group palaeognaths. Though Nesbitt and Clarke [64] were unable to achieve any resolution within Lithornithidae, lithornithid monophyly received relatively high support. However, the authors acknowledge the need for future analy ses assimilating additional lithornithid character sets to further test the monophyly and phylogenetic position of lithornithids.

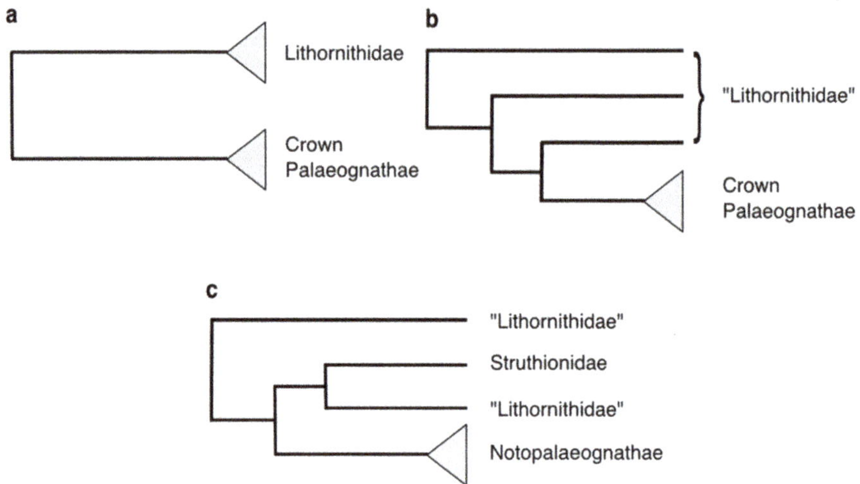

Figure 8. Possible relationships of Lithornithidae to the remainder of Palaeognathae. (**a**) Scenario A shows a monophyletic Lithornithidae, (**b**) Scenario B shows a paraphyletic Lithornithidae, and (**c**) Scenario C shows a polyphyletic Lithornithidae.

A strict consensus tree using parsimony constrained to match recent molecular phylogenetic topologies recovered a monophyletic Lithornithidae sister to Tinamidae, but when the molecular constraint was removed and replaced with constraints enforcing sister group relationships between Palaeognathae + Neognathae and Neoaves + Galloanserae, Lithornithidae instead resolved sister to a *Dinornis* + *Dromaius* + *Struthio* clade to the exclusion of tinamous [136]. In an analysis of this same dataset with new characters added and increased taxon sampling, Bayesian analysis placed lithornithids as stem palaeognaths, and a maximum parsimony analysis of this dataset with cranial characters weighted more strongly found strong support for a monophyletic Lithornithidae in this same position [137]. When characters were unweighted in the maximum parsimony analysis but constrained to a molecular backbone, a monophyletic Lithornithidae was once again sister to Tinamidae [137]. Almeida, et al. [42] also recovered lithornithids as sister to crown Palaeognathae in their Bayesian topology, but sister to tinamous in their maximum parsimony and maximum likelihood trees. Maximum likelihood trees inferred using characters exhibiting low homoplasy also supported a position on the palaeognath stem for Lithornithidae [49], though the monophyly of the clade was dependent on the matrix used. Ten non-homoplastic characters from Houde [62] yielded a paraphyletic Lithornithidae, while 92 non-homoplastic characters from Worthy, et al. [136] supported them as a monophyletic group. The authors considered their results as supportive of the hypothesis that all extant palaeognaths evolved independently from *Lithornis*-like birds [42]. Given lingering uncertainties regarding the monophyly and phylogenetic position of lithornithids, a careful revaluation of character states and species limits within the group would be timely, though this is beyond the scope of the present review.

2.2. African and Eurasian Palaeognaths: Struthioniformes

Two ostrich species are extant. The Common Ostrich *Struthio camelus* inhabits open areas across much of sub-Saharan Africa, and the Somali Ostrich *Struthio molybdophanes* of Eastern Africa was once considered conspecific with *S. camelus* but is now given species status [17,138]. While the two extant species of ostrich are now confined to Africa, their range extended into Asia during the Holocene. Ostriches may have persisted as far east as Mongolia until 7500 years ago based on Carbon-14 dating of eggshells [139] (though see Khatsenovich, et al. [140] regarding uncertainties surrounding the dating of ostrich eggs from Mongolia and Siberia), and ostriches of the subspecies *S. c. syriacus*, whose native range stretched from the Arabian Peninsula to Syria and Iraq, did not become extinct until 1966 [17]. Ostriches are arguably the most cursorial of all birds, able to run at speeds in excess of 70 km per hour [67]. Their extreme cursoriality is evinced by their unique foot morphology: ostriches are the only extant didactyl birds, an anatomical configuration that may be the result of similar selective pressures as those that drove digit reduction in horses [77]. The fossil record of ostrich eggshell is rich, and although the present review focuses only on skeletal remains, we note that the occurrence of palaeognath eggshells in the early Miocene of China 17 million years ago [77,141] supports the theory that struthionids either originated outside of Africa or else underwent rapid range expansion after their emergence. For a thorough review of the ostrich eggshell record, see Mikhailov and Zelenkov [78].

2.2.1. Eurasian Stem Struthionids

Our understanding of palaeognath evolution and particularly the transition to flight-lessness in ratites has been hampered by a lack of recognizable stem group representatives of extant palaeognath lineages. Fortunately, recent research advances have provided a valuable window into the nature of early stem struthionids, which were previously unknown prior to the Miocene. The flightless palaeognaths *Palaeotis weigelti* and *Remiornis heberti* have long been known from the Paleogene of Europe [76,142–145], but their relation to the remainder of Palaeognathae was unclear [76,142]. *Palaeotis*, the better-known of the two taxa, has been variably recovered as the sister taxon to rheids [146], sister to a clade including Struthionidae, Rheidae, and Casuariidae [147], and sister to a clade comprised of lithornithids and tinamous [33]. The unconstrained analysis of Nesbitt and Clarke [64] recovered *Palaeotis* outside a *Struthio* + *Dromaius* + *Rhea* clade. When relationships of living palaeognaths were constrained to match those recovered by molecular phylogenies, the same authors recovered *Palaeotis* as the sister taxon of extant palaeognaths (to the exclusion of lithornithids). Mayr [142] noted the resemblance of the skull of *Palaeotis* to that of lithornithids, and that the scapulocoracoid differs from all extant ratites, but was unable to find a well-supported placement for *Palaeotis* and proposed that it may represent yet another independent acquisition of ratite features among palaeognaths. The phylogenetic position of *Remiornis heberti* was also challenging to estimate with confidence. Mayr [76] considered that it may belong with Palaeotididae before amending this hypothesis based upon the lack of a supratendinal bridge and extensor sulcus in *Remiornis*, both of which are present in *Palaeotis* [148].

Without information on its palatal anatomy, it would be extremely difficult to recognize *Palaeotis* as a palaeognath on the basis of its postcranial skeleton, as several aspects of its hindlimb morphology, such as a notch in the distal rim of the medial condyle of the tibiotarsus and intratendinous ossifications on the tarsometatarsus, are unusual for palaeognaths and are more reminiscent of Gruiformes [148]. Recently, Mayr [148] transferred *Galligeranoides boriensis* from the stem gruiform clade Geranoididae [149] to Palaeotididae. *G. boriensis* had been described on the basis of leg bones from the early Eocene of France [150]. Its initial assignment to Geranoididae was notable, as this clade was only known from the Eocene of North America [76,149,151]. The transfer of *G. boriensis* from Geranoididae to Palaeotididae raises the possibility that additional records of early palaeognaths could be hiding in plain sight in museum collections, misidentified due to their lack of obvious palaeognath synapomorphies.

This scenario was indeed the case with Eogruidae, a group of crane-sized birds known primarily from hindlimb elements from Central Asia. Since the remainder of the skeleton of eogruids was virtually unknown, these taxa were difficult to place phylogenetically. Eocene eogruids show a trend towards reduction in the size of the inner toe as a possible adaptation for cursoriality [152], and later eogruids of the subclade Ergilornithidae take this trend even further, to the point where the inner toe is vestigial or absent [148,152]. This feature led several earlier authors to hypothesize a placement for Eogruidae as stem struthionids [153–155]. However, this hypothesis was not widely accepted, and eogruids were generally viewed as representatives of Gruiformes (either as sister to a clade containing Aramidae and Gruidae [156] or sister to Gruidae [149]), implying that the didactyly of some eogruids was convergent with Struthionidae.

A previously undescribed partial skull PIN 3110–170 from the latest Eocene locality of Khoer Dzan, Mongolia has rendered the hypothesis of eogruids as gruiforms untenable [6]. Although the palate is missing, the skull preserves an articular surface for the otic capitulum of the quadrate, but apparently does not exhibit an articular surface for the squamosal capitulum of the quadrate. Both articular surfaces would be expected for a gruiform, and indeed for most neognaths, which have a bipartite otic process of the quadrate. Instead, the skull appears to genuinely exhibit only one articular facet for the quadrate, a condition seen only in palaeognaths [157]. This feature, in combination with the reduction and eventual loss of the inner toe, strongly indicate a stem struthioniform placement for Eogruidae. If taxa with greater toe reduction are more closely related to crown struthionids, eogruids would form a paraphyletic grade along the ostrich stem lineage [6] (Figure 9).

With the reassignment of Eogruidae, there is now a clear record of stem Struthionidae in Eurasia well before the first crown struthionids appear in the Miocene of Africa. It now appears likely that this iconic clade of extant African birds first arose outside the continent. In addition to recognizing eogruids as stem struthionids, Mayr and Zelenkov [6] also hypothesized that *Palaeotis* represents a total-clade struthionid based upon similarities in the shape of its skull with the newly described specimen. With palaeotidids interpreted as stem struthionids, the case for a Eurasian origin of Struthioniformes is strengthened even further (Figure 9).

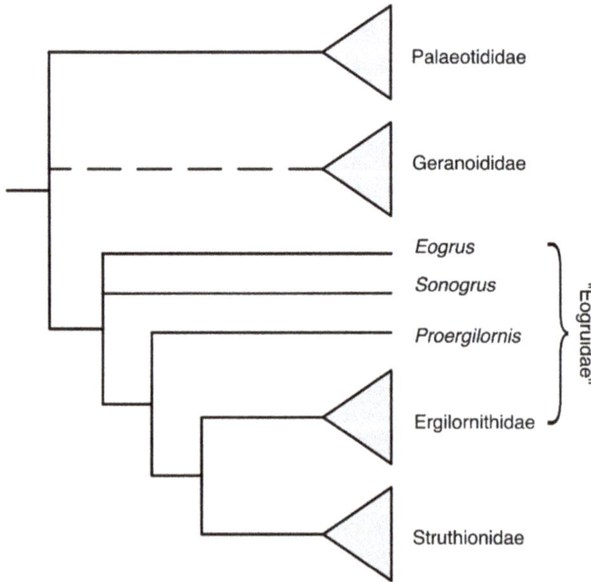

Figure 9. Relationships within Struthioniformes as hypothesized by Mayr and Zelenkov [6]. "Eogruidae" is here estimated to be a paraphyletic grade of crownward stem struthioniforms, and Geranoididae is tentatively inferred to be a clade of early stem struthioniforms.

The oldest flightless, non-lithornithid palaeognaths in Eurasia belong to Palaeotididae. *Galligeranoides boriensis* is now the oldest known probable palaeotidid, found in rocks ranging between the ages of 56 to 51 Ma [158]. It is known from a right tibiotarsus, a distal portion of a left tibiotarsus, and an incomplete right tarsometatarsus [150]. The nominate and best known palaeotidid, *Palaeotis weigelti*, was initially interpreted as a bustard [145] and subsequently as a crane [159] before it was finally recognized as a palaeognath by Houde and Haubold [143], who hypothesized that it was as a stem ostrich despite its lack of obvious cursorial adaptations, an assessment that, in light of the recent work discussed above, has gained robust support. *P. weigelti* is known from six specimens from the middle Eocene of the Messel and Geisel Valley sites of Germany (Table 2). One of these specimens is a complete two-dimensionally preserved skeleton. It stood slightly under 1 m tall, and was more gracile than the older *Remiornis* [76].

Table 2. Fossil record of stem struthioniforms.

Continent	Geological Unit	Location	Epoch	Stage	Age Reference	Taxa	Institutions	Reference
Europe	Châlons-sur-Vesles Formation	Cernay and Berru, Marne, France	late Paleocene	Thanetian	Buffetaut and Angst [160]	*Remiornis heberti*	MNHN	Lemoine [144]; Martin [161]; Mayr [76]
	Sables de Bracheux Formation	Rivecourt, France	late Paleocene	Thanetian	Smith, et al. [162]	*Remiornis heberti*	MV	Buffetaut and de Ploëg [163]
	Argiles rutilantes d'Issel et de Saint-Papoul	Saint-Papoul, France	early Eocene	Ypresian	Laurent, et al. [164]; Danilo, et al. [165]	*Galligeranoides boriensis*	MHNT	Bourdon, et al. [150]; Mayr [148]
	Messel Formation	Messel, Germany	middle Eocene	Ypresian-Lutetian	Franzen and Haubold [108]; Schaal and Ziegler [109]; Lenz, et al. [110]	*Palaeotis weigelti*	HLMD	Peters [146]; Houde and Haubold [143]; Mayr [142]
	Geiseltal brown coal	Geisel Valley lignite pits, Germany	middle Eocene	Lutetian	Franzen and Haubold [108]	*Palaeotis weigelti*	GMH	Lambrecht [145]; Houde and Haubold [143]; Mayr [142]; Mayr [148]
	unlisted	Kolkotova Balka, Tiraspol, Moldova	late Miocene	Tortonian-Messinian	Zelenkov and Kurochkin [166]	*Urmiornis ukrainus*	PIN	Zelenkov and Kurochkin [166]
	unlisted	Hrebeniki, Odessa Oblast, Ukraine	late Miocene	Tortonian-Messinian	Zelenkov and Kurochkin [166]	*Urmiornis ukrainus*	NNPM	Zelenkov and Kurochkin [166]
	unlisted	Morozovka, Odessa Oblast, Ukraine	late Miocene	Tortonian-Messinian	Zelenkov and Kurochkin [166]	*Urmiornis ukrainus*	NNPM	Zelenkov and Kurochkin [166]
	unlisted	Armavir, Krasnodar Krai, Russia	late Miocene	Tortonian-Messinian	Zelenkov and Kurochkin [166]	*Urmiornis ukrainus*	Armavir Regional Museum	Zelenkov and Kurochkin [166]
	unlisted	Samos, Greece	late Miocene	Tortonian	Zelenkov, et al. [167]	*Ampipelargus majori*	NHMUK	Lydekker [168]; Zelenkov, et al. [167]
	Triglia Formation	Kryopigi, Chalkidiki, Greece	late Miocene	Tortonian-Messinian	Tsoukala and Bartsiokas [169]; Lazaridis and Tsoukala [170]	*?Ampipelargus* sp.	AUG	Boev, et al. [171]; Zelenkov, et al. [167]
Asia	Irdin Manha Formation	Shara Murun region, Inner Mongolia, China	middle Eocene	Lutetian	Li [172]	*Eogrus aeola*	AMNH, PIN	Wetmore [173]; Kurochkin [152]; Zelenkov and Kurochkin [166]
	Khaichin Formation	Omnogvi Province, Mongolia	middle Eocene	Lutetian	Zelenkov and Kurochkin [166]	*Eogrus aeola*	PIN	Zelenkov and Kurochkin [166]

Table 2. *Cont.*

Continent	Geological Unit	Location	Epoch	Stage	Age Reference	Taxa	Institutions	Reference
	Obayla Formation	Kalmakpai River, East Kazakstan	late Eocene	Priabonian	Clarke, et al. [156]	*Eogrus turanicus*	PIN	Bendukidze [174]; Zelenkov and Kurochkin [166]
	unlisted	Tsagan Khutel, Bayanhongor Province, Mongolia	late Eocene	Priabonian	Russell and Zhai [175]	*Eogrus crudus*	PIN	Kurochkin [176]; Zelenkov and Kurochkin [166]
	unlisted	Alag Tsav, Dornogovi Province, Mongolia	late Eocene	Priabonian	Dashzèvèg [177]; Clarke, et al. [156]	Eogruidae incertae sedis	IGM	Clarke, et al. [156]
	Kustovskaya Formation	East Kazakstan	late Eocene	Priabonian	Musser, et al. [178]	*Eogrus* sp.	PIN	Kozlova [179]; Kurochkin [176]; Musser, et al. [178]
	Ergilin Dzo Formation	Dornogovi Province, Mongolia	latest Eocene-earliest Oligocene	Priabonian-Rupelian	Dashzèvèg [177]	*Eogrus* sp., *Ergilornis rapidus*, *Ergilornis minor*, *Ergilornis* sp., Ergilornithidae incertae sedis, *Sonogrus gregalis*	PIN	Wetmore [173]; Kozlova [179]; Kurochkin [152]; Kurochkin [176]; Zelenkov and Kurochkin [166]; Mayr and Zelenkov [6]
	unlisted	Mynsualmas, Kazakstan	early Miocene	Aquitanian-Burdigalian	Karhu [180]; Zelenkov and Kurochkin [166]	*Urmiornis brodkorbi*	PIN	Karhu [180]; Zelenkov and Kurochkin [166]
	Upper Aral Formation	Altynshokysu, Kazakstan	early Miocene	Aquitanian-Burdigalian	Karhu [180]; Zelenkov and Kurochkin [166]	*Urmiornis brodkorbi*	PIN	Karhu [180]; Zelenkov and Kurochkin [166]
	Tunggur Formation	Shara Murun region, Inner Mongolia, China	middle Miocene	Serravallian	Wang, et al. [181]	*Eogrus wetmorei*	AMNH	[173]; Brodkorb [182]; Cracraft [183]
	unlisted	Sharga, Govi-Altai Province, Mongolia	middle Miocene	Serravallian	Musser, et al. [178]	*Ergilornis* sp.		Zelenkov, et al. [167]; Musser, et al. [178]
	Nagri and Chinji Formations	Gilgit-Baltistan, Pakistan	late middle- early late Miocene	Serravallian-Tortonian	Barry, et al. [184]	? *Urmiornis cracrafti*		Harrison and Walker [185]; Musser, et al. [178]
	unlisted	Maragheh, Iran	late Miocene	Tortonian-Messinian	Musser, et al. [178]	*Urmiornis maraghanus*	MNHN	Mecquenem [186]
	Lower Pavlodar Formation	Pavlodar, Kazakstan	late Miocene	Tortonian-Messinian	Zelenkov and Kurochkin [166]	*Urmiornis* sp.	PIN	Kurochkin [176]; Zelenkov and Kurochkin [166]

Table 2. *Cont.*

Continent	Geological Unit	Location	Epoch	Stage	Age Reference	Taxa	Institutions	Reference
	Karabulak Formation	Kalmakpai, Zaisan, East Kazakstan	late Miocene	Tortonian-Messinian	Zelenkov and Kurochkin [166]	*Urmiornis orientalis*	PIN	Kurochkin [176]; Zelenkov and Kurochkin [166]
	Liushu Formation	Zhuangeji town, Gansu, China	late Miocene	Messinian	Fang, et al. [187]	*Sinoergilornis guanheensis*	IVPP	Musser, et al. [178]
	Khirgis-Nur Formation	Khirgis-Nur, Sunur Province, Mongolia	late Miocene	Messinian	Zelenkov and Kurochkin [166]	*Urmiornis* sp.	PIN	Kurochkin [176]; Zelenkov and Kurochkin [166]
	Khirgis-Nur Formation	Chono-Khariakh, Kobdos Province, Mongolia	early Pliocene	Zanclean	Zelenkov and Kurochkin [166]	*Urmiornis dzabghanensis*	PIN	Kurochkin [188]; Zelenkov and Kurochkin [166]
	Khirgis-Nur Formation	Dzagso-Khairkhan-Obo, Ubsunur Province, Mongolia	early Pliocene	Zanclean	Zelenkov and Kurochkin [166]	*Urmiornis dzabghanensis*	PIN	Kurochkin [188]; Zelenkov and Kurochkin [166]

Eogruids are younger than Palaeotididae, occurring from the middle Eocene to the early Pliocene, and comprise fifteen named species in six genera (Table 2). The oldest species, *Eogrus aeola*, has been collected from the middle Eocene of Inner Mongolia and Mongolia's Omnogvi Province [152,166,173] (Table 2). Like nearly all eogruids, it is known only from hindlimb elements. Other members of this genus from the late Eocene include *Eogrus crudus* from central Mongolia [176], and *Eogrus turanicus* from Eastern Kazakhstan [174] (Table 2).

Outcrops of the latest Eocene-earliest Oligocene Ergilin Dzo Formation in Dorngovi Province, Mongolia have produced an enormous wealth of eogruid fossils. It is in this formation that Ergilornithidae first appear. Once recognized as a separate family [179], they are now considered a subclade of Eogruidae [156,167]. Ergilornithids recovered from this formation include *Ergilornis rapidus* [179], *Ergilornis minor* [176,179], and *Sonogrus gregalis* [176] (Table 2). The partial skull PIN 3110–170 was collected from the latest Eocene Sevkhul member of this formation [6,155]. As the Sevkhul member has produced huge quantities of hindlimb material belonging to *Sonogrus gregalis* and *Ergilornis minor* and no other large birds, the skull was presumed to belong to one of the two species [6].

We were unable to find any documented occurrences of this clade for the remainder of the Oligocene. The ergilornithid genus *Urmiornis* first appears in the early Miocene, with two occurrences of *Urmiornis brodkorbi* in western Kazakhstan [180]. The latest occurrence of the genus *Eogrus* is in the middle Miocene of Inner Mongolia with *Eogrus wetmorei* [173,182,183]. By the late Miocene, eogruids had expanded their range outside of Central Asia and reached their greatest generic diversity, with *Amphipelargus majori* occurring on Samos island [167,168] and another member of the same genus on the Greek mainland [167,171], *Urmiornis ukrainus* occurring in Ukraine, Moldova, and southwestern Russia [166,176], *Urmiornis maraghanus* in Iran [183,186,189], *?Urmiornis cracrafti* in the Siwaliks of northern Pakistan [185], and *Sinoergilornis guangheensis* in Gansu, China [178] (Table 2). Although Kurochkin [176] noted differences between *U. ukrainus* and *U. maraghanus*, the validity of *U. ukrainus* requires further conformation and *U. maraghanus* would take nomenclatural priority if they are shown to be the same species [166]. The group continued to thrive in their Central Asian stronghold, with *Urmiornis orientalis* found near Zaisan, Kazakhstan [166,176] and *Urmiornis* sp. in the Sunur province of Mongolia and Pavlodar, Kazakhstan [166,176]. The youngest species, *Urmiornis dzabghanensis*, was found in the early Pliocene Khirgis-Nur Formation of Mongolia [166,188] (Table 2).

The possibility that the eogruids were flightless has been proposed by several authors [152,173], though others contend that such a conclusion is premature based on existing evidence [156,178]. The trochlea for the second toe is vestigial or entirely absent in *Ergilornis*, *Sinoergilornis*, *Urmiornis*, and *Ampipelargus* [6,166,176,178], which is indicative of a highly cursorial lifestyle as seen in extant struthionids. In addition, a proximal humerus PIN 3110–60 from the Ergilin Dzo Formation attributed to *Ergilornis* has a greatly reduced deltopectoral crest (the portion of the humerus serving as the major insertion point for major flight muscles), and from this it was assumed that at least this taxon was flightless [152]. If some eogruids were volant, it could imply that multiple transitions to flightlessness occurred among stem struthionids, following the phylogeny of Mayr and Zelenkov (Figure 9) [6].

That the North American Geranoididae may also be struthioniforms has been suggested on several occasions, but unlike Eogruidae no strong evidence for such a placement has yet been found [6,148,155]. Geranoidids share several derived features with Palaeotididae, including an elongated tarsometatarsus, a pronounced extensor sulcus along the dorsal surface of the tarsometatarsus, a proximodistally elongated hypotarsus that forms a long medial crest, and a notched distal rim of the medial condyle of the tibiotarsus [148]. With the recent reassignment of *G. boriensis* (discussed above), an investigation into possible palaeognath affinities for fossils assigned to the remaining members of this clade is clearly merited. *Eogeranoides campivagus* from the Wilwood Formation of Wyoming has a deep extensor sulcus along the dorsal surface of the tarsometatarsus, a feature it shares with *Palaeotis* [142,148]. Considering that North American and European avifaunas were generally similar during the Eocene [114,148], and that certain flightless bird taxa such as Gastornithidae occurred on both sides of the Atlantic [76,77], the possibility that palaeotidids existed in North America is plausible. A clade uniting Palaeotididae, Geranoididae, Eogruidae, and Struthionidae is supported by the following characters highlighted by Mayr and Zelenkov [6]: a very long and narrow tarsometatarsus, a short trochlea for digits II and IV, a tubercle adjacent to the supratendinal bridge, and a shortening of all non-ungual phalanges on pedal digit IV.

Also uncertain is the placement of *Remiornis heberti* [144] from the late Paleocene of France [161] (Table 2). It is known from several isolated elements belonging to different individuals that include a tibiotarsus, tarsometatarsus, and fragmentary associated remains [76,161,163]. It appears to have been recognized as a palaeognath based on its overall resemblance to *Palaeotis*, as the two genera share a deep furrow on the dorsal surface of the tarsometatarsus and a similar configuration of the distal trochleae [76]. Mayr [148] excluded it from Palaeotididae based on its lack of an ossified supratendinal bridge and extensor sulcus, and Mayr and Zelenkov did not include *Remiornis* at all in their new hypothesis of struthioniform interrelationships [6]. However, in light of the variability exhibited by the supratendinal bridge, extensor sulcus, and hypotarsus among palaeognaths, rejecting a struthioniform affinity for *Remiornis* may be premature. An ossified supratendinal bridge of the tibiotarsus is present in Tinamidae and Dinornithidae and is variably present in Apterygidae, but is missing from all other crown palaeognaths [137,148]. Worthy et al. [137] note that given its variability in clades including crown Palaeognathae and Cariamiformes, the presence or absence of this feature should not be viewed to negate potential sister relationships. The extensor sulcus of the tibiotarsus is also variably present in palaeognaths. It is narrow in Lithornithidae, Apterygidae, Tinamidae, and Dinornithidae, and absent in Struthionidae, Casuariidae, Rheidae, and Aepyornithidae [148]. Eogruids have a hypotarsal canal, while all other palaeognaths lack this feature [148]. The putative gruid *Palaeogrus princeps* [190] from the middle Eocene of Italy also shares similarities in the distal tibiotarsus with *Palaeotis* and could represent yet another record of this clade [148].

Several other taxa that deserve further revision of their taxonomic placement are listed here, though it is far less likely that they belong within Palaeognathae. *Eleutherornis cotei* [191,192] from the middle Eocene of Switzerland and France is known from a partial pelvis and hindlimb elements and was originally assumed to be a ratite due to its large size, but was reinterpreted as a phorusrhacoid [193]. *Eremopezus eocaenus* [194] is known from hindlimb elements from the late Eocene Fayum Formation of Egypt [76,195]. Rasmussen, et al. [195] suggest that it could represent a non-palaeognathous endemic African group that independently became large and flightless. More material will be needed to firmly rule out palaeognathous affinities for this taxon [76]. Whether or not these species are indeed palaeognaths, we expect that further revaluation of Paleogene fossil collections is bound to reveal more palaeognaths from a critical time period that may capture their transitions to flightlessness.

2.2.2. African and Eurasian Crown Struthionids

As shown in Table 3.

Table 3. Crown struthionid fossil record.

Continent	Geological Unit	Location	Epoch	Stage	Age Reference	Taxa	Institutions	Reference
Africa	Elisabethfeld silts	Northern Sperrgebiet, Namibia	early Miocene	Aquitanian	Pickford and Senut [196]	*Struthio coppensi*		Mourer-Chauviré, et al. [197]; Mourer-Chauviré [198]
	unlisted	Kadianga West, Kenya	middle Miocene	Langhian	Pickford [199]	*Struthio sp.*	KNM	Leonard, et al. [200]
	unlisted	Central Nyanza, Kenya	middle Miocene	Serravallian	Pickford [199]	*Struthio sp.*	KNM	Leonard, et al. [200]
	unlisted	Ngorora, Kenya	middle Miocene	Serravallian	Pickford [199]	*Struthio sp.*	KNM	Leonard, et al. [200]
	Beglia Formation	Bled el Douarah, Tunisia	late Miocene	Tortonian	Werdelin [201]	*Struthio sp.*		Rich [202]
	Varswater Formation	Langebaanweg, South Africa	early Pliocene	Zanclean	Roberts, et al. [203]	*Struthio cf. asiaticus*		Rich [204]; Manegold, et al. [205], but see Mikhailov and Zelenkov [78]
	unlisted	Ahl al Oughlam, Casablanca, Morocco	late Pliocene	Piacenzian	Geraads [206]	*Struthio asiaticus*		Mourer-Chauviré and Geraads [207], but see Mikhailov and Zelenkov [78]
	Olduvai series	Olduvai Gorge Bed I, Tanzania	early Pliestocene	Gelasian	Hay [208]	*Struthio oldawayi*		Lowe [209]; Leakey [210]
	unlisted	Aïn Boucherit, Algeria	early Pleistocene	Gelasian	Werdelin [201]	*Struthio barbarus*		Arambourg [211]; Mikhailov and Zelenkov [78]
Asia	Turgut strata	Çandir, Turkey	middle Miocene	Langhian	Becker-Platen, et al. [212]	*Struthio cf. brachydactylus*	BGR	Sauer [213]
	unlisted	Maragha, Iran	late Miocene	Tortonian		*Palaeostruthio karatheodoris*		Mecquenem [189]; Lambrecht [214]; Mikhailov and Zelenkov [78]
	Baynunah Formation	United Arab Emirates	late Miocene	Tortonian		*Palaeostruthio karatheodoris*		Louchart, et al. [215]
	unlisted	Pavlodar, Kazakhstan	late Miocene	Messinian (?)		*Palaeostruthio karatheodoris*		Tugarinov [216]; Kurochkin [188]; Mikhailov and Zelenkov [78]
	Liushu Formation	Gansu province, China	late Miocene	Tortonian-Messinian	Deng, et al. [217]	*Struthio (Orientornis) linxiaensis*		Hou, et al. [218]

Table 3. *Cont.*

Continent	Geological Unit	Location	Epoch	Stage	Age Reference	Taxa	Institutions	Reference
	unlisted	Baode county, China	late Miocene	Messinian	Kaakinen, et al. [219]	*Struthio wimani*		Lowe [220]; Mikhailov and Zelenkov [78]
	Dhok Pathan Formation?, Siwalik series	Siwalik Hills, India	late Miocene-early Pliocene	Messinian-Zanclean	Sahni, et al. [221]; Sahni, et al. [222]; Stern, et al. [223]; Patnaik, et al. [224]	*Struthio asiaticus*		Davies [225]; Lydekker [226]; Mikhailov and Zelenkov [78]
	unlisted	Çalta, Ankara, Turkey	early Pliocene	Zanclean	Ginsburg, et al. [227]; Sen [228]; Janoo and Sen [229]	*Struthio sp.*		Janoo and Sen [229]
	unlisted	Pavlodar, Kazakhstan	early Pliocene	Zanclean		*Struthio chersonensis*		Beliaeva [230]
	upper Issykulian Formation	Akterek, Kyrgyzstan	late Pliocene	Piacenzian	Sotnikova, et al. [231]	*Pachystruthio transcaucasius*		Sotnikova, et al. [231]
	Nihewan Formation	Nihewan Basin, China	early Pleistocene	Gelasian	Cai, et al. [232]	*Pachystruthio indet.*	MNHN	Buffetaut and Angst [233]
	unlisted	Zhoukoudian, China	middle-late Pleistocene	Calabrian-Chibanian		*"Struthio anderssoni"*		Hou [234]
Europe	unlisted	Varnitsa, Moldova	late Miocene	Tortonian	Vangengeim and Tesakov [235]	*Struthio orlovi*		Kurochkin and Lungu [236]
	unlisted	Pikermi, Greece	late Miocene	Tortonian	Solounias, et al. [237]	*Palaeostruthio cf. karatheodoris*		Bachmayer and Zapfe [238]; Michailidis, et al. [239]
	Nikiti Formation	Nikiti, Greece	late Miocene	Tortonian		*Palaeostruthio cf. karatheodoris*		Koufos, et al. [240]
	unlisted	Hadzhidimovo, Bulgaria	late Miocene	Tortonian	Spassov [241]	*Palaeostruthio karatheodoris*	NMNHS	Boev and Spassov [242]
	unlisted	Novoelizavetovka, Ukraine	late Miocene	Tortonian-Messinian	Vangengeim and Tesakov [235]	*Struthio novorossicus*	ONU	Aleksejev [243]; Mikhailov and Zelenkov [78]
	unlisted	Kuyal'nik, Ukraine	late Miocene	Tortonian-Messinian		*Struthio sp.*		Burchak-Abramovich [244]; Mikhailov and Zelenkov [78]
	unlisted	Samos, Greece	late Miocene	Tortonian-Messinian		*Palaeostruthio karatheodoris*	MGL	Forsyth Major [245]; Mikhailov and Zelenkov [78]

Table 3. *Cont.*

Continent	Geological Unit	Location	Epoch	Stage	Age Reference	Taxa	Institutions	Reference
	Strumyani Genetic Lithocomplex	Kamimantsi, Bulgaria	late Miocene	Tortonian-Messinian	Tzankov, et al. [246]; Spassov, et al. [247]	*Palaeostruthio cf. karatheodoris*	NMNHS	Boev and Spassov [242]
	unlisted	Kerassia, Greece	late Miocene	Tortonian-Messinian	Theodorou, et al. [248]	*Palaeostruthio karatheodoris*		Kampouridis, et al. [249]
	unlisted	Grebeniki, Ukraine	late Miocene	Tortonian	Vangengeim and Tesakov [235]	*Palaeostruthio karatheodoris, Struthio brachydactylus*		Burchak-Abramovich [250]; Mikhailov and Zelenkov [78]
	Odessa Catacombs	Odessa, Ukraine	early Pliocene	Zanclean		*Struthio sp.* "Odessa Ostrich"	ONU	Burchak-Abramovich [244]; Mikhailov and Zelenkov [78]
	unlisted	Kvabebi, Georgia	late Pliocene	Piacenzian		*Pachystruthio transcaucasius*		Burchak-Abramovich and Vekua [251]; Mikhailov and Zelenkov [78]
	Khapry Formation	Liventzovka, Rostovskaya Oblast, Russia	early Pleistocene	Gelasian	Tesakov [252]; Tesakov, et al. [253]	*Struthio sp.* "Odessa Ostrich"		Kurochkin and Lungu [236]
	Sésklo basin sedimentary fill	Sésklo, Thessaly, Greece	early Pleistocene	Gelasian		*Struthio cf. chersonensis*		Athanassiou [254]
	unlisted	Dmanisi, Georgia	early Pleistocene	Gelasian	Ferring, et al. [255]	*Pachystruthio dmanisensis*		Burchak-Abramovich and Vekua [256]; Mikhailov and Zelenkov [78]
	Taurida Cave	Taurida, Crimea	early Pleistocene	Gelasian	Lopatin, et al. [257]	*Pachystruthio dmanisensis*		Lopatin, et al. [257]; Zelenkov, et al. [258]
	unlisted	Kisláng, Hungary	early-middle Pleistocene	Gelasian-Calabrian	Mayhew [259]	*Pachystruthio pannonicus*	GMB	Kretzoi [260]; Mikhailov and Zelenkov [78]

The body fossil record of crown ostriches begins 21 million years ago in the early Miocene of Africa with *Struthio coppensi* (Figure 7, Table 3), named on the basis of the shaft and distal part of a left tibiotarsus, proximal left femur, distal left tarsometatarsus, right tarsometatarsus shaft, and a left fibula from the early Miocene of the Northern Sperrgebiet, Namibia [197]. As noted by Mourer-Chauviré [198], it was smaller and more gracile than *S. camelus*, and a vestigial trochlea metatarsi II shows this early ostrich was didactyl [197,198]. A late middle Miocene ostrich from western Kenya assigned to *Struthio* also had a didactyl foot and was smaller than extant ostriches, though still larger than *S. coppensi* [200]. Other Kenyan middle Miocene ostrich fossils have been discovered, but they remain undescribed [78,261]. A distal tarsometatarsus was found from the middle-late Miocene boundary in Tunisia [201,202], indicating their presence in North Africa. The size of this bone is roughly comparable with that of the extant *S. camelus* [78].

No late Miocene ostrich body fossils have yet been found from sub-Saharan Africa, but they are relatively common in Eurasia during this period (Figure 7, Table 3) [78]. A pedal phalanx from the middle Miocene of Turkey is the oldest body fossil of crown struthionids outside Africa [213]. From the late Miocene onwards, this clade occupied an enormous geographical range, from the Balkans to northeastern China and eastern Siberia, and south to India. The oldest ostrich from Eastern Europe, *Struthio orlovi*, was found in the early late Miocene of Moldova [236]. Late Miocene Southern and Eastern European ostrich species limits are somewhat contentious. *S. karatheodoris* [245] was larger than extant ostriches [78], and many specimens from the Balkans have been referred to this taxon [238–240,242,249]. A large pelvis from the late Miocene of the United Arab Emirates was assigned to this species based on its size [215], and sacral vertebrae of a very large ostrich found in the terminal Miocene of northern Kazakhstan [188,216] may also belong to *S. karatheodoris* [78]. *S. novorossicus* [243] is considered a *nomem dubium* by Mikhailov and Zelenkov [78], as it cannot be distinguished from *S. asiaticus*. Koufos, et al. [240] suggested that *S. brachydactylus* [250] may be a junior synonym of *S. karatheodoris*, but Mikhailov and Zelenkov [78] consider them separate taxa, as *S. brachydactylus* was roughly the size of *S. camelus* and therefore much smaller than *S. karatheodoris*. Mikhailov and Zelenkov [78] refer *Palaeostruthio sternatus* [244] to *S. karatheodoris*, creating the new combination *Palaeostruthio karatheodoris*.

Struthio ("*Orientornis*") *linxiaensis* from the late Miocene of Gansu province, China is one of the oldest East Asian ostriches [77,218,262]. Slightly larger than *S. camelus*, Mikhailov and Zelenkov [78] argued that it likely belongs in its own genus, but tentatively treat it as *Struthio*. Other late Miocene Asian ostriches include *S. wimani*, known from a fragmentary pelvis from China [220], and *S. asiaticus* [263] from the Siwalik series in North India and Pakistan. The latter species has been treated as somewhat of a wastebasket taxon, with eggshell fragments attributed to it from sediments as young as the late Pleistocene of the Baikal region [264], and body fossils from as far away as South Africa [204,205] (Table 3). Ostrich eggshells ranging in age from 11 to 1.3 Ma are known from the Siwalik series [223]. However, the distribution, temporal range, and taxonomic identifications of these specimens are in need of revision.

Several large ostriches are known from the Pliocene. *S. transcaucasius* is known from a pelvis from the late Pliocene of Georgia [251] and was recently assigned to the genus *Pachystruthio* [258]. Many others have not been assigned to a species level taxon. It is evident from hindlimb fragments that a large ostrich existed in the lower Pliocene of South Africa, which was referred to *Struthio* cf. *asiaticus* [204,205]. Pliocene fossils from Ahl al Oughlam, Casablanca, Morocco, were also attributed to S. asiaticus [207]. Another large ostrich is known from the early Pliocene of Central Turkey [229]. An ostrich from Odessa, Ukraine, also from the early Pliocene, has only been assigned to *Struthio* [78,244].

Multiple species of large ostriches persisted through the Pleistocene. *Struthio oldawayi* of the early Pleistocene of Tanzania was similar to the extant *S. camelus*, though considerably larger [209,220]. Large Pleistocene ostrich bones from Kenya's Olduvai Gorge site may also belong to this species [210]. A large ostrich from the early Pleistocene of

Algeria was assigned to *S. barbarus* [201,211], and a middle Pleistocene cervical vertebra from the Nefud desert in northeastern Saudi Arabia bears a close resemblance to the extant *S. molybdophanes* [265]. Two giant Eurasian ostriches of the early Pleistocene, *Pachystruthio pannonicus* and *Struthio dmanisensis,* may be one species [258]. These birds were truly massive; a femur from the lower Pleistocene Taurida Cave of Crimea yields a mass estimate of 450 kg [258] using the equation of Field, et al. [266]. A 1.8-million-year-old right femur from Nihewan, North China may also belong to *Pachystruthio.* Assigned to *Pachystruthio indet.,* its estimated mass is a smaller, though still enormous 300 kg [233]. *S. anderssoni* of the late Pleistocene of eastern China [234] was 1.5 times the size of *S. camelus,* at about 270 kg based on estimates from its minimum femur circumference [267]. Why ostriches disappeared across Eurasia remains a mystery. One hypothesis is that their decline was at least partially linked to climatic cooling throughout the Cenozoic [77]. However, fossil eggshells indicating the possible persistence of ostriches in Mongolia well into the Holocene [139] (though again, see Khatsenovich, et al. [140]) would seem to negate such an explanation, and a stronger explanation for their disappearance is needed.

2.3. South American Palaeognaths: Rheiformes and Tinamiformes

South America is notable for being the only continent to host two family-level palaeognath clades that have persisted to the present day. Two species belong to Rheidae, the Greater Rhea *Rhea americana* and the Lesser Rhea or Darwin's Rhea *Rhea pennata* (alternatively *Pterocnemia pennata* in certain taxonomies). Both species are cursorial and inhabit open areas, with the Greater Rhea's range covering much of eastern and southern South America while the Lesser Rhea is found in Patagonia and the Altiplano region [68,268,269]. The Lesser Rhea was formerly placed in its own genus, *Pterocnemia,* but genetic studies suggest it is closely related to the Greater Rhea, with which it can hybridize [268,270]. There is some debate surrounding species limits among Lesser Rheas populations, as some consider the Altiplano subspecies *R. p. garleppi* and *R. p. tarapacensis* to form a separate species from the nominate Patagonian subspecies, *R. p. pennata* [268].

Tinamous (Tinamidae) are by far the most speciose extant palaeognath clade, and occupy a wide range of habitats in Central and South America [14]. The clade is divided into two major subclades, the forest-adapted Tinaminae which contains 29 species in the genera *Tinamus, Crypturellus,* and *Nothocercus,* and the open and arid habitat-dwelling Nothurinae, with 17 species in the genera *Taoniscus, Nothura, Nothoprocta, Rhynchotus, Eudromia,* and *Tinamotis* [14,42,271,272]. Like many ground-dwelling birds, tinamous have short wings relative to their body size which results in high wing loading [273]. High wing loading is associated with rapid flight but makes flight energetically costly [273], therefore tinamous tend to escape from threats on foot unless flight is necessary [61]. The pectoral muscles in tinamids are enormous relative to their body size, and allow for rapid takeoff to escape potential predators [273,274].

2.3.1. Rheid Fossil Record

The oldest named ratite, *Diogenornis fragilis,* provides a key minimum-bound age estimate for the evolution of larger body size and flightlessness among palaeognaths. The type specimen was found in the middle-late Paleocene of Itaboraí, Brazil and consists of limb bones, vertebrae, and the tip of a premaxilla deriving from several individuals [76,275]. The precise age of the Itaboraí fauna has been subject to debate, and an early Eocene age has also been suggested [276]. However, the distal end of a right tibiotarsus missing most of its lateral condyle from the even older middle Paleocene Rio Chico Formation of Argentina was also referred to this genus [277]. It was about two-thirds the size of the Greater Rhea, and its wings were less reduced [77]. For biogeographical reasons, *Diogenornis* is often presumed to be a stem rheiform [77,275]. However, Alvarenga [278] reported casuariid affinities for *Diogenornis,* and [277] also noted dissimilarities between the referred tibiotarsus and those of rheids. The cranial end of the medial condyle in medial view is larger and projects further distally than the caudal portion, which optimizes as a synapomorphy of casuariids [5,29].

While we consider it unlikely that *Diogenornis* represents a casuariiform, the phylogenetic affinities of these fossils remain somewhat uncertain. We conservatively treat *D. fragilis* as a total-clade rheid (Figure 7, Table 4). Another possible Paleogene rheid is represented by pedal phalanges from the middle Paleocene of Patagonia [279].

Table 4. Rheid fossil record.

Continent	Geological Unit	Location	Epoch	Stage	Age Reference	Taxa	Institutions	Reference
South America	Itaboraí Formation	São José, Brazil	late Paleocene	Selandian	Pascual and Ortiz-Jaureguizar [280]	*Diogenornis fragilis*		Alvarenga [275]
	Rio Chico Formation	Chubut province, Argentina	late Paleocene	Thanetian	Raigemborn, et al. [281]	*Diogenornis sp.*, Rheidae indet.	MACN	Tambussi [279]; Agnolín [277]
	Kolt el Kaike Formation	El Gauchito, Chubut province, Argentina	late Paleocene	Thanetian	Krause and Bellosi [282]	gen. et sp. indet.	MLP	Agnolín [277]
	Sarmiento Formation	Chubut province, Argentina	middle Eocene to early Miocene	unknown	Paredes, et al. [283]	gen. et sp. Indet.	MACN	Agnolín [277]
	Chichinales Formation	Río Negro province, Argentina	early Miocene	Burdigalian	Kramarz, et al. [284]	*Opisthodactylus horacioperezi*	MPCN	Agnolín and Chafrat [285]
	Santa Cruz Formation	Santa Cruz province, Argentina	early Miocene	Burdigalian-Langhian	Marshall and Patterson [286]; Fleagle, et al. [287]; Blisniuk, et al. [288]; Perkins, et al. [289]; Cuitiño, et al. [290]	*Opisthodactylus patagonicus*	NHMUK, MPM, YPM, MNHN	Ameghino [291]; Buffetaut [292]; Diederle and Noriega [293]
	Aisol Formation	Mendoza province, Argentina	early Miocene	Burdigalian-Langhian	Forasiepi, et al. [294]	*Pterocnemia cf. mesopotamica*	FMNH	Agnolín and Noriega [295]
	Level 13 of Ganduglia (1977)	Río Negro province, Argentina	middle Miocene	Langhian	Ganduglia [296]	gen et sp. indet.	MLP	Agnolín [277]
	Ituzaingó Formation	Entre Ríos province, Argentina	late Miocene	Messinian	Cione, et al. [297]	*Pterocnemia mesopotamica*, *Pterocnemia sp.*, Rheidae indet.	MACN, MASP, CICYTTP	Agnolín and Noriega [295]
	Cerro Azul Formation	La Pampa province, Argentina	late Miocene	Messinian	Cerdeño and Montalvo [298]; Verzi, et al. [299]	*Pterocnemia sp.*	GHUNLP	Cenizo, et al. [300]
	Andalhuala Formation	Tucumán province, Argentina	late Miocene-early Pliocene	Messinian-Zanclean	Marshall and Patterson [286]; Bossi and Muruaga [301]; Reguero and Candela [302]	*Opisthodactylus kirchneri*	MUFYCA	Noriega, et al. [303]
	Monte Hermoso Formation	Buenos Aires province, Argentina	early Pliocene	Zanclean	Deschamps, et al. [304]; Tomassini, et al. [305]	*Heterorhea dabbenei*, *Hinasuri nehuensis*	MLP	Rovereto [306]; Tambussi [279]

Other apparent ratite fossils from South America whose relations to modern palaeognaths are unclear are an incomplete right tibiotarsus from the middle Paleocene Koluel Kaike Formation of Argentina [277], a pedal phalanx from a poorly dated portion of the Sarmiento Formation that could be anywhere between middle Eocene and early Miocene in age [283], and a distal end of a tibiotarsus from the late Miocene of Patagonia [277]. By the late Miocene there was a marked increase in aridity across the continent, in contrast with the paratropical and warm temperate forests that stretched all the way south into Patagonia before this time [307]. Agnolín [277] puts forth the idea that this environmental change could have led to the extinction of hypothetical forest-adapted non-rheid ratites in South America, while favouring the open-habitat adapted rheids. Due to the high degree of anatomical homoplasy among the various ratite lineages, we may never know the true affinities of *Diogenornis* and these other unnamed ratite-like fossils with certainty, and can only hope that further fossil material will be found that can shed light on their proper phylogenetic placement and ecological habits.

Eocene bird records from South America are unfortunately rare in general [308]. The next oldest rheid fossils are significantly younger, dating from the Miocene (Figure 7, Table 4). *Pterocnemia mesopotamica* was found in the late Miocene of the Mesopotamia region of Argentina [295], and an isolated tarsometatarsus referred to *Pterocnemia cf. mesopotamica* could extend the temporal range of this species back to the middle Miocene [295]. *Opisthodactylus kirchneri*, another rheid from the late Miocene, was described on the basis of a right femur, a right and left tibiotarsus, left and right tarsometatarsi, and pedal phalanges [303]. The robust rheid *Hinasuri nehuensis* is known from a single left femur from the early Pliocene of Buenos Aires province, Argentina [309]. Extant rheid species appear in the Pleistocene, with *Rhea anchorenensis* [310] and *Rhea pampeana* [311] of the Pleistocene of Argentina reassigned to the extant Greater Rhea (*Rhea americana*) [312,313].

2.3.2. Tinamid Fossil Record

The oldest fossils belonging to crown group Tinamidae appear in the early Miocene Pinturas and Santa Cruz Formations of southern Patagonia (Figure 7, Table 5) [314–316]. This apparently abrupt appearance is most likely an artefact of the region's limited Eocene record. Molecular divergence time estimates suggest that the origin of crown Tinamidae occurred in the late Eocene or early Oligocene, concurrent with large-scale cooling and the emergence of open habitat in South America that led to turnover of the region's mammalian fauna [42,317]. Most of these early Miocene fossils are fragmentary and cannot be identified at a generic level, though phylogenetic analyses placed them within the open habitat-specialised tinamid subclade Nothurinae [42,315]. A left humerus from the Santa Cruz Formation was described as a new species, *Crypturellus reai* (*Crypturellus* is an extant genus within the tinamid subclade Tinaminae, which is sister to Nothurinae [316]). Fragmentary remains from the late Miocene were assigned to the extant genera *Eudromia* and *Nothoprocta* [300], both of which belong to Nothurinae. Only two species have been assigned to genera that are no longer extant: *Roveretornis intermedius* and *Tinamisornis parvulus*, both from the early Pliocene Monte Hermoso Formation [306,318], and *Tinamisornis* was later referred to the extant genus *Eudromia* [319]. The extinct *Eudromia olsoni* was also described from the same formation [320], and *Nothura parvula* was found alongside the extant *Nothura darwinii* and *Eudromia elegans* in the late Pliocene Chapadmalal Formation [308,321,322]. More recently, *Nothura parvula* was placed as sister to a *Nothura* + *Taoniscus* + *Rynchotus* + *Nothoprocta* clade [42]. As-yet undiscovered representatives of the Tinamidae stem group, which will likely be Eocene in age, are sorely needed to better understand the evolutionary history of this group, and whether the ancestors of crown tinamids were adapted for flight styles other than the highly specialized burst flight seen in tinamous today.

Table 5. Tinamid fossil record.

Continent	Geological Unit	Location	Epoch	Stage	Age Reference	Taxa	Institutions	Reference
South America	Pinturas Formation	Santa Cruz province, Argentina	early Miocene	Burdigalian	Fleagle, et al. [287]	Tinamidae gen. et sp. indet	MACN	Bertelli and Chiappe [315]
	Santa Cruz Formation	Santa Cruz province, Argentina	early Miocene	Burdigalian	Marshall and Patterson [286]; Fleagle, et al. [287]; Blisniuk, et al. [288]; Perkins, et al. [289]; Cuitiño, et al. [290]	*Crypturellus reai*, Tinamidae gen. et sp. indet	MPM, MACN, AMNH	Bertelli and Chiappe [315]; Chandler [316]
	Cerro Azul Formation	La Pampa province, Argentina	late Miocene	Messinian	Cerdeño and Montalvo [298]; Verzi, et al. [299]	*Eudromia* sp., *Nothura* sp.	MLP, GHUNLP	Cenizo, et al. [300]
	Monte Hermoso Formation	Buenos Aires province, Argentina	early Pliocene	Zanclean	Deschamps, et al. [304]; Tomassini, et al. [305]	*Eudromia olsoni*, *Eudromia cf. elegans*, *Roveretornis intermedius*, *Tinamisornis parvulus*	MACN	Brodkorb [318]; Tambussi and Tonni [320]; Tomassini, et al. [305]
	Chapadmalal Formation	Buenos Aires province, Argentina	late Pliocene	Zanclean-Placenzian	Marshall, et al. [323]; Deschamps, et al. [304]	*Eudromia elegans*, *Eudromia* sp., *Nothura parvula*, *Nothura darwinii*	MLP	Tambussi and Noriega [324]; Tambussi and Degrange [308]

2.4. Australian Ratites: Casuariiformes

Both the cursorial emu and the graviportal cassowary belong to the family-level clade Casuariidae [325]. The Emu *Dromaius novaehollandiae* is the only member of its genus, with the recently extinct dwarf Kangaroo Island Emu *D. baudinianus* [326], King Island Emu *D. minor* [327], and Tasmanian Emu *D. diemenensis* [328] now considered to be subspecies of *D. novaehollandiae* [329–331]. Emu are found across most of continental Australia, with the exception of areas of sandy desert and dense forest [332]. Cassowaries have an extremely distinctive appearance, with a casque on the head and wattles on the neck. Unlike Emu, cassowaries typically inhabit dense rainforest habitats. Three cassowary species are currently accepted: the Southern Cassowary *Casuarius casuarius*, the Dwarf Cassowary *Casuarius bennetti*, and the Northern Cassowary *Casuarius unappendiculatus* [66]. All three species inhabit the island of New Guinea, and the Southern Cassowary's range extends into northeastern Queensland, Australia, and some adjacent islands. No casuariiform fossils are known before the Late Oligocene [333], and thus far there is no indication that any other palaeognath lineage has ever been present in Australia (Figure 7, Table 6).

Table 6. Casuarid fossil record.

Continent	Geological Unit	Location	Epoch	Stage	Age Reference	Taxa	Institutions	Reference
Australia	Etadunna Formation	Lake Palankarinna, South Australia, Australia	late Oligocene	Chattian	Woodburne, et al. [334]; Megirian, et al. [335]	*Emuarius guljaruba*	SAM	Boles [333]
	Wipajiri Formation	Etadunna Station, South Australia, Australia	latest Oligocene-early Miocene	Chattian-Aquitanian	Woodburne, et al. [334]; Megirian, et al. [335]	*Emuarius gidju*	SAM, AM	Patterson and Rich [336]; Boles [337]
	Riversleigh faunal zones A-C	Riversleigh, Queensland, Australia	latest Oligocene-middle Miocene	Chattian-Langhian	Archer, et al. [338]; Travouillon, et al. [339]; Megirian, et al. [335]	*Emuarius gidju*	AM, QM	Boles [337]; Boles [340]; Worthy, et al. [341]
	Camfield beds	Bullock Creek, Northern Territory, Australia	middle Miocene	unknown	Woodburne, et al. [342]	*Dromaius* sp.		Rich [343]; Rich and Van Tets [344]
	Waite Formation	Alcoota, Northern Territory, Australia	late Miocene	unknown	Rich [343]	*Dromaius* sp.	QM, UCMP	Woodburne [345]; Stirton, et al. [346]; Rich [343]; Rich and Van Tets [344]
	Chinchilla Sands	Chinchilla, Queensland, Australia	early Pliocene	Zanclean	Rich and Van Tets [344]	*Dromaius novaehollandiae*	QM	Woods [347]; Stirton, et al. [346]; Rich and Van Tets [344]
	Tirari Formation	Lake Palankarinna, South Australia, Australia	late Pliocene-early Pleistocene	Piacenzian-Gelasian	Stirton, et al. [348]; Rich and Van Tets [344]	*Dromaius ocypus*	UCMP	Miller [349]; Rich [343]; Rich and Van Tets [344]
New Guinea	Otibanda Formation	Morobe, Papua New Guinea	late Pliocene	Piacenzian	Hoch and Holm [350]	*Casuarius* sp.	UCMP	Plane [351]; Rich and Van Tets [344]
	Cave deposits	unknown	Pleistocene?	Unknown	Lydekker [168]; Miller [352]	*Casuarius lydekkeri*	AM	Lydekker [168]; Rothschild [353]; Miller [352]; Worthy, et al. [341]
	Pleistocene swamp deposits	Pureni, Papua New Guinea	late Pleistocene	Chibanian	Williams, et al. [354]	*Casuarius lydekkeri*	CPC	Rich, et al. [355]

One of these earlyfossil Casuariiformes, *Emuarius gidju* [337], had a temporal range spanning from approximately 24 Ma to 15 Ma and is known from a large number of specimens [341]. *E. gidju* was first described on the basis of a distal tibiotarsus, proximal tarsometatarsus and shaft, and a complete pes from the Lake Ngapakaldi Leaf Locality of the Wipajiri Formation in South Australia [336]. Two more specimens were found in late Miocene deposits in Alcoota, Northern Territory [336,356], and even more from formations spanning the late Oligocene to early late Miocene of Riversleigh, Queensland [337,340]. The genus *Emuarius* differs from *Dromaius* in its retention of a cassowary like-femur, while the tibiotarsus and tarsometatarsus have cursorial modifications and are emu-like [337,340]. The pedal phalanges are of an intermediate morphology between the extant emu and cassowary, being more dorsoventrally compressed than those of cassowaries but less than those of emu [337,341]. This taxon is frequently used to calibrate molecular divergence dates between *Casuarius* and *Dromaius*, and a phylogenetic analysis of morphological characters provided robust confirmation for *E. gidju* and *Dromaius* being sister taxa [341]. The derived tibiotarsus and tarsometatarsus of *Emuarius* and *Dromaius* likely evolved after the emu-cassowary split as the emu lineage began to evolve towards a more cursorial mode of life [337,341]. The humerus is less reduced than in *Dromaius*, which may represent the plesiomorphic state of a bird less removed in time from its volant ancestors than extant Emu and cassowaries are [341]. *E. gidju* was smaller than the extant *D. novaehollandiae*, with an estimated weight of 19–21 kg [340] compared with 30–55 kg in emus [332]. Smaller orbits than *Dromaius* indicates *Emuarius* had smaller eyes relative to its skull, and this feature combined with the limited extent of its cursorial specialisations have been interpreted as being representative of the less open habitats present in Australia before the continent underwent extensive aridification beginning in the latter half of the Miocene [341,357].

Emuarius guljaruba, from the 24.1 Ma late Oligocene Etadunna Formation [333–336], is known from a single complete left tarsometatarsus [333]. It is larger than *E. gidju* and most likely a separate species, but its allocation to *Emuarius* remains provisional because no femur has yet been discovered. The extant genus *Dromaius* first appears in the middle Miocene Camfield beds of the Northern Territory [336,343]. *Dromaius arleyekweke* from the late Miocene Waite Formation in the Alcoota scientific reserve, Northern Territory [358] is the oldest named species in this genus. Small and gracile, it is notable in that it exhibits extreme cursorial adaptation, with the tarsometatarsus even more elongated than in *D. novaehollandiae* [358]. It was a small emu, with an estimated body mass based on tibiotarsus least shaft circumference using the algorithm of Campbell and Marcus [359] between 16 and 17.2 kg [358]. Derived features including a distally flattened external condyle of the distal tibiotarsus, the elongated tarsometatarsus, a reduced trochlea metatarsi II as compared with trochlea metatarsi IV, and a shallow median sulcus of the distal trochlea metatarsi II indicate a close affinity with *Dromaius* rather than *Emuarius* [358]. The oldest occurrence of the extant *Dromaius novaehollandiae* is in the early to middle Pliocene-aged Chinchilla Sands of Queensland [336,346,347]. Another species, *Dromaius ocypus*, is known from a tarsometatarsus from the Pliocene Tirari Formation of Lake Palankarinna, South Australia [349]. *D. arleyekweke* was found as the sister taxon of *D. ocypus* and *D. novaehollandiae* [358]. With *D. ocypus* interpreted as less cursorial than either *D. arleyekweke* or *D. novaehollandiae*, this relationship implies an independent acquisition of cursoriality in *D. arleyekweke* or a loss in *D. ocypus*, which may complicate the traditional view of emu evolutionary history as having involved a trend towards increasing cursorial specialisation [358].

The cassowary fossil record is very poor, likely owing to the clade's preference for tropical forest habitats in which fossils are unlikely to form or be found. Phalanges found from the late Pliocene-aged Otibanda Formation of Papua New Guinea most closely match the extant *C. bennetti* in size but do not appear similar enough to justify being considered conspecific [351]. *Casuarius lydekkeri* [353] is known from a distal right tibiotarsus that is likely Pleistocene in age. The provenance of this fossil is debated [355], and may be from Darling Downs, Queensland based on its preservation [331,341]. Worthy, et al. [341]

assessed the *C. lydekkeri* type material and concluded that its placement within *Casuarius* is likely correct, but there are significant differences between it and the extant *C. bennetti* and *C. casuarius*. A partial skeleton from swamp deposits dating to the late Pleistocene of Pureni, Papua New Guinea was assigned to *C. lydekkeri*, and it was noted to be smaller than any extant cassowary, with a more gracile femur [355]. Unfortunately, no elements from this specimen overlap with those from the Otibanda Formation specimen [355], so the relationship between the only known fossil cassowaries remains a mystery. Naish and Perron [360] speculated that crown cassowaries may be a relatively young clade that evolved in post-Pliocene Australia, with movement into New Guinea occurring during the Pleistocene with the appearance of land bridges between the two landmasses. Of course, this scenario will remain purely speculative until more of these elusive fossils come to light.

2.5. New Zealand Ratites: Apterygiformes and Dinornithiformes

Until just a few centuries ago, New Zealand hosted two ratite lineages: Apterygiformes (kiwi) and Dinornithiformes (moa). Without mammalian competition, kiwi and moa filled the niches of small terrestrial insectivorous and large browsing mammals respectively. Five extant species of kiwi (Apterygidae) are currently recognized, all in the same genus: the Southern Brown Kiwi *Apteryx australis*, the North Island Brown Kiwi *Apteryx mantelli*, the Great Spotted Kiwi *Apteryx haastii*, the Little Spotted Kiwi *Apteryx owenii*, and the Okarito Brown Kiwi *Apteryx rowi* [10]. Convergence between kiwi and small ground mammals is often noted, and is indeed remarkable [361]: kiwi are relatively small-bodied and nocturnal, with hair-like plumage and a superb sense of smell that compensates for their poor vision. Their long bills are used to probe the soil and leaf litter for invertebrates. Their eggs, which are the largest relative to body size of any bird, are laid in burrows [10]. Additionally, they are unique in that they are the only known crown birds with two functioning ovaries [362]. All five species face serious threats from introduced mammalian predators, and introduction of kiwi to predator-free offshore islands has been key to their continued survival [363]. Because of their sedentary nature, substantial local diversity exists, and a study examining thousands of mtDNA loci found 16 to 17 genetically distinct lineages within the five extant kiwi species [364].

Moa took the trend of forelimb reduction in flightless birds to the furthest possible extreme by losing the forelimbs entirely. There is no indication of a humeral articular facet on the scapulocoracoid, which itself is highly reduced and, along with the sternum, is the only vestige of the pectoral girdle [77]. A vestigial furcula is present in the genus *Dinornis* but is absent in all other moa [77]. Curiously, the forelimb-specific gene *tbx5* that is essential for the induction of forelimb development appears to have been fully functional in moa, suggesting that other developmental pathways were responsible for the loss of their wings [365]. The moa clade exhibited an extreme degree of reverse sexual dimorphism that for some time led to confusion regarding the number of known species-level taxa. The accepted number of recent taxa based on ancient DNA is nine species in three families: Dinornithidae, containing *Dinornis robustus* and *Dinornis novaezealandiae*, Megalapterygidae containing the monotypic *Megalapteryx didinus*, and Emeidae, containing *Anomalopteryx didiformis*, *Emeus crassus*, *Euryapteryx curtus*, *Pachyornis geranoides*, *Pachyornis elephantopus*, and *Pachyornis australis* [11]. In the largest-bodied genus, *Dinornis*, females could be up to three times larger than males, and it required a study of ancient sex-linked DNA sequences to reveal that individuals of the previously recognized *D. struthoides* actually represented the much smaller males of *D. giganteus* and *D. novaezealandiae* [366]. The extinction of moa is believed to have occurred extremely rapidly, within 200 years of human settlement approximately 600 years BP [367]. Evidence of their existence remains in New Zealand's flora, some of which retains anachronistic defenses against browsing by moa [368,369]. Moa coprolites and preserved gizzard contents indicate that they were generalist herbivores, though some degree of species-specific dietary niche partitioning existed [370].

How and when moa and kiwi arrived in New Zealand is still unknown [371], as unfortunately neither group has a clear fossil record from before the Pliocene [372]. Molecular phylogenetic evidence generally supports the hypothesis that moa and tinamous are sister taxa [371], suggesting that moa and kiwi colonised New Zealand and became flightless independently. Depending on the timing of their arrival, both clades may have been greatly affected by the Oligocene drowning of New Zealand, which culminated 25 Mya [373,374]. Coincidentally, this time frame appears to have been a key interval for the emergence of recognizable crown group representatives of other palaeognath clades on different landmasses (Tables 3–6).

Debates regarding how much of Zealandia was above water during the Oligocene drowning episode, and how this event impacted the origins of New Zealand's endemic flora and fauna continue [375,376]. Cooper and Cooper [377] postulate that only 18% of the present land area was above sea level during peak inundation as a low-lying archipelago. Trewick, et al. [376] and Landis, et al. [374] proposed that the islands were inundated completely, meaning that the entirety of New Zealand's terrestrial flora and fauna must have arrived in the past 22 million years. An increasing amount of biological evidence suggests at least some land must have remained above sea level during this period and has shifted the consensus against a total inundation [372]. Divergence dating of taxa with poor dispersal ability including frogs of the genus *Leiopelma* [378], *Craterostigmus* centipedes [379], mite harvestmen [380], and zopherid beetles [381] indicates that taxa within these groups diverged well before the drowning event, suggesting that all of them would have needed to independently disperse to New Zealand post-flooding had it been fully submerged. Wallis and Jorge [382] reviewed 248 published divergence dates between New Zealand lineages and their closest relatives elsewhere and found evidence for 74 lineages that diverged before 23 Mya, and of those, 25 lineages dated back before Zealandia split from Australia, making them of true Gondwanan vicariant origin. Interestingly, they found no evidence for a spike in extinctions or new arrivals around the time of the transgression. No study has yet presented unequivocal geological evidence for complete submergence [376,383], and clastic sediments deposited during the Waitakian stage in the southern Taranaki Basin suggests a nearby terrestrial sediment source [384].

Cooper and Cooper [377] examined mitochondrial genetic diversity in kiwi, moa, and acanthisitid wrens and found it to be unusually low compared to other ratites and other avian taxa, and interpreted this as evidence for a bottleneck effect due to the Oligocene drowning. They estimated that re-radiation of these endemic New Zealand lineages began 19–24 Mya. Could this be evidence that moa and kiwi survived the drowning in situ on small islands, or that small volant founding populations arrived afterwards? The apparent survival through the drowning event by other New Zealand taxa means the first scenario is certainly possible. If absence of volant non-tinamid palaeognaths after the middle Eocene is not an artifact of the fossil record, then the ancestral founding populations that ultimately gave rise to kiwi and moa must have arrived before the drowning of New Zealand. Ultimately, only new fossil discoveries from before the drowning event are likely to be able to resolve this question completely.

2.5.1. Apterygid Fossil Record

The oldest kiwi and moa fossils are from the St. Bathans terrestrial vertebrate faunal assemblage from the early Miocene of St. Bathans, in the central Otago region of the South Island (Figure 7, Table 7). The site is dated to 19–16 Ma [385,386], and has provided a rare glimpse at New Zealand's Neogene fauna just after the drowning of New Zealand. The earliest known kiwi, *Proapteryx micromeros*, was described on the basis of a right femur missing its distal condyles [387]. The only referred specimen is also fragmentary, consisting of a left quadrate missing the orbital process anterior to the pterygoid condyle and much of the lateral mandibular condyle [387]. Based on the femur circumference, the estimated body mass of *P. micromeros* was between 234.1 and 377 g, making it only slightly larger than the smallest extant kiwi, *A. owenii* [387]. If this species is representative of size of the earliest total-clade apterygids, its size would seem to refute the hypothesis that kiwi are phyletic dwarfs. The classic explanation for the extremely large eggs of kiwi was that kiwi evolved from a large-bodied ancestor, and body size decreased over time while egg size remained the same [361,388,389]. Instead, it may be more likely to have arisen as a novel feature related to producing highly precocial young [387,390]. Based on the gracile shape of the femur, the authors went as far as to propose that *P. micromeros* may have been volant, though that hypothesis is impossible to assess on the basis of presently known fossil material. If *P. micromeros* was volant, it would represent the only known example of a volant stem member of an extant ratite lineage, and would indicate that kiwi may have arrived in New Zealand after the drowning event. Recently, a 1-million-year-old kiwi fossil from the North Island [391] was identified as a new species *Apteryx littoralis* [392]. No other fossils of intermediate age are yet known between the St. Bathans fauna and the Holocene, making it difficult to trace the origins of crown kiwi.

Table 7. Apterygid fossil record.

Continent	Geological Unit	Location	Epoch	Stage	Age Reference	Taxa	Institutions	Reference
New Zealand	Bannockburn Formation	Otago, South Island, New Zealand	late early Miocene	Burdigalian	Mildenhall and Pocknall [385]; Pole and Douglas [386]	*Proapteryx micromeros*	NMNZ	Worthy, et al. [387]
	Kaimatira Pumice	Marton, North Island, New Zealand	middle Pleistocene	Calabrian	Worthy [393]	*Apteryx littoralis*	NMNZ	Tennyson and Tomotani [392]

Thus far, the only molecular studies that sample multiple *Apteryx* species yield alternative estimates of the timescale over which species-level diversification within *Apteryx* took place. Using concatenated sequences of nuclear and mitochondrial DNA, Grealy, et al. [41] estimated that *Apteryx mantelli* diverged from other kiwi approximately 13 MYA, whereas *A. haastii* and *A. owenii* diverged at about 4 MYA. The phylogenomic time tree produced by Yonezawa, et al. [49] included nuclear and mitochondrial sequences from all five extant kiwi species, and is in agreement with those divergence time estimates, inferring an origin of crown group kiwi at approximately 12 MYA. By contrast, Weir, et al. [364] inferred a much younger origin of the kiwi crown group at 3.85 MYA using mitochondrial DNA from a large sample of individuals. This was interpreted as evidence that the kiwi radiation coincided with the last glacial period when populations were isolated in glacial refugia, particularly those on the South Island [364].

2.5.2. Dinornithid Fossil Record

The St. Bathans fauna also provides a window into moa evolution (Figure 7, Table 8), though the moa fossils known from this locality are even more fragmentary than those of kiwi. Eggshell fragments found at the site suggest at least two species of moa were present [372,394,395]. Several large avian bone fragments have been found, including one that was identified as a portion of the proximal shaft of a right tibiotarsus [395]. Other large New Zealand landbirds such as flightless adzebills and giant geese existed at the time, but the fibular and outer cnemial crests are separated further on this tibiotarsus fragment than they would be in those groups, and instead resemble those of palaeognaths most closely [395]. One can only hope that the St. Bathans site yields bones that can be more conclusively identified as belonging to early representatives of the moa lineage. Many late Pleistocene-Holocene moa fossils are known [391,396], but Pliocene-Pleistocene moa fossils are much scarcer, and very few are known from before the Otira glaciation event which began ~75,000 years ago [397]. A tibiotarsus assigned to *Euryapteryx* was found in marine mudstone reported to be Pliocene in age [397], and *Dinornis* was present on the North Island at least two million years ago [397]. A tibiotarsus and tarsometatarsus fragments belonging to *Anomalopteryx didiformis* were found in a clay bed below a basalt [398], and if they are indeed older than the basalt and not fissure-fill, they would be about 2.5 million years old [397].

Table 8. Dinornithid fossil record.

Continent	Geological Unit	Location	Epoch	Stage	Age Reference	Taxa	Institutions	Reference
New Zealand	Bannockburn Formation	Otago, South Island, New Zealand	late early Miocene	Burdigalian	Mildenhall and Pocknall [385]; Pole and Douglas [386]	Dinornithidae indet.	NMNZ	Tennyson, et al. [395]
	unlisted	Timaru, South Island, New Zealand	early Pleistocene	Gelasian	Mathews and Curtis [399]	*Anomalopteryx didiformis*		Forbes [398]; Worthy, et al. [397]
	unlisted	Hawke's Bay, North Island, New Zealand	early Pleistocene?	Gelasian?	Beu and Edwards [400]	*Eurapteryx curtus*	AIM	Worthy, et al. [397]
	unlisted	Wairapa, North Island, New Zealand	early Pleistocene?	Gelasian?	Oliver [401]; Beu and Edwards [400]	*"Eurapteryx geranoides"*	NMNZ	Worthy, et al. [397]
	Tewkesbury Formation	Wanganui, North Island, New Zealand	early Pleistocene	Calabrian	Beu and Edwards [400]	*Dinornis novaezealanidae, Emeidae* indet.	NMNZ	Marshall [402]; Worthy, et al. [397]

As with kiwi, molecular time trees have yielded divergent hypotheses regarding the timing of the moa radiation. Bunce, et al. [11] found evidence for the radiation being relatively recent. The deepest divergence (between Megalapterygidae and the remaining family-level moa taxa) was estimated at 5.8 MYA, within the same time frame as rapid mountain formation on the South Island during the Miocene-Pliocene [11]. Indeed, the uplift of the Southern Alps would have led to greater habitat diversity [403], and may have spurred the diversification of moa. Interestingly, Haddrath and Baker [38] placed this earliest moa divergence much earlier, at 19 MYA, which roughly coincides with the end of the Oligocene drowning event. Regardless of when the earliest phylogenetic divergence within the moa clade occurred, the fossil record suggests moa crossed onto the North Island via a land bridge 1.5–2 million years ago, which may have led to even greater species diversity as the land bridge reappeared and disappeared during Pleistocene glacial cycles [11]. Whether kiwi were similarly restricted to the South Island before the Pleistocene is unknown, and more fossils from sediments of intermediate age between the Miocene and Pleistocene are needed to make any further advances.

2.6. Malagasy Ratites: Aepyornithiformes

Extremely little is known of the evolutionary history of Madagascar's giant elephant birds. The island's Cenozoic terrestrial vertebrate record is notoriously poor, and thus far all fossil finds are restricted to the last 80,000 years [404–406]. What little we do know comes from subfossil bones and eggshells, the latter of which are extremely abundant in some areas. Detailed records of late Pleistocene and Holocene aepyornithid subfossils are beyond the scope of this paper, but can be found in Angst and Buffetaut [407]. Isotopic analysis of eggshells from southern Madagascar reveals that the birds that laid them mainly browsed on non-succulent trees and shrubs [408], some of which retain anachronistic defenses against ratite browsing similar to plants in New Zealand [369]. Palaeoneurological evidence shows that elephant birds had extremely reduced optic lobes, presumably associated with a predominantly nocturnal or crepuscular lifestyle [409].

Even the number of elephant bird species that existed into the Holocene is not known with certainty. Morphometric analysis of subfossil limb bones by Hansford and Turvey [12] recovered evidence for four species-level taxa: *Mullerornis modestus, Aepyornis hildebrandti, Aepyornis maximus,* and the heaviest bird ever discovered, *Vorombe titan. M. modestus, A. maximus,* and *V. titan* were found to be sympatrically distributed across much of Madagascar, while *A. hildebrandti* was restricted to the central highlands [12]. Molecular studies are needed to evaluate this morphology-based taxonomic scheme, as well as additional fossil collecting in other regions of Madagascar, as most known specimens come from the south of the island and the central highlands [12]. Nuclear and mitochondrial DNA recovered from eggshells suggested that *Aepyornis* and *Mullerornis* diverged approximately 27.6 MYA [41]. A divergence at 3.3 MYA between *A. hildebrandti* and *A. maximus* had previously been estimated [45]. The third genus found by Hansford and Turvey [12] appears not to have been sampled, highlighting the need to extract aDNA from additional eggshells and subfossil specimens.

Unraveling the decline and eventual demise of elephant birds in Madagascar is less straightforward than for moa, which went extinct within a brief window of time following human arrival in New Zealand [367]. Debate as to how long humans have been present on Madagascar, and thus for how long they coexisted with the island's endemic megafauna, is ongoing. Based on rare findings of stone tools and butcher marks on elephant bird bones, humans may have arrived early, between 10,000 and 4000 years BP [410,411]. Some anthropologists advocate a more recent arrival, between 1600 and 1000 BP [412], while an intermediate arrival time between 2000 and 1600 BP is supported by ^{14}C data associated with human activity [413]. If humans and elephant birds indeed coexisted for a long period of time, their extinction cannot be easily attributed to the rapid overkill of a naïve population as with moa [411,414]. Instead, a more complex scenario for the extinction of the Malagasy megaherbivores, which also included giant lemurs and tortoises, as well as dwarf hippopotami, has been proposed. Instead of overhunting, the key factor in their decline may have been the introduction of livestock such as Zebu cattle and a shift towards pastoralism. The introduction of large herbivores by humans coincides with the time frame of Malagasy megafaunal extinction, and under this scenario a combination of resource competition with introduced herbivores, alteration of the landscape by humans to suit the needs of livestock, and increased bushmeat hunting due to the expanding human population could have led to the demise of the Malagasy megafauna [414]. Whatever the direct cause or causes, the extinction of Aepyornithidae occurred roughly 1000 years BP according to radiometric data [415], concurrent with the drastic decline and extinction of the remainder of the endemic megafauna of the island [416], though some colonial records suggest they may have survived in isolated areas into the 17th century [407,417].

2.7. Antarctic Ratites

Antarctica was once a very different place from the frozen continent we recognize today. The formation of a continental ice sheet did not occur until the Eocene—Oligocene boundary [418]. Up until this time, the continent boasted thriving flora and fauna that were isolated from large mammalian predators—an ideal environment for flightless birds to evolve. Palynological records from sediment cores dated to 53.6–51.9 MYA from the eastern Antarctic Wilkes Land coast reveal that a diverse paratropical rainforest with frost-free winters existed during the early Eocene climatic optimum [419,420]. Sparse pollen from more cold-tolerant trees such as *Nothofagus* (southern beech) and *Araucaria* ("monkey puzzle") trees suggest temperate rainforests further inland [419,420]. By the middle Eocene, cores from 49.3–46 MYA indicate species diversity had decreased [420] and that cool temperate *Nothofagus*-dominated forests had taken over [419,420]. As a point of comparison, petrified wood samples from King George island in the South Shetland Islands aged 49–43 MYA (Middle Eocene) indicate a forest similar in composition to the cold temperate Valdivian rainforest of Chile [421], which is not dissimilar to the temperate rain forests of New Zealand that moa once inhabited.

There is fossil evidence of large terrestrial birds in Antarctica during this time, but they are too fragmentary to allow firm diagnoses (Table 9). A distal fragment of a right tarsometatarsus purported to be a ratite was found in the middle Eocene of the La Meseta Formation of Seymour Island, just off the Antarctic peninsula [422]. Unfortunately, there is no evidence for its ratite affinities other than its large size. Its unusually large trochlea for the second toe is different from that of all other known ratites [76], and it bears consideration that misattribution of large bones to ratites is not uncommon [423]. An anterior part of a premaxilla originally attributed to a phorusrhacid, also from the La Meseta Formation [424–428], was recently suggested to belong to a palaeognath [429,430]. The presence of ratites on Seymour Island would not be surprising given the environmental conditions at the time, as evidenced by abundant petrified conifer wood from the La Meseta Formation [431]. Confirmation of their existence will have to await more complete specimens, but remains a tantalizing possibility.

Table 9. Putative Antarctic ratites.

Continent	Geological Unit	Location	Epoch	Stage	Age Reference	Taxa	Institutions	Reference
Antarctica	La Meseta Formation	Seymour Island	late Eocene	Lutetian-Priabonian	Amenábar, et al. [432]	"ratititae"	MLP, UCR	Tambussi, et al. [422]; Cenizo [429]; Acosta Hospitaleche, et al. [430]

The majority of Cenozoic Antarctic bird fossils belong to penguins and other marine birds, but Seymour Island was also host to a thriving terrestrial fauna during the Eocene. The stem falconid *Antarctoboenus carlinii* [433,434] was named from a distal end of tarsometatarsus from the early Eocene portion of the La Meseta Formation [430]. Small mammals were abundant, and included the extinct and highly enigmatic sudamericid gondwanatheres [435,436] as well as didelphimorphid, polydolopimorphid, and microbiotheriid marsupials [436–440]. Seymour Island also hosted South American meridiungulates [436,441–445], and a large sparnotheriodont with an estimated body mass of 395–440 kg [446] indicates the ecosystem was fully capable of sustaining large herbivores. The presence of meridiungulates also indicates that overland dispersal from South America was possible, and there is no reason why South American ratites could not have made the journey as well. The Drake passage between South America and the Antarctic Peninsula did not begin to open until approximately 41 Ma [447], meaning these faunas lived during an era where biotic interchange was possible. Such interchange with Australia was also hypothetically possible for a brief window during the Paleocene and early Eocene, as dinocyst assemblages indicate the flow of ocean water across the Tasman gateway by 50–49 Ma [448]. It is also possible for a unique ratite lineage to have arisen on Antarctica, though—as with all other ideas regarding Antarctic palaeognaths—this will remain highly speculative until more fossils are recovered. Regardless of whether the Antarctic terrestrial fauna included ratites, the complete glaciation of the continent in the Oligocene would have doomed them to extinction.

3. Molecular Phylogenetic Hypotheses of Palaeognath Interrelationships

Interpreting phylogenetic relationships among extant and fossil palaeognaths was historically challenging due to morphological homoplasy, and although molecular phylogenetic approaches have yielded some consensus on palaeognath interrelationships, areas of disagreement remain. Thus far, all recent molecular phylogenetic studies of palaeognaths have recovered ostriches as the sister taxon to the rest of the clade, yielding congruent support for a reciprocally monophyletic clade called Notopalaeognathae comprising rheas, tinamous, kiwi, moa, and elephant birds [36–41,44–46,48–50,54–58,449,450]. Limited morphological evidence has also been found in support of a monophyletic Notopalaeognathae [33,77]. In addition, all molecular phylogenetic studies investigating ancient DNA from palaeognath subfossils have strongly supported elephant birds as sister to kiwi [41,45,46,49,57], and tinamous as sister to moa [38,40,41,44–46,49,50,57].

The internal relationships of Notopalaeognathae remain controversial, particularly in regard to the position of rheids. The internal branches at the base of Notopalaeognathae appear to be very short, indicating that the clade may have undergone relatively rapid diversification early in its history, which may have led to incomplete lineage sorting and limited phylogenetically informative character acquisition along deep internodes [38,39,56]. This may have pushed Notopalaeognathae into an empirical anomaly zone in which the most common gene trees from molecular phylogenetic analyses do not match the species tree [56]. Rheids are most often recovered in one of two phylogenetic positions:

1. As the sister taxon of the remaining notopalaeognaths [36,37,39,41,42,44–46,48,49,54–56], though this position is generally weakly supported (Figure 10) [41,44,49].

2. As sister to a casuariid + apterygid + aepyornithid clade ("Novaeratitae") [38,43,48,50,56–58,450] (Figure 11). Several alternative topologies in addition to these have been recovered that place rheas sister to the tinamid-dinornithid clade [37,39,449] or sister to casuariids [38].

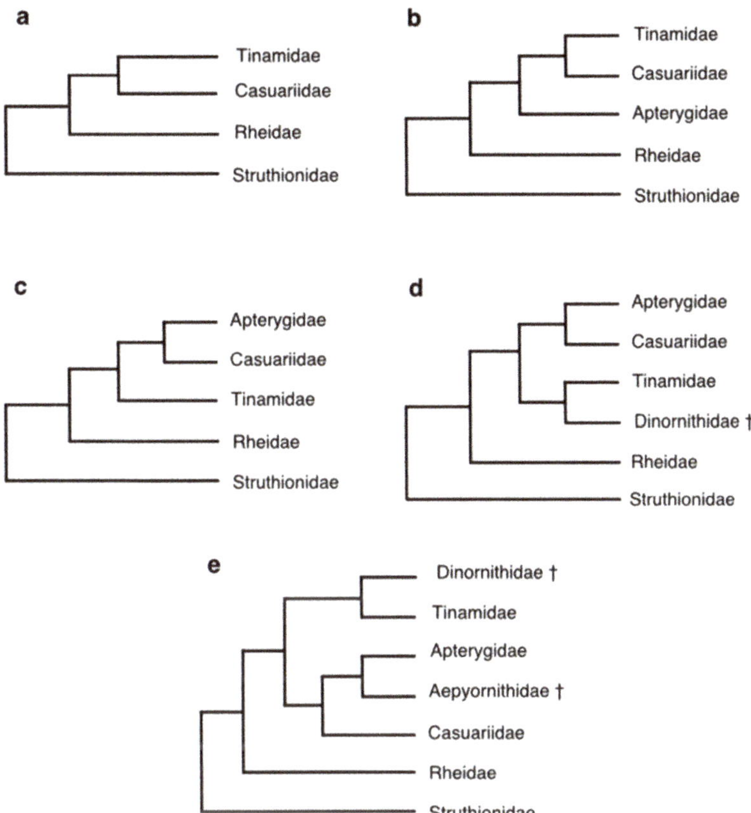

Figure 10. A summary of recent molecular phylogenetic studies that recover Rheidae as sister to the remaining notopalaeognaths. Extinct clades are indicated by †. (**a**) Smith, et al. [39] primary concordance and total evidence tree. (**b**) Prum, et al. [48] concatenated dataset; Kuhl, et al. [54]. (**c**) Hackett, et al. [36]; Harshman, et al. [37] maximum likelihood and Bayesian tree; Claramunt and Cracraft [55]. (**d**) Phillips, et al. [44]; Cloutier, et al. [56] concatenated dataset. (**e**) Mitchell, et al. [45]; Grealy, et al. [41]; Yonezawa, et al. [49], Urantówka, et al. [46], Almeida, et al. [42].

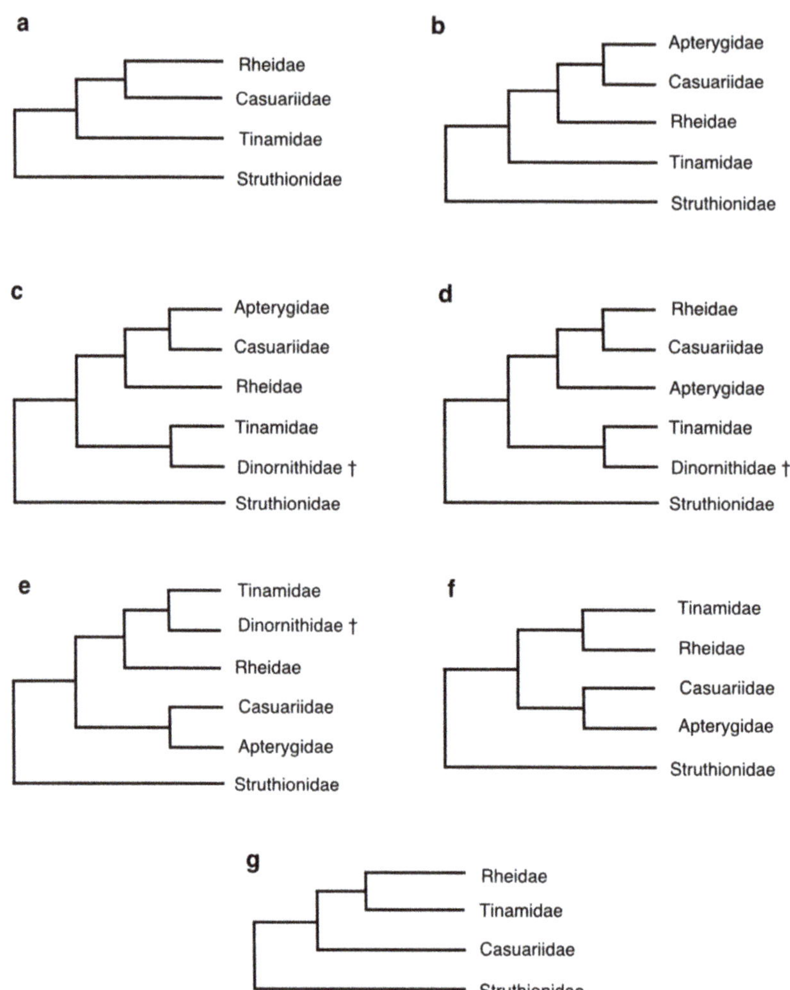

Figure 11. A summary of recent molecular phylogenetic studies that do not recover Rheidae as sister to the remaining notopalaeognaths. Extinct clades are indicated by †. (**a**) Kimball, et al. [450]. (**b**) Prum, et al. [48] binned ASTRAL analysis; Reddy, et al. [58]; Sackton, et al. [50]; Feng, et al. [43] maximum likelihood analysis of avian growth hormone gene copies. (**c**) Haddrath and Baker [38] 10 and 27 gene concatenated dataset, 27 gene consensus tree; Baker, et al. [40]; Cloutier, et al. [56] total evidence consensus tree. (**d**) Haddrath and Baker [38] 10 gene consensus tree. (**e**) Smith, et al. [39] maximum likelihood reanalysis of Phillips, et al. [44]; (**f**) Harshman, et al. [37] maximum parsimony and RY coded maximum likelihood analysis; Wang, et al. [449]; (**g**) Smith, et al. [39] using 40 loci.

Determining why these discrepancies exist could be key to finally resolving the internal branching order of Notopalaeognathae. In their attempt to address this question using genome-wide datasets of conserved nonexonic elements, introns, and ultraconserved elements, Cloutier, et al. [56] found that the consensus species tree building methods MP-EST and ASTRAL-II placed rheids sister to the casuariid-apterygid-aepyornithid clade with maximal bootstrap support from MP-EST for all three datasets. Their concatenated supermatrix dataset recovered rheids as sister to all other notopalaeognaths, but with weaker statistical support. In general, concatenated analyses have often yielded different results to consensus tree building methods regarding the interrelationships of Notopalaeognathae, with concatenated data more frequently recovering rheids as sister to all other notopalaeognaths [56]. Sackton, et al. [50] found similar results and claim that their genome-wide approach is more robust to incomplete lineage sorting than concatenation, which is what leads to discrepancies between studies. "Novaeratitae", a proposed clade that places casuariids sister to an elephant bird + kiwi clade, received high bootstrap support when mitochondrial and genomic data were combined but not when each were analysed individually [41]. In order to finally resolve the messy internal relationships of notopalaeognaths, a greater number of faster-evolving retrotransposons and introns may need to be analysed [41], and the models of sequence evolution employed must fit the type of genomic data being investigated [58].

Molecular Divergence Time Estimates

The vast majority of molecular divergence time analyses have recovered an estimate for the palaeognath-neognath divergence in the Cretaceous Period, preceding the K-Pg extinction event (e.g., [38,41,42,44,45,47–49,54,55,449]), an estimate that is consistent with the known (yet sparse) fossil record of Mesozoic neornithines [72]. However, estimates of the age of the neornithine root vary enormously, ranging from 131 Ma [38] to 63.2 Ma [42]. Importantly, the oldest published divergence time estimates do not invalidate Gondwanan vicariance as a potential driver of crown palaeognath divergences [38]. The enormous temporal breadth of deep neornithine divergence time estimates have stimulated discussion about the role of model misspecification in driving erroneously ancient divergence time estimates [451]. Hypothesized selection for reduced body size across the end-Cretaceous mass extinction event could have transiently increased molecular substitution rates along the deepest branches within neornithine phylogeny, which would be expected to drive overestimates of node ages around the neornithine root [452]. Indeed, simulations suggest that 40 million years' worth of age disparity for the neornithine root node can plausibly be explained by the effect of body size on nucleotide substitution rates [452]. Importantly, the palaeognath stem lineage is inferred to have exhibited high nucleotide substitution rates, consistent with ancestral palaeognaths having been small-bodied (the last common ancestor of crown palaeognaths was estimated to have weighed approximately 2.9 kg) [452]. With smaller body sizes and shorter generation times than other extant palaeognaths, tinamous exhibit anomalously high nucleotide substitution rates compared with other palaeognaths [37,449], which may additionally drive erroneously ancient divergence time estimates near the neornithine root [45,453].

Lingering uncertainty regarding the phylogenetic divergence times of crown palaeognaths complicates attempts to place lithornithids within the broader context of palaeognath evolution. Since most palaeognath divergence time estimates pre-date the earliest well corroborated lithornithid fossils [41,45,49,449] (with the possible exception of the ~66 million year old isolated scapula from the Hornerstown Formation [63]), the hypothesis that at least some lithornithids represent early stem group representatives of major palaeognath subclades is temporally viable. However, Prum, et al. [48] estimated the origin of the palaeognath crown group at 51 Ma, during the Ypresian stage of the early Eocene. In this temporal scenario, most lithornithid fossils predate the crown palaeognath radiation, in which case nearly all lithornithids with the exception of those found in the younger Messel Formation could only represent stem palaeognaths. This relatively young age for the palaeognath crown group would also imply that early Paleogene remains such as *Diogenornis*, Palaeotididae, and the Middle Paleocene fossils identified as belonging to a stem rheid fall outside the palaeognath crown group

4. Key Gaps in the Palaeognath Fossil Record

4.1. Cretaceous Stem Palaeognaths

Virtually no examples of Cretaceous stem palaeognaths have yet been identified, despite consensus—on the basis of divergence time estimates as well as the presence of fossil total-clade neognaths—that they must have existed at this time. This is perhaps the most glaring gap in the known palaeognath fossil record, but is perhaps an unsurprising one given the general scarcity of well-supported Cretaceous neornithines at present. A probable example of a Cretaceous total-clade neognath is *Vegavis iaai*, recovered from the late Maastrictian of Vega Island, Antarctica [69]. This fossil taxon shows apparent specialisations for foot-propelled diving, and has been variably placed within Anatoidea [69], as a stem neognath, or even outside of Neornithes altogether [72,454]. *Asteriornis maastrichtensis*, from the Maastrichtian of Belgium, is another probable Cretaceous total-clade neognath. At 66.7–66.8 million years old, *Asteriornis* is slightly older than *Vegavis*, and therefore the oldest well-corroborated neornithine yet discovered [72]. A relatively small bird (estimated to have weighed roughly 490 grams), *Asteriornis* was identified as a total-clade galloanseran [72], although a recent study raised the (weakly supported) hypothesis that it instead represents a total-clade palaeognath [8]. The presence of probable total-clade neognaths from before the K-Pg mass extinction, such as *Vegavis* and *Asteriornis*, implies that the palaeognath-neognath split must have occurred even earlier in the Cretaceous (though, as described above, molecular divergence dates do not agree on the true antiquity of the basal neornithine phylogenetic divergence).

Longstanding biogeographic hypotheses held that Neornithes originated in Gondwana [26,55], partly on the basis that there are far more extant endemic bird clades on the southern continents of South America, Africa, and Australia than there are on the northern continents of North America and Eurasia [455]. However, the discovery of *Asteriornis* in Europe indicates that deeply diverging crown bird lineages have a long evolutionary history in the Northern Hemisphere [72]. More broadly, many clades that are currently restricted to tropical latitudes have fossil stem group representatives in the Paleocene and Eocene of the Northern Hemisphere (e.g., [70,74,120,124,455–457]), implying far more widespread geographic distributions early in these clades' evolutionary histories. Given the generally dispersive capacity of birds, as well as the fact that hothouse climatic conditions predominated throughout the early Paleogene and led to the expansion of paratropical forests into high latitudes, the present-day geographic distributions of many extant tropical clades may not reliably indicate their ancestral areas of origin [74]. In light of these considerations, determining the most likely fossil localities for revealing the first evidence of a Cretaceous stem palaeognath is challenging, and it would seem equally probable that an early palaeognath could derive from Late Cretaceous deposits of either the northern or the southern hemisphere.

4.2. Stem Group Representatives of Extant Palaeognath Subclades

If contemporary hypotheses of ratite paraphyly and dispersal are accurate, small volant palaeognaths should have been present on landmasses where extant palaeognaths are found during the Paleocene or Eocene [45]. However, the timing of each independent palaeognath transition to large body size and flightlessness is uncertain. Transitions to complete flightlessness among island-dwelling birds typically necessitate few terrestrial predators and a food source that does not require flight [458,459]. If these conditions are met, flightlessness may be advantageous because it allows for energy conservation through reduction in the size of the pectoral musculature [460]. Indeed, the basal metabolic rates of flightless rails are lower than those of closely related flighted rails [460]. Given the right circumstances, transitions to flightlessness and large body size can apparently arise quite rapidly. The extinct giant flightless Hawaiian goose *Branta rhuax* is nested within the Canada Goose *Branta canadensis* species complex, and its presence on the main island of Hawai'i means it must have become large and flightless in less than 500,000 years [461].

Most geologically recent transitions to avian flightlessness occurred on oceanic islands in the absence of predation and competition from terrestrial mammals [458,459]. Were these conditions met on continents in the wake of the K-Pg mass extinction event, allowing multiple lineages of ratites to evolve flightlessness and large body sizes before mammalian predators and competitors could evolve? These conditions appear to have been met on at least some landmasses, as even 10 million years after the extinction event most mammals remained relatively small and unspecialized [462]. The Corral Bluffs site in Colorado suggests that the mammalian fauna in the immediate aftermath of the K-Pg was dominated by small omnivores and insectivores [463], and generally there was a dearth of specialized mammalian carnivores in the early Paleocene [76,464,465]. The makeup of terrestrial mammalian faunas at the time could well have favoured the evolution of flightlessness in birds that could obtain food on the ground, and other large flightless Paleogene bird clades such as Gastornithidae, Phorusrhacidae, and Dromornithidae may have followed a similar pattern along with ratites [76]. In particular, the lack of placental carnivores in South America through most of the Cenozoic may have contributed to the diversity of flightless birds on that continent, which also included Phorusrhacoidea and the giant anseriform *Brontornis* [76].

If volant stem group representatives of various palaeognath subclades evolved into large-bodied, flightless forms during a relatively narrow temporal window in the early Paleogene, the chances of finding direct fossil evidence of these small-bodied ancestral forms might be relatively low. Indeed, short internodes near the root of Notopalaeognathae indicate a rapid diversification of palaeognath lineages during the Paleogene [41,56]. However, if some transitions to flightlessness were protracted, the chances of identifying informative fossils documenting such transitions would be more likely. With their recent reassignment to total clade Struthionidae, eogruids are a superb example of previous unrecognised stem group representatives of an extant ratite lineage, though better data on their wing apparatus are needed in order to assess whether all known taxa were flightless. If some taxa were volant, Eogruidae could provide an illuminating window into the relative timing of transitions to cursoriality, large body size, and loss of flight in a ratite lineage.

A further challenging aspect of reconstructing the early evolutionary history of the various ratite lineages is that, if flightlessness and large body size arose numerous independent times, confidently assigning a given volant palaeognath fossil from the Paleogene to the correct palaeognath subclade may prove difficult due to convergence. However, the ongoing exploration of certain localities may yield further insight into transitions to flightlessness among certain ratite lineages—for example, additional finds from the St. Bathans fauna could shed more light on the origins of moa and kiwi, and help elucidate whether the stem kiwi *Proapteryx* was indeed small and volant as initially hypothesized [387].

5. Reconstructing the Most Recent Common Ancestor of Palaeognaths

Understanding the nature of the most recent common ancestor (MRCA) of extant palaeognaths will reveal much about palaeognath macroevolution, and neornithine macroevolution more broadly. For instance, insight into the flight apparatus of the crown palaeognath MRCA will help explain how the geographic distributions of extant palaeognaths arose. Moreover, stem palaeognaths (along with stem galloanserans and stem neoavians) are inferred to have survived the end-Cretaceous mass extinction event [41,48,71,72], while all non-neornithine birds appear to have perished [73]. Strong evidence regarding the morphology and ecology of early palaeognaths may also help clarify ecological factors that may have favoured the survivorship of crown birds with respect to non-neornithine avialans—one of the more contentious questions in contemporary palaeornithology [71,77]. Inevitably, given that the palaeognath-neognath split is the deepest divergence within crown birds, a better understanding of the nature of the palaeognath MRCA will in turn shed light on the common ancestral condition of all extant birds. Although much remains to be learned, there are several inferences that can be made regarding the nature of the most recent common ancestor (MRCA) of palaeognaths based upon the information currently available.

5.1. The Flight Apparatus of the Crown Palaeognath MRCA

Due to the relaxation of stabilizing selection, significant polymorphism exists in the wing musculature of ratites [466], complicating attempts to infer features of the ancestral crown palaeognath wing. As the only extant flighted palaeognaths, tinamids presumably provide the best source of data on the muscular anatomy of the wings of early flighted palaeognaths. Nearly all flight muscles present in neognaths are found in tinamids, with the exception of the biceps slip [274,467,468]. Extant phylogenetic bracketing [469] therefore indicates that the same suite of muscles would be expected to be present in both the crown palaeognath and crown neornithine MRCAs. Of course, tinamids are specialized for burst flight over relatively short distances, and as such are probably imperfect analogues of the ancestral crown palaeognaths that must have colonized distant landmasses in the early Cenozoic [470]. Subsequent losses of dispersal capacity, and the extinction of dispersive ancestral lineages, can leave the inaccurate impression that poorly dispersive taxa underwent oceanic dispersal via stochastic events. For example, the phasianid galliforms *Margaroperdix* (Madagascar) and *Anurophasis* (New Guinea) are poor dispersers, yet are found on isolated islands [470]. Phylogenomic analyses revealed that these taxa are nested within *Coturnix* quails and likely evolved from a dispersive *Coturnix*-like ancestor. Both taxa apparently independently evolved towards a non-dispersive partridge-like morphotype, reminiscent of how the ratite condition appears to have repeatedly evolved in palaeognaths [470]. As discussed in this review, some lithornithids appear to have been reasonably capable fliers and could provide more accurate insight into the nature of dispersive ancestral crown palaeognaths.

5.2. Inferred Ecology of the Palaeognath MRCA and K-Pg Survivorship

Non-neornithine avialans thrived throughout the Cretaceous and remained diverse through the Maastrichtian, before suddenly disappearing at the K-Pg boundary [73]. Until this point, Enantiornithes were the dominant Mesozoic avialan clade with more than 60 known species and a worldwide distribution [471–473]. Why did they become extinct, while neornithines survived? The answer may be associated with their ecology and habitat preferences. The K-Pg impact was devastating to the world's forests and resulted in significant species turnover [71,77,474–478]. Palynology of K-Pg boundary sections across the globe indicates that ground cover following the impact consisted primarily of ferns. This "fern spike" is interpreted as evidence of a disaster flora following the destruction of forests worldwide [71,464,474–476] by widespread fires ignited by the impact and subsequent cold and darkness [479,480]. This fern spike persisted for approximately 1000 years, and closed-canopy forests appear to have remained generally rare during this interval [481]. Indeed, it may have taken as long as 1.4 Ma for floral diversity hotspots to reappear [482]. This widespread habitat destruction would have been a powerful agent of selection against the mostly arboreal Enantiornithes, though this hypothesis does not explain the extinction of contemporaneous marine avialans such as Ichthyornithes and Hesperornithes. Instead, the demise of these marine piscivorous taxa may have been part of a broader collapse of marine food chains in the aftermath of the Chicxulub impact [77,81,483–486]. Importantly, ancestral state reconstructions of crown birds predict that the MRCAs of crown birds and the deepest crown bird subclades (Neornithes, Palaeognathae, Neognathae, and Neoaves) were all non-arboreal [71]. As such, the ancestors of palaeognaths may have made it through this mass extinction event partly by virtue of having exhibited terrestrial non-arboreal lifestyles.

As the most stemward palaeognaths known [49,64], lithornithids provide the best opportunity to draw fossil-informed inferences about the nature of the crown paleognath MRCA. Vibrotactile bill tips in *Lithornis promiscuus* and *Paracathartes howardae* may have been associated with probe-feeding in the ground, an interpretation congruent with the hypothesis of predominant K-Pg survivorship among non-arboreal taxa. A vibrotactile bill tip organ composed of mechanoreceptors known as Herbst corpuscles embedded within the bone was hypothesized to be a plesiomorphy of Neornithes by du Toit, et al. [80], which would support the neornithine MRCA and its immediate descendants as having been ground-foraging birds. Such organs are found in palaeognathous and neognathous probe-foragers, enabling them to locate prey buried in substrate through vibration detection [487,488]. In non-probe-foraging palaeognaths, the vibrotactile bill tip organ is vestigial [80,489]. The hypothesis that lithornithids and the palaeognath MRCA were probe-feeders agrees with ideas put forth by Houde [62], who suggested that lithornithids may have preferred to live near water and probed for food using their long beaks, noting the similarity of their jaw apparatus to those of kiwi. Additionally, the genus *Lithornis* appears to have had relatively large olfactory lobes, similar to olfactory foraging taxa including Procellariiformes and kiwi [490]. Since ground feeding birds are more likely to become flightless than arboreal taxa, a volant, non-arboreal, probe-feeding taxon would seem to be a provide a reasonable expectation for the ecology of the MRCA of crown palaeognaths.

6. Conclusions

Our understanding of palaeognath evolution has progressed markedly over the past two decades thanks to the development and application of sophisticated molecular phylogenetic approaches and the continued interrogation of the fossil record; however, many fundamental questions about the origins of extant palaeognath diversity remain unanswered. The present review affirms that the palaeognath crown group has a reasonably thorough fossil record from the late Oligocene-early Miocene onwards, with the exception of early elephant birds and early representatives of the New Zealand ratites, whose fossil record remains sparse until the Pleistocene [392,397,409]. However, the fossil record still fails to clearly illuminate how and when independent transitions to large body size and flightlessness arose among the multiple lineages of "ratites". As yet, volant stem members of these extant flightless clades remain unknown (besides the possible exception of Proapteryx [387]), leaving the early evolutionary history of crown group palaeognaths shrouded in mystery. Lithornithids currently provide the best insight into the nature of the earliest total-clade palaeognaths, and their relatively small size, probable non-arboreal ecology, and apparent capacity for sustained flight may make them useful models for understanding the nature of avian survivors of the end-Cretaceous mass extinction event. In the coming years, we anticipate increased consensus on both the evolutionary relationships and age of Palaeognathae and its major subclades, and hope that such advances are accompanied by the recognition of new fossil total-group palaeognaths from the Mesozoic and early Cenozoic. Such advances will be necessary to fill the many gaps in the palaeognath fossil record identified in this review, and to shed light on the repeated independent origins of "ratites"—one of the most striking examples of convergent evolution in birds, or indeed any other vertebrate clade.

Author Contributions: Conceptualization, K.W. and D.J.F.; methodology, K.W. and D.J.F. investigation, K.W.; data curation, K.W.; writing—original draft preparation, K.W.; writing—review and editing, K.W. and D.J.F.; visualization, K.W.; supervision, D.J.F.; funding acquisition, K.W. and D.J.F. All authors have read and agreed to the published version of the manuscript.

Funding: This research was funded by UKRI Future Leaders Fellowship, grant number MR/S032177/1 to D.J.F.

Institutional Review Board Statement: Not applicable.

Acknowledgments: We thank E. Buffetaut and D. Angst for the opportunity to contribute to this Special Issue, K. Welch for proofreading, and G. Mayr as well as an anonymous reviewer for constructive comments on our manuscript.

Conflicts of Interest: The authors declare no conflict of interest. The funders had no role in the design of the study; in the collection, analyses, or interpretation of data; in the writing of the manuscript, or in the decision to publish the results.

References

1. Billerman, S.; Keeney, B.; Rodewald, P.; Schulenberg, T. Birds of the World. Available online: https://birdsoftheworld.org/bow/home (accessed on 30 March 2021).
2. Pycraft, W.P. On the Morphology and Phylogeny of the Palæognathæ (Ratitæ and Crypturi) and Neognathæ (Carinatæ). *Trans. Zool. Soc. Lond.* **1900**, *15*, 149–290. [CrossRef]

3. Bock, W. The cranial evidence for ratite affinities. In Proceedings of the XIII International Ornithological Congress, Ithaca, NY, USA, 17–24 June 1962; pp. 39–54.

4. Parkes, K.C.; Clark, G.A., Jr. An Additional Character Linking Ratites and Tinamous, and an Interpretation of their Monophyly. *Condor* **1966**, *68*, 459–471. [CrossRef]

5. Cracraft, J. Phylogeny and evolution of the ratite birds. *Ibis* **1974**, *116*, 494–521. [CrossRef]

6. Mayr, G.; Zelenkov, N. Extinct crane-like birds (Eogruidae and Ergilornithidae) from the Cenozoic of Central Asia are indeed ostrich precursors. *Ornithology* **2021**, *138*, 1–15. [CrossRef]

7. Huxley, T.H. On the Classification of Birds and on the Taxonomic Value of the Modifications of Certain of the Cranial Bones Observed in that Class. In *Proceedings of the Zoological Society of London*; Zoological Society of London: London, UK, 1867.

8. Torres, C.R.; Norell, M.A.; Clarke, J.A. Bird neurocranial and body mass evolution across the end-Cretaceous mass extinction: The avian brain shape left other dinosaurs behind. *Sci. Adv.* **2021**, *7*, eabg7099. [CrossRef]

9. Field, D.J.; Benito, J.; Kuo, P.; Jagt, J. Mesozoic Fossil Insight into the Palaeognath-Neognath Anatomical Dichotomy. In Proceedings of the The Society of Vertebrate Paleontology Annual Meeting, Vurtual Meeting, 1–5 November 2021; p. 278.

10. Winkler, D.W.; Billerman, S.M.; Lovette, I.J. Kiwis (Apterygidae), version 1.0. *Birds World* **2020**. [CrossRef]

11. Bunce, M.; Worthy, T.H.; Phillips, M.J.; Holdaway, R.N.; Willerslev, E.; Haile, J.; Shapiro, B.; Scofield, R.P.; Drummond, A.; Kamp, P.J.J.; et al. The evolutionary history of the extinct ratite moa and New Zealand Neogene paleogeography. *Proc. Natl. Acad. Sci. USA* **2009**, *106*, 20646–20651. [CrossRef]

12. Hansford, J.P.; Turvey, S.T. Unexpected diversity within the extinct elephant birds (Aves: Aepyornithidae) and a new identity for the world's largest bird. *R. Soc. Open Sci.* **2018**, *5*, 181295. [CrossRef]

13. Fromm, A.; Meiri, S. Big, flightless, insular and dead: Characterising the extinct birds of the Quaternary. *J. Biogeogr.* **2021**, 1–10. [CrossRef]

14. Winkler, D.; Billerman, S.; Lovette, I. Tinamous (Tinamidae). *Birds World* **2020**. [CrossRef]

15. Cabot, J.; Christie, D.A.; Jutglar, F.; Sharpe, C.J. Dwarf Tinamou (*Taoniscus nanus*), version 1.0. *Birds World* **2020**. [CrossRef]

16. Cabot, J.; Jutglar, F.; Garcia, E.F.J.; Boesman, P.F.D.; Sharpe, C.J. Gray Tinamou (*Tinamus tao*), version 1.0. *Birds World* **2020**. [CrossRef]

17. Folch, A.; Christie, D.A.; Jutglar, F.; Garcia, E.F.J. Common Ostrich (*Struthio camelus*), version 1.0. *Birds World* **2020**. [CrossRef]

18. Davies, S.J.J.F. Moas. In *Grzimek's Animal Life Encyclopedia. Birds I: Tinamous and Ratites to Hoatzins*, 2nd ed.; Hutchins, M., Ed.; Gale Group: Farmington Hills, MI, USA, 2003; Volume 8, pp. 95–98.

19. Szabo, M.J. South Island Giant Moa. Available online: http://nzbirdsonline.org.nz/species/south-island-giant-moa (accessed on 18 April 2021).

20. McDowell, S. The Bony Palate of Birds. Part I The Palaeognathae. *Auk* **1948**, *65*, 520–549. [CrossRef]

21. Fürbringer, M. *Untersuchungen zur Morphologie und Systematik der Vögel: Zugleich ein Beitrag zur Anatomie der Stütz-und Bewegungsorgane*; T. van Holkema: Amsterdam, The Netherlands, 1888; Volume 15.

22. Allen, G.M. *Birds and Their Attributes*; Marshall Jones Company: Boston, MA, USA, 1925.

23. Stresemann, E. Aves in: Kükenthal-Krumbach, Handbuch der Zoologie. *Zool. VII* **1927**, *7*, 2.

24. Verheyen, R. Outline of procedure in basic avian systematics. *Gerfaut* **1960**, *1960*, 50.

25. Cracraft, J. Continental drift, paleoclimatology, and the evolution and biogeography of birds. *J. Zool.* **1973**, *169*, 455–543. [CrossRef]

26. Cracraft, J. Avian evolution, Gondwana biogeography and the Cretaceous-Tertiary mass extinction event. *Proc. R. Soc. B* **2001**, *268*, 459–469. [CrossRef] [PubMed]

27. Roff, D. The evolution of flightlessness: Is history important? *Evol. Ecol.* **1994**, *8*, 639–657. [CrossRef]

28. Merrem, B. Tentamen systematis naturalis avium. *Abh. Der Königlichen Akad. Der Wiss. Berl.* **1813**, 237–259.

29. Lee, K.; Feinstein, J.; Cracraft, J. Chapter 7—The Phylogeny of Ratite Birds: Resolving Conflicts between Molecular and Morphological Data Sets. In *Avian Molecular Evolution and Systematics*; Mindell, D.P., Ed.; Academic Press: San Diego, CA, USA, 1997; pp. 173–209.

30. Livezey, B.C.; Zusi, R.L. Higher-order phylogeny of modern birds (Theropoda, Aves: Neornithes) based on comparative anatomy. II. Analysis and discussion. *Zool. J. Linn. Soc.* **2007**, *149*, 1–95. [CrossRef] [PubMed]

31. Bourdon, E.; De Riqles, A.; Cubo, J. A new Transantarctic relationship: Morphological evidence for a Rheidae–Dromaiidae–Casuariidae clade (Aves, Palaeognathae, Ratitae). *Zool. J. Linn. Soc.* **2009**, *156*, 641–663. [CrossRef]

32. Elzanowski, A. Cretaceous birds and avian phylogeny. *Cour. Forsch. Senckenberg* **1995**, *181*, 37–53.

33. Johnston, P. New morphological evidence supports congruent phylogenies and Gondwana vicariance for palaeognathous birds. *Zool. J. Linn. Soc.* **2011**, *163*, 959–982. [CrossRef]

34. Owen, R. *The Anatomy of Vertebrates, Volume 2: Birds and Mammals*; Longmans, Green and Company: London, UK, 1866.

35. Chojnowski, J.L.; Kimball, R.T.; Braun, E.L. Introns outperform exons in analyses of basal avian phylogeny using clathrin heavy chain genes. *Gene* **2008**, *410*, 89–96. [CrossRef]

36. Hackett, S.J.; Kimball, R.T.; Reddy, S.; Bowie, R.C.K.; Braun, E.L.; Braun, M.J.; Chojnowski, J.L.; Cox, W.A.; Han, K.; Harshman, J.; et al. A Phylogenomic Study of Birds Reveals Their Evolutionary History. *Science* **2008**, *320*, 1763–1768. [CrossRef]

37. Harshman, J.; Braun, E.L.; Braun, M.J.; Huddleston, C.J.; Bowie, R.C.K.; Chojnowski, J.L.; Hackett, S.J.; Han, K.; Kimball, R.T.; Marks, B.D.; et al. Phylogenomic evidence for multiple losses of flight in ratite birds. *Proc. Natl. Acad. Sci. USA* **2008**, *105*, 13462–13467. [CrossRef]

38. Haddrath, O.; Baker, A. Multiple nuclear genes and retroposons support vicariance and dispersal of the palaeognaths, and an Early Cretaceous origin of modern birds. *Proc. R. Soc. B: Biol. Sci.* **2012**, *279*, 4617–4625. [CrossRef]
39. Smith, J.V.; Braun, E.L.; Kimball, R.T. Ratite Nonmonophyly: Independent Evidence from 40 Novel Loci. *Syst. Biol.* **2012**, *62*, 35–49. [CrossRef]
40. Baker, A.J.; Haddrath, O.; McPherson, J.D.; Cloutier, A. Genomic Support for a Moa–Tinamou Clade and Adaptive Morphological Convergence in Flightless Ratites. *Mol. Biol. Evol.* **2014**, *31*, 1686–1696. [CrossRef]
41. Grealy, A.; Phillips, M.; Gifford, M.; Gilbert, M.T.P.R.; Jean-Marie Lambert, D.; Bunce, M.; Haile, J. Eggshell palaeogenomics: Palaeognath evolutionary history revealed through ancient nuclear and mitochondrial DNA from Madagascan elephant bird (*Aepyornis* sp.) eggshell. *Mol. Phylogenetics Evol.* **2017**, *109*, 151–163. [CrossRef] [PubMed]
42. Almeida, F.C.; Porzecanski, A.L.; Cracraft, J.L.; Bertelli, S. The evolution of tinamous (Palaeognathae: Tinamidae) in light of molecular and combined analyses. *Zool. J. Linn. Soc.* **2021**. [CrossRef]
43. Feng, S.; Stiller, J.; Deng, Y.; Armstrong, J.; Fang, Q.; Reeve, A.H.; Xie, D.; Chen, G.; Guo, C.; Faircloth, B.C.; et al. Dense sampling of bird diversity increases power of comparative genomics. *Nature* **2020**, *587*, 252–257. [CrossRef] [PubMed]
44. Phillips, M.J.; Gibb, G.C.; Crimp, E.A.; Penny, D. Tinamous and Moa Flock Together: Mitochondrial Genome Sequence Analysis Reveals Independent Losses of Flight among Ratites. *Syst. Biol.* **2009**, *59*, 90–107. [CrossRef]
45. Mitchell, K.J.; Llamas, B.; Soubrier, J.; Rawlence, N.J.; Worthy, T.H.; Wood, J.; Lee, M.S.Y.; Cooper, A. Ancient DNA reveals elephant birds and kiwi are sister taxa and clarifies ratite bird evolution. *Science* **2014**, *344*, 898–900. [CrossRef]
46. Urantówka, A.D.; Kroczak, A.; Mackiewicz, P. New view on the organization and evolution of Palaeognathae mitogenomes poses the question on the ancestral gene rearrangement in Aves. *BMC Genom.* **2020**, *21*, 874. [CrossRef]
47. Jarvis, E.D.; Mirarab, S.; Aberer, A.J.; Li, B.; Houde, P.; Li, C.; Ho, S.Y.W.; Faircloth, B.C.; Nabholz, B.; Howard, J.T.; et al. Whole-genome analyses resolve early branches in the tree of life of modern birds. *Science* **2014**, *346*, 1320–1331. [CrossRef]
48. Prum, R.O.; Berv, J.S.; Dornburg, A.; Field, D.J.; Townsend, J.P.; Lemmon, E.M.; Lemmon, A.R. A comprehensive phylogeny of birds (Aves) using targeted next-generation DNA sequencing. *Nature* **2015**, *526*, 569–573. [CrossRef]
49. Yonezawa, T.; Segawa, T.; Mori, H.; Campos, P.F.; Hongoh, Y.; Endo, H.; Akiyoshi, A.; Kohno, N.; Nishida, S.; Wu, J. Phylogenomics and morphology of extinct paleognaths reveal the origin and evolution of the ratites. *Curr. Biol.* **2017**, *27*, 68–77. [CrossRef]
50. Sackton, T.B.; Grayson, P.; Cloutier, A.; Hu, Z.; Liu, J.S.; Wheeler, N.E.; Gardner, P.P.; Clarke, J.A.; Baker, A.J.; Clamp, M.; et al. Convergent regulatory evolution and loss of flight in paleognathous birds. *Science* **2019**, *364*, 74–78. [CrossRef]
51. Norberg, U.L. Evolution of flight in animals. *Flow Phenom. Nat.* **2007**, *1*, 36–48.
52. Livezey, B.C. Evolution of flightlessness in rails (Gruiformes, Rallidae): Phylogenetic, Ecomorphological, and Ontogenetic Perspectives. *Ornithol. Monogr.* **2003**, *53*, iii–654. [CrossRef]
53. Gaspar, J.; Gibb, G.C.; Trewick, S.A. Convergent morphological responses to loss of flight in rails (Aves: Rallidae). *Ecol. Evol.* **2020**, *10*, 6186–6207. [CrossRef] [PubMed]
54. Kuhl, H.; Frankl-Vilches, C.; Bakker, A.; Mayr, G.; Nikolaus, G.; Boerno, S.T.; Klages, S.; Timmermann, B.; Gahr, M. An Unbiased Molecular Approach Using 3′-UTRs Resolves the Avian Family-Level Tree of Life. *Mol. Biol. Evol.* **2020**, *38*, 108–127. [CrossRef] [PubMed]
55. Claramunt, S.; Cracraft, J. A new time tree reveals Earth history's imprint on the evolution of modern birds. *Sci. Adv.* **2015**, *1*, e1501005. [CrossRef]
56. Cloutier, A.; Sackton, T.B.; Grayson, P.; Clamp, M.; Baker, A.J.; Edwards, S.V. Whole-Genome Analyses Resolve the Phylogeny of Flightless Birds (Palaeognathae) in the Presence of an Empirical Anomaly Zone. *Syst. Biol.* **2019**, *68*, 937–955. [CrossRef]
57. Kimball, R.; Oliveros, C.; Wang, N.; White, N.D.; Barker, F.K.; Field, D.J.; Ksepka, D.; Chesser, R.; Moyle, R.; Braun, M.; et al. A Phylogenomic Supertree of Birds. *Diversity* **2019**, *11*, 109. [CrossRef]
58. Reddy, S.; Kimball, R.T.; Pandey, A.; Hosner, P.A.; Braun, M.J.; Hackett, S.J.; Han, K.; Harshman, J.; Huddleston, C.J.; Kingston, S.; et al. Why Do Phylogenomic Data Sets Yield Conflicting Trees? Data Type Influences the Avian Tree of Life more than Taxon Sampling. *Syst. Biol.* **2017**, *66*, 857–879. [CrossRef]
59. Faux, C.; Field, D.J. Distinct developmental pathways underlie independent losses of flight in ratites. *Biol. Lett.* **2017**, *13*. [CrossRef]
60. Farlie, P.G.; Davidson, N.M.; Baker, N.L.; Raabus, M.; Roeszler, K.N.; Hirst, C.; Major, A.; Mariette, M.M.; Lambert, D.M.; Oshlack, A.; et al. Co-option of the cardiac transcription factor Nkx2.5 during development of the emu wing. *Nat. Commun.* **2017**, *8*, 132. [CrossRef]
61. Forshaw, J.M. *Encyclopedia of Animals: Birds*; Merehurst Limited: London, UK, 1991.
62. Houde, P. *Paleognathous Birds from the Early Tertiary of the Northern Hemisphere*; Paynter, J.R.A., Ed.; Nuttall Ornithological Club: Cambridge, MA, USA, 1988; Volume 22.
63. Parris, D.; Hope, S. New interpretations of birds from the Navesink and Hornerstown formations, New Jersey, USA (Aves: Neornithes). In Proceedings of the 5th Symposium of the Society of Avian Paleontology and Evolution, Beijing, China, 1–4 June 2002; pp. 113–124.
64. Nesbitt, S.J.; Clarke, J.A. The anatomy and taxonomy of the exquisitely preserved Green River formation (early Eocene) Lithornithids (Aves) and the relationships of Lithornithidae. *Bull. Am. Mus. Nat. Hist.* **2016**, 1–91. [CrossRef]
65. Torres, C.R.; Norell, M.A.; Clarke, J.A. Estimating Flight Style of Early Eocene Stem Palaeognath Bird *Calciavis grandei* (Lithornithidae). *Anat. Rec.* **2020**, *303*, 1035–1042. [CrossRef] [PubMed]

66. Winkler, D.W.; Billerman, S.M.; Lovette, I.J. Cassowaries and Emu (Casuariidae), version 1.0. *Birds World* **2020**. [CrossRef]
67. Winkler, D.W.; Billerman, S.M.; Lovette, I.J. Ostriches (Struthionidae), version 1.0. *Birds World* **2020**. [CrossRef]
68. Winkler, D.W.; Billerman, S.M.; Lovette, I.J. Rheas (Rheidae), version 1.0. *Birds World* **2020**. [CrossRef]
69. Clarke, J.A.; Tambussi, C.P.; Noriega, J.I.; Erickson, G.M.; Ketcham, R.A. Definitive fossil evidence for the extant avian radiation in the Cretaceous. *Nature* **2005**, *433*, 305–308. [CrossRef]
70. Ksepka, D.T.; Stidham, T.A.; Williamson, T.E. Early Paleocene landbird supports rapid phylogenetic and morphological diversification of crown birds after the K-Pg mass extinction. *Proc. Natl. Acad. Sci. USA* **2017**, *114*, 8047–8052. [CrossRef]
71. Field, D.J.; Bercovici, A.; Berv, J.S.; Dunn, R.; Fastovsky, D.E.; Lyson, T.R.; Vajda, V.; Gauthier, J.A. Early Evolution of Modern Birds Structured by Global Forest Collapse at the End-Cretaceous Mass Extinction. *Curr. Biol.* **2018**, *28*, 1825–1831.e2. [CrossRef]
72. Field, D.J.; Benito, J.; Chen, A.; Jagt, J.; Ksepka, D. Late Cretaceous neornithine from Europe illuminates the origins of crown birds. *Nature* **2020**, *579*, 397–401. [CrossRef]
73. Longrich, N.; Tokaryk, T.; Field, D.J. Mass extinction of birds at the Cretaceous Paleogene (K-Pg) boundary. *Proc. Natl. Acad. Sci.* **2011**, *108*, 15253–15257. [CrossRef]
74. Saupe, E.E.; Farnsworth, A.; Lunt, D.J.; Sagoo, N.; Pham, K.V.; Field, D.J. Climatic shifts drove major contractions in avian latitudinal distributions throughout the Cenozoic. *Proc. Natl. Acad. Sci. USA* **2019**, *116*, 12895–12900. [CrossRef] [PubMed]
75. Crouch, N.M.A.; Clarke, J.A. Body size evolution in palaeognath birds is consistent with Neogene cooling-linked gigantism. *Palaeogeogr. Palaeoclimatol. Palaeoecol.* **2019**, *532*, 109224. [CrossRef]
76. Mayr, G. *Paleogene Fossil Birds*; Springer: Berlin/Heidelberg, Germany, 2009; p. 262.
77. Mayr, G. *Avian Evolution – The Fossil Record of Birds and Its Paleobiological Significance*; Wiley Blackwell: Chichester, UK, 2017; p. 289.
78. Mikhailov, K.; Zelenkov, N. The late Cenozoic history of the ostriches (Aves: Struthionidae), as revealed by fossil eggshell and bone remains. *Earth-Sci. Rev.* **2020**, *208*, 103270. [CrossRef]
79. Houde, P.; Olson, S. Paleognathous Carinate Birds from the Early Tertiary of North America. *Science* **1981**, *214*, 1236–1237. [CrossRef] [PubMed]
80. Du Toit, C.J.; Chinsamy, A.; Cunningham, S. Cretaceous origins of the vibrotactile bill-tip organ in birds. *Proc. R. Soc. B* **2020**, *287*, 20202322. [CrossRef]
81. Clarke, J. Morphology, Phylogenetic Taxonomy, and Systematics of *Ichthyornis* and *Apatornis* (Avialae: Ornithurae). *Bull. Am. Mus. Nat. Hist.* **2004**, *286*, 1–179. [CrossRef]
82. Benito, J.; Chen, A.; Wilson, L.E.; Bhullar, B.A.S.; Burnham, D.; Field, D.J. 40 new specimens of *Ichthyornis* provide unprecedented insight into the postcranial morphology of crownward stem group birds. *bioRxiv* **2022**. [CrossRef]
83. Scotese, C.R. PALEOMAP PaleoAtlas for GPlates and the PaleoData Plotter Program. Available online: http://www.earthbyte.org/paleomap-paleoatlas-for-gplates/ (accessed on 3 April 2021).
84. Müller, R.D.; Cannon, J.; Qin, X.; Watson, R.J.; Gurnis, M.; Williams, S.; Pfaffelmoser, T.; Seton, M.; Russell, S.H.J.; Zahirovic, S. GPlates: Building a Virtual Earth Through Deep Time. *Geochem. Geophys. Geosystems* **2018**, *19*, 2243–2261. [CrossRef]
85. Olson, S.L.; Parris, D.C. The Cretaceous birds of New Jersey. *Smithson. Contrib. Paleobiol.* **1987**, *63*, 1–22. [CrossRef]
86. Staron, R.; Grandstaff, D.; Grandstaff, B.; Gallagher, W. Mosasaur taphonomy and geochemistry implications for a KT bone bed in the New Jersey coastal plain. *J. Vertebr. Paleontol.* **1999**, *19*, 78A.
87. Lofgren, D.L.; Lillegraven, J.; Clemens, W.; Gingerich, P.; Williamson, T.; Woodburne, M.O. Paleocene Biochronology: The Puercan Through Clarkforkian Land Mammal Ages. In *Late Cretaceous and Cenozoic Mammals of North America*; Woodburne, M., Ed.; Columbia University Press: New York, NY, USA, 2004.
88. Stidham, T.A.; Lofgren, D.; Farke, A.A.; Paik, M.; Choi, R. A lithornithid (Aves: Palaeognathae) from the Paleocene (Tiffanian) of southern California. *PaleoBios* **2014**, *31*. [CrossRef]
89. Lofgren, D.L.; Honey, J.G.; McKenna, M.C.; Zondervan, R.L.; Smith, E.E.; Wang, X.; Barnes, L. Paleocene primates from the Goler Formation of the Mojave Desert in California. In *Geology and Vertebrate Paleontology of Western and Southern North America, Contributions in Honor of David P. Whistler: National History Museum of Los Angeles County Science Series*; National History Museum of Los Angeles County: Los Angeles, CA, USA, 2008; Volume 41, pp. 11–28.
90. Albright, L.I.; Lofgren, D.; McKenna, M. Magnetostratigraphy, mammalian biostratigraphy, and refined age assessment of the Goler Formation (Paleocene), California. *Mus. North Ariz. Bull.* **2009**, *65*, 259–278.
91. Lofgren, D.; Mckenna, M.; Honey, J.; Nydam, R.; Wheaton, C.; Yokote, B.; Henn, L.; Hanlon, W.; Manning, S.; Mcgee, C. New Records of Eutherian Mammals from the Goler Formation (Tiffanian, Paleocene) of California and Their Biostratigraphic and Paleobiogeographic Implications. *Am. Mus. Novit.* **2014**, *2014*, 1–57. [CrossRef]
92. Smith, M.E.; Carroll, A.R.; Singer, B.S. Synoptic reconstruction of a major ancient lake system; Eocene Green River Formation, western United States. *Geol. Soc. Am. Bull.* **2008**, *120*, 54–84. [CrossRef]
93. Murphey, P.C.; Evanoff, E. Paleontology and stratigraphy of the middle Eocene Bridger Formation, southern Green River basin, Wyoming. In Proceedings of the 9th Conference on Fossil Resources, Kemmerer, WY, USA, 26–28 April 2011; pp. 83–109.
94. Smith, T.; Smith, R. Terrestrial mammals as biostratigraphic indicators in upper Paleocene-lower Eocene marine deposits of the southern North Sea Basin. *Geol. Soc. Am. Spec. Pap.* **2003**, *369*, 513–520. [CrossRef]
95. De Bast, E.; Steurbaut, E.; Smith, T. New mammals from the marine Selandian of Maret, Belgium, and their implications for the age of the Paleocene continental deposits of Walbeck, Germany. *Geol. Belg.* **2013**, *16*, 4.

96. Mayr, G.; Smith, T. New Paleocene bird fossils from the North Sea Basin in Belgium and France. *Geol. Belg.* **2019**, *22*, 35–46. [CrossRef]

97. Aguilar, J.; Augustı, J.; Alexeeva, N.; Antoine, P.; Antunes, M.; Archer, M. Syntheses and correlation tables. In Proceedings of the Actes du Congres BiochroM, Montpellier, France, 14–17 April 1997; pp. 769–805.

98. Mayr, G. The Birds from the Paleocene Fissure Filling of Walbeck (Germany). *J. Vertebr. Paleontol.* **2007**, *27*, 394–408. [CrossRef]

99. Steurbaut, E. High-resolution holostratigraphy of Middle Paleocene to Early Eocene strata in Belgium and adjacent areas. *Palaeontogr. Abt. A: Palaozool. Stratigr.* **1997**, *247*, 91–156.

100. Moreau, F.; Mathis, S. Les élasmobranches du Thanétien (Paléocène) du Nord de la France, des carrières de Templeuve et de Leforest. *Cossmanniana* **2000**, *7*, 1–18.

101. Heilmann-Clausen, C.; Schmitz, B. The late Paleocene thermal maximum δ13C excursion in Denmark? *Gff* **2000**, *122*, 70. [CrossRef]

102. Bourdon, E.; Lindow, B.E.K. A redescription of *Lithornis vulturinus* (Aves, Palaeognathae) from the Early Eocene Fur Formation of Denmark. *Zootaxa* **2015**, *4032*, 493–514. [CrossRef] [PubMed]

103. Chambers, L.; Pringle, M.; Fitton, G.; Larsen, L.; Pedersen, A.; Parrish, R. Recalibration of the Palaeocene-Eocene boundary (PE) using high precision U-Pb and Ar-Ar isotopic dating. In Proceedings of the EGS-AGU-EUG Joint Assembly, Nice, France, 6–11 April 2003; p. 9681.

104. Leonard, L.; Dyke, G.; van Tuinen, M. A New Specimen of the Fossil Palaeognath *Lithornis* from the Lower Eocene of Denmark. *Am. Mus. Novit.* **2005**, *3491*, 1–11. [CrossRef]

105. King, C. The stratigraphy of the London Clay and associated deposits. *Tert. Res. Spec. Pap.* **1981**, *6*, 1–158.

106. Ellison, R.A.; Woods, M.A.; Allen, D.J.; Forster, A.; Pharaoh, T.C.; King, C. Palaeogene–Eocene. In *Geology of London*; British Geological Survey: London, UK, 2004; pp. 44–54.

107. Friedman, M.; Beckett, H.T.; Close, R.A.; Johanson, Z. The English Chalk and London Clay: Two remarkable British bony fish Lagerstätten. *Geol. Soc. Lond. Spec. Publ.* **2016**, *430*, 165–200. [CrossRef]

108. Franzen, J.; Haubold, H. The middle Eocene of European mammalian stratigraphy. Definition of the Geiseltalian. *Mod. Geol.* **1986**, *10*, 159–170.

109. Schaal, S.; Ziegler, W. *Messel: Ein Schaufenster in die Geschichte der Erde und des Lebens*; Waldemar Kramer: Frankfurt, Germany, 1988.

110. Lenz, O.K.; Wilde, V.; Mertz, D.F.; Riegel, W. New palynology-based astronomical and revised 40Ar/39Ar ages for the Eocene maar lake of Messel (Germany). *Int. J. Earth Sci.* **2015**, *104*, 873–889. [CrossRef]

111. Mayr, G. First substantial Middle Eocene record of the Lithornithidae (Aves): A postcranial skeleton from Messel (Germany). *Ann. De Paleontol.* **2008**, *94*, 29–37. [CrossRef]

112. Mayr, G. Towards the complete bird—The skull of the middle Eocene Messel lithornithid (Aves, Lithornithidae). *Bull. De L'Institute R. Des. Sci. Nat. De Belg.* **2009**, *79*, 167–173.

113. Blondel, J.; Mourer-Chauviré, C. Evolution and history of the western Palaearctic avifauna. *Trends Ecol. Evol.* **1998**, *13*, 488–492. [CrossRef]

114. Mayr, G. Birds–The most species-rich vertebrate group in Messel. *Senckenberg Ges. Für Nat. Frankf. Am Main* **2018**, 169–214.

115. Harrison, C. A new cathartid vulture from the lower Eocene of Wyoming. *Tert. Res.* **1979**, *5*, 7–10.

116. Grande, L. *Paleontology of the Green River Formation, with a Review of the Fish Fauna*, 2nd ed.; University of Wyoming: Laramie, WY, USA, 1984.

117. Wetmore, A. Fossil birds from the Green River deposits of eastern Utah. *Ann. Carnegie Mus.* **1926**, *16*, 391–497.

118. Brodkorb, P. An eocene puffbird from Wyoming. *Rocky Mt. Geol.* **1970**, *9*, 13–15.

119. Olson, S. A Lower Eocene Frigatebird from the Green River Formation of Wyoming (Pelecaniformes: Fregatidae). *Smithson. Contrib. Paleobiol.* **1977**, 35. [CrossRef]

120. Olson, S. An early Eocene oilbird from the Green River Formation of Wyoming (Caprimulgiformes: Steatornithidae). In Proceedings of the Table Ronde internationale du CNRS, Lyon, France, 18–21 September 1985; pp. 57–69.

121. Olson, S. A new family of primitive landbirds from the early Eocene Green River Formation of Wyoming. *Pap. Avian Paleontol. Honor. Pierce Brodkorb* **1992**, *36*, 127–136.

122. Mayr, G.; Daniels, M. A new short-legged landbird from the Early Eocene of Wyoming and contemporaneous European sites. *Acta Palaeontol. Pol.* **2001**, *46*, 393–402.

123. Olson, S.; Matsuoka, H. New specimens of the early Eocene frigatebird *Limnofregata* (Pelecaniformes: Fregatidae), with the description of a new species. *Zootaxa* **2005**, *1046*, 1–15. [CrossRef]

124. Ksepka, D.T.; Clarke, J.A. Affinities of *Palaeospiza bella* and the Phylogeny and Biogeography of Mousebirds (Coliiformes). *Auk* **2009**, *126*, 245–259. [CrossRef]

125. Ksepka, D.T.; Clarke, J.A. *Primobucco mcgrewi* (Aves: Coracii) from the Eocene Green River Formation: New anatomical data from the earliest constrained record of stem rollers. *J. Vertebr. Paleontol.* **2010**, *30*, 215–225. [CrossRef]

126. Weidig, I. New birds from the Lower Eocene Green River Formation, North America. In Proceedings of the VII International Meeting of the Society of Avian Paleontology and Evolution, ed. W.E. Boles and T.H. Worthy. *Rec. Aust. Mus.* **2010**, *62*, 29–44.

127. Ksepka, D.T.; Clarke, J.A.; Grande, L. Stem Parrots (Aves, Halcyornithidae) from the Green River Formation and a Combined Phylogeny of Pan-Psittaciformes. *J. Paleontol.* **2011**, *85*, 835–852. [CrossRef]

128. Smith, N.D.; Grande, L.; Clarke, J.A. A new species of Threskiornithidae-like bird (Aves, Ciconiiformes) from the Green River Formation (Eocene) of Wyoming. *J. Vertebr. Paleontol.* **2013**, *33*, 363–381. [CrossRef]

129. Owen, R. Description of the remains of a mammal, a bird, and a serpent from the London Clay. *Proc. Geol. Soc. Lond.* **1840**, *3*, 162–166.

130. Owen, R. Description of the Fossil Remains of a Mammal (*Hyracotherium leporinum*) and of a Bird (*Lithornis vulturinus*) from the London Clay. *Trans. Geol. Soc.* **1841**, *2*, 203–208. [CrossRef]

131. Harrison, C.J.O.; Walker, C.A. Birds of the British Lower Eocene. *Tert. Res. Spec. Pap.* **1977**, *3*, 1–52.

132. Harrison, C.J.O. Rail-like cursorial birds of the British Lower Eocene, with descriptions of two new species. *Lond. Nat.* **1984**, *63*, 14–23.

133. Bledsoe, A.H. A Phylogenetic Analysis of Postcranial Skeletal Characters of the Ratite Birds. *Ann. Carnegie Mus.* **1988**, *57*, 73–90.

134. Mayr, G.; Clarke, J. The deep divergences of neornithine birds: A phylogenetic analysis of morphological characters. *Cladistics* **2003**, *19*, 527–553. [CrossRef]

135. Clarke, J.; Zhou, Z.; Zhang, F. Insight into the evolution of avian flight from a new clade of Early Cretaceous ornithurines from China and the morphology of *Yixianornis grabaui*. *J. Anat.* **2006**, *208*, 287–308. [CrossRef]

136. Worthy, T.H.; Mitri, M.; Handley, W.D.; Lee, M.S.Y.; Anderson, A.; Sand, C. Osteology supports a stem-galliform affinity for the giant extinct flightless bird *Sylviornis neocaledoniae* (Sylviornithidae, Galloanseres). *PLoS ONE* **2016**, *11*, e0150871. [CrossRef]

137. Worthy, T.H.; Degrange, F.J.; Handley, W.D.; Lee, M.S.Y. The evolution of giant flightless birds and novel phylogenetic relationships for extinct fowl (Aves, Galloanseres). *R. Soc. Open Sci.* **2017**, *4*, 170975. [CrossRef] [PubMed]

138. Del Hoyo, J.; Collar, N.; Garcia, E.F.J. Somali Ostrich (*Struthio molybdophanes*), version 1.0. *Birds World* **2020**. [CrossRef]

139. Janz, L.; Feathers, J.K.; Burr, G.S. Dating surface assemblages using pottery and eggshell: Assessing radiocarbon and luminescence techniques in Northeast Asia. *J. Archaeol. Sci.* **2015**, *57*, 119–129. [CrossRef]

140. Khatsenovich, A.; Rybin, E.; Gunchinsuren, B.; Bolorbat, T.; Odsuren, D.; Angaragdulguun, G.; Margad-Erdene, G. Human and Struthio asiaticus: One page of Paleolithic art in the eastern part of Central Asia. *Irkutsk State Univ. Bull. Ser. Geoarcheol. Ethnol. Anthropol.* **2017**, *21*, 80–106.

141. Wang, S.; Hu, Y.; Wang, L. New ratite eggshell material from the Miocene of Inner Mongolia, China. *Chin. Birds* **2011**, *2*, 18–26. [CrossRef]

142. Mayr, G. The middle Eocene European "ratite" *Palaeotis* (Aves, Palaeognathae) restudied once more. *Paläontologische Z.* **2015**, *89*, 503–514. [CrossRef]

143. Houde, P.; Haubold, H. *Palaeotis weigelti* restudied; a small middle Eocene ostrich (Aves: Struthioniformes). *Palaeovertebrata* **1987**, *17*, 27–42.

144. Lemoine, V. *Recherches sur les Oiseaux Fossiles des Terrains Tertiaires Inférieurs des Environs de Reims, Deuxième Partie*; Matot-Braine: Reims, France, 1881.

145. Lambrecht, K. *Palaeotis weigelti* n. g. n. sp., eine fossile Trappe aus der mitteleozänen Braunkohle des Geiseltales. *Jahrb. Des Halleschen Verb. Für Die Erforsch. Mitteldtsch. Bodenschätze* **1928**, *7*, 1–11.

146. Peters, D.S. Ein vollständiges Exemplar von *Palaeotis weigelti* (Aves, Palaeognathae). *Cour. Forsch. Senckenberg* **1988**, *107*, 223–233.

147. Dyke, G.J. The fossil record and molecular clocks: Basal radiations within Neornithes. In *Telling the Evolutionary Time—Molecular Clocks and the Fossil Record*; Smith, P., Donoghue, P., Eds.; Taylor and Francis: London, UK, 2003; pp. 263–278.

148. Mayr, G. Hindlimb morphology of *Palaeotis* suggests palaeognathous affinities of the Geranoididae and other crane-like birds from the Eocene of the Northern Hemisphere. *Acta Palaeontol. Pol.* **2019**, *64*, 669–678. [CrossRef]

149. Mayr, G. On the taxonomy and osteology of the Early Eocene North American Geranoididae (Aves, Gruoidea). *Swiss J. Palaeontol.* **2016**, *135*, 315–325. [CrossRef]

150. Bourdon, E.; Mourer-Chauviré, C.; Laurent, Y. Early Eocene Birds from La Borie, Southern France. *Acta Palaeontol. Pol.* **2016**, *61*, 175–190. [CrossRef]

151. Cracraft, J. Systematics and evolution of the Gruiformes (Class, Aves). 1. The Eocene family Geranoididae and the early history of the Gruiformes. *Am. Mus. Novit.* **1969**, *2388*, 1–41.

152. Kurochkin, E. A survey of the Paleogene birds of Asia. *Smithson. Contrib. Paleobiol.* **1976**, *27*, 75–86.

153. Burchak-Abramovich, N.I. Urmiornis (*Urmiornis maraghanus* Mecq.) strausopodobnaya ptitsa gipparionovoi fauny Zakavkaz'a i yuzhnoi Ukrainy [Urmiornis (*Urmiornis maraghanus* Mecq.), an ostrich-like bird from the Hipparion fauna of Transcaucasia and southern Ukraine]. *Izv. Akedemii Nauk Azerbaidzhanskoi SSR* **1951**, *6*, 83–94.

154. Feduccia, A. *The Age of Birds*; Harvard University Press: Cambridge, MA, USA, 1980.

155. Olson, S.L. The fossil record of birds. *Avian Biol.* **1985**, 79–238.

156. Clarke, J.A.; Norell, M.A.; Dashzévëg, D. New Avian Remains from the Eocene of Mongolia and the Phylogenetic Position of the Eogruidae (Aves, Gruoidea). *Am. Mus. Novit.* **2005**, 1–17. [CrossRef]

157. Elzanowski, A.; Paul, G.S.; Stidham, T.A. An avian quadrate from the Late Cretaceous Lance Formation of Wyoming. *J. Vertebr. Paleontol.* **2001**, *20*, 712–719. [CrossRef]

158. Vandenberghe, N.; Hilgen, F.J.; Speijer, R.P.; Ogg, J.G.; Gradstein, F.M.; Hammer, O.; Hollis, C.J.; Hooker, J.J. Chapter 28—The Paleogene Period. In *The Geologic Time Scale*; Gradstein, F.M., Ogg, J.G., Schmitz, M., Ogg, G., Eds.; Elsevier: Amsterdam, The Netherlands, 2012; pp. 855–921.

159. Lambrecht, K. Drei neue Vogelformen aus dem Lutétian des Geiseltales. *Nova Acta Leopold.* **1935**, *3*, 361–367.

160. Buffetaut, E.; Angst, D. Stratigraphic distribution of large flightless birds in the Palaeogene of Europe and its palaeobiological and palaeogeographical implications. *Earth-Sci. Rev.* **2014**, *138*, 394–408. [CrossRef]

161. Martin, L.D. The status of the Late Paleocene birds *Gastornis* and *Remiornis*. *Nat. Hist. Mus. Los Angeles Cty. Sci. Ser.* **1992**, *36*, 97–108.

162. Smith, T.; Quesnel, F.; De Plöeg, G.; De Franceschi, D.; Métais, G.; De Bast, E.; Solé, F.; Folie, A.; Boura, A.; Claude, J.; et al. First Clarkforkian equivalent Land Mammal Age in the latest Paleocene basal Sparnacian facies of Europe: Fauna, flora, paleoenvironment and (bio)stratigraphy. *PLoS ONE* **2014**, *9*, e86229. [CrossRef] [PubMed]

163. Buffetaut, E.; de Ploëg, G. Giant Birds from the Uppermost Paleocene of Rivecourt (Oise, Northern France). *Bol. Do Cent. Port. De Geo-História E Pré-História* **2020**, *2*, 1.

164. Laurent, Y.; Adnet, S.; Bourdon, E.; Corbalan, D.; Danilo, L.; Duffaud, S.; Fleury, G.; Garcia, G.; Godinot, M.; Le Roux, G. La Borie (Saint-Papoul, Aude): Un gisement exceptionnel dans l'Éocène basal du Sud de la France. *Bull. De La Société D'histoire Nat. De Toulouse* **2010**, *146*, 89–103.

165. Danilo, L.; Remy, J.A.; Vianey-Liaud, M.; Marandat, B.; Sudre, J.; Lihoreau, F. A new Eocene locality in southern France sheds light on the basal radiation of Palaeotheriidae (Mammalia, Perissodactyla, Equoidea). *J. Vertebr. Paleontol.* **2013**, *33*, 195–215. [CrossRef]

166. Zelenkov, N.; Kurochkin, E.N. *Class. Aves, Iskopaemye Pozvonochnye Rossii i Sopredel'nyh Stran. Iskopaemye Reptilii i Ptitsy. Chast'3*; Kurochkin, E., Lopatin, A., Zelenkov, N., Eds.; GEOS: Moscow, Russia, 2015.

167. Zelenkov, N.; Boev, Z.; Lazaridis, G. A large ergilornithine (Aves, Gruiformes) from the Late Miocene of the Balkan Peninsula. *Paläontologische Z.* **2016**, *90*, 145–151. [CrossRef]

168. Lydekker, R. *Catalogue of the Fossil Birds in the British Museum (Natural History)*; Order of the Trustees: London, UK, 1891.

169. Tsoukala, E.; Bartsiokas, A. New *Mesopithecus pentelicus* specimens from Kryopigi, Macedonia, Greece. *J. Hum. Evol.* **2008**, *54*, 448–451. [CrossRef] [PubMed]

170. Lazaridis, G.; Tsoukala, E. *Hipparion phlegrae*, sp. nov. (Mammalia, Perissodactyla): A new species from the Turolian locality of Kryopigi (Kassandra, Chalkidiki, Greece). *J. Vertebr. Paleontol.* **2014**, *34*, 164–178. [CrossRef]

171. Boev, Z.; Lazaridis, G.; Tsoukala, E. *Otis hellenica* sp. nov., a new Turolian bustard (Aves: Otididae) from Kryopigi (Chalkidiki, Greece). *Geol. Balc.* **2013**, *42*, 59–64. [CrossRef]

172. Li, Q. Eocene fossil rodent assemblages from the Erlian Basin (Inner Mongolia, China): Biochronological implications. *Palaeoworld* **2016**, *25*, 95–103. [CrossRef]

173. Wetmore, A. Fossil Birds from Mongolia and China. *Am. Mus. Novit.* **1934**, *711*, 1–16.

174. Bendukidze, O. Novyj prestavitel' semeist-va Geranoididae (Aves, Gruiformes) iz eotsenovykh otlozhenij Zaisan. *Soobtzhenija Akad. Nauk Gruz. SSSR* **1971**, *63*, 749–751.

175. Russell, D.; Zhai, R. The Paleogene of Asia: Mammals and stratigraphy. *Mémoires Du Muséum Natl. D'histoire Nat. Série C* **1987**, *52*, 1–488.

176. Kurochkin, E.N. New representatives and evolution of two archaic gruiform families in Eurasia. *Tr. Sovmest. Sov. Mong. Paleontol. Ekspeditsija* **1981**, *15*, 59–85.

177. Dashzèvèg, D. Some carnivorous mammals from the Paleogene of the eastern Gobi Desert, Mongolia, and the application of Oligocene carnivores to stratigraphic correlation. *Am. Mus. Novit.* **1996**, *3179*, 1–14.

178. Musser, G.; Li, Z.; Clarke, J.A. A new species of Eogruidae (Aves: Gruiformes) from the Miocene of the Linxia Basin, Gansu, China: Evolutionary and climatic implications. *Auk* **2019**, *137*. [CrossRef]

179. Kozlova, E.V. Novye iskopaemye ptitsy iz yugovostochnoi Gobi. *Tr. Probl. I Temat. Soveshanii ZIN* **1960**, *9*, 323–329.

180. Karhu, A. A new species of *Urmiornis* (Gruiformes: Ergilornithidae) from the early Miocene of western Kazakhstan. *Paleontol. J.* **1997**, *31*, 102–107.

181. Wang, X.; Qiu, Z.; Opdyke, N.D. Litho-, bio-, and magnetostratigraphy and paleoenvironment of Tunggur Formation (Middle Miocene) in central Inner Mongolia, China. *Am. Mus. Novit.* **2003**, *2003*, 1–31. [CrossRef]

182. Brodkorb, P. Catalogue of fossil birds. Part 3 (Ralliformes, Ichthyornithiformes, Charadriiformes). *Bull. Fla. State Mus. Biol. Sci. II* **1967**, *3*, 1–220.

183. Cracraft, J. Systematics and evolution of the Gruiformes (class Aves). 3. Phylogeny of the suborder Grues. *Bull. Am. Mus. Nat. Hist.* **1973**, *151*, 1–128.

184. Barry, J.C.; Morgan, M.E.; Flynn, L.J.; Pilbeam, D.; Behrensmeyer, A.K.; Raza, S.M.; Khan, I.A.; Badgley, C.; Hicks, J.; Kelley, J. Faunal and environmental change in the late Miocene Siwaliks of northern Pakistan. *Paleobiology* **2002**, *28*, 1–71. [CrossRef]

185. Harrison, C.; Walker, C.A. Fossil birds from the Upper Miocene of northern Pakistan. *Tert. Res.* **1982**, *4*, 53–69.

186. Mecquenem, R. Contribution à l'étude du gisement des vertébrés de Maragha et de ses environs. *Ann. D'histoire Nat.* **1908**, *1*, 27–29.

187. Fang, X.; Wang, J.; Zhang, W.; Zan, J.; Song, C.; Yan, M.; Appel, E.; Zhang, T.; Wu, F.; Yang, Y.; et al. Tectonosedimentary evolution model of an intracontinental flexural (foreland) basin for paleoclimatic research. *Glob. Planet. Chang.* **2016**, *145*, 78–97. [CrossRef]

188. Kurochkin, E.N. Birds of the central Asia in Pliocene. *Trans. Jt. Sov. Mong. Paleontol. Exped.* **1985**, *26*, 1–120.

189. Mecquenem, R. Oiseaux. Contribution à l'étude des fossiles de Maragha. *Ann. De Paléontologie* **1925**, *14*, 54–56.

190. Portis, A. Contribuzioni alla ornitolitologia italiana. *Mem. Della R. Accad. Delle Sci. Di Torino* **1884**, *36*, 361–384.

191. Schaub, S. Ein Ratitenbecken aus dem Bohnerz von Egerkingen. *Eclogae Geol. Helv.* **1940**, *33*, 274–284.

192. Gaillard, C. Un oiseau géant dans les dépôts éocènes du Mont-d'Or lyonnais. *Compte Rendus De L'académie Des Sci.* **1936**, *202*, 965–967.

193. Angst, D.; Buffetaut, E.; Lécuyer, C.; Amiot, R. "Terror Birds" (Phorusrhacidae) from the Eocene of Europe imply trans-Tethys dispersal. *PLoS ONE* **2013**, *8*, e80357. [CrossRef] [PubMed]

194. Andrews, C.W. On the Pelvis and Hind-limb of *Mullerornis betsilei* M. Edw. & Grand.; with a Note On the Occurrence of a Ratite Bird in the Upper Eocene Beds of the Fayurn, Egypt. *Proc. Zool. Soc. Lond.* **1904**, *74*, 163–171.

195. Rasmussen, D.T.; Simons, E.L.; Hertel, F.; Judd, A. Hindlimb of a giant terrestrial bird from the upper Eocene, Fayum, Egypt. *Palaeontology* **2001**, *44*, 325–337. [CrossRef]

196. Pickford, M.; Senut, B. Miocene Palaeobiology of the Orange River Valley, Namibia. *Mem. Geol. Surv. Namib.* **2003**, *19*, 1–22.

197. Mourer-Chauviré, C.; Senut, B.; Pickford, M.; Mein, P. The oldest representative of the genus *Struthio* (Aves: Struthionidae), *Struthio coppensi* n. sp., from the Lower Miocene of Namibia. *Comptes Rendus De L'académie Des Sci. Sér. 2 Fasc. A Sci. De La Terre Des Planètes* **1996**, *322*, 325–332.

198. Mourer-Chauviré, C. Birds (Aves) from the Early Miocene of the Northern Sperrgebiet, Namibia. *Mem. Geol. Surv. Namib.* **2009**, *20*, 147–167.

199. Pickford, M. Preliminary Miocene mammalian biostratigraphy for western Kenya. *J. Hum. Evol.* **1981**, *10*, 73–97. [CrossRef]

200. Leonard, L.M.; Dyke, G.J.; Walker, C.A. New specimens of a fossil ostrich from the Miocene of Kenya. *J. Afr. Earth Sci.* **2006**, *45*, 391–394. [CrossRef]

201. Werdelin, L. Chronology of Neogene mammal localities. In *Cenozoic Mammals of Africa*; Werdelin, L., Sanders, W.J., Eds.; University of California Press: Berkeley, CA, USA, 2010; pp. 27–43.

202. Rich, P.V. A fossil avifauna from the Upper Miocene Beglia Formation of Tunisia. *Notes De Serv. Géologique Du Tunis* **1972**, *35*, 29–66.

203. Roberts, D.L.; Matthews, T.; Herries, A.I.; Boulter, C.; Scott, L.; Dondo, C.; Mtembi, P.; Browning, C.; Smith, R.M.; Haarhoff, P. Regional and global context of the Late Cenozoic Langebaanweg (LBW) palaeontological site: West Coast of South Africa. *Earth-Sci. Rev.* **2011**, *106*, 191–214. [CrossRef]

204. Rich, P. Preliminary report on the fossil avian remains from late Tertiary sediments at Langebaanweg (Cape Province), South Africa. *South. Afr. J. Sci.* **1980**, *76*, 166–170.

205. Manegold, A.; Louchart, A.; Carrier, J.; Elzanowski, A. The early Pliocene avifauna of Langebaanweg (South Africa): A review and update. In Proceedings of the 8th International Meeting of the Society of Avian Paleontology and Evolution, Vienna, Austria, 11–16 June 2012; pp. 135–152.

206. Geraads, D. Carnivores du Pliocène terminal de Ahl al Oughlam (Casablanca, Maroc). *Geobios* **1997**, *30*, 127–164. [CrossRef]

207. Mourer-Chauviré, C.; Geraads, D. The Struthionidae and Pelagornithidae (Aves: Struthioniformes, Odontopterygiformes) from the late Pliocene of Ahl al Oughlam, Morocco [Les Struthionidae et les Pelagornithidae (Aves, Struthioniformes et Odontopterygiformes) du Pliocène final d'Ahl al Oughlam, Maroc]. In Proceedings of the 6ème Symposium international de la Society for Avian Paleontology and Evolution (SAPE), Quillan, France, 28 September–3 October 2004; pp. 169–194.

208. Hay, R.L. *Geology of the Olduvai Gorge*; University of California Press: Berkeley, CA, USA, 1976.

209. Lowe, P.R. XLI.—On Some Struthious Remains:—1. Description of some Pelvic Remains of a large Fossil Ostrich, *Struthio oldawayi*, sp. n., from the Lower Pleistocene of Oldaway (Tanganyika Territory); 2. Egg-shell Fragments referable to Psammornis and other Struthiones collected by Mr. St. John Philby in Southern Arabia. *Ibis* **1933**, *75*, 652–658. [CrossRef]

210. Leakey, L.S.B. *Olduvai Gorge. 1951–1961. Vol 1. A Preliminary Report on the Geology and Fauna*; Cambridge University Press: Cambridge, UK, 1967.

211. Arambourg, C. *Vertébrés villafranchiens d'Afrique du Nord. (Artiodactyles, Carnivores, Primates, Reptiles, Oiseaux)*; Fondation Singer-Polignac: Paris, France, 1979; p. 141.

212. Becker-Platen, J.D.; Benda, L.; Steffens, P. *Litho-und Biostratigraphische Deutung radiometrischer Altersbestimmungen aus dem Jungtertiär der Türkei*; Schweizerbart'sche Verlagsbuchhandlung: Sttutgart, Germany, 1977.

213. Sauer, E.G. A Miocene Ostrich from Anatolia. *Ibis* **1979**, *121*, 494–501. [CrossRef]

214. Lambrecht, K. *Handbuch der Palaeornithologie*; Gebrüder Borntraeger: Berlin, Germany, 1933.

215. Louchart, A.; Bibi, F.; Stewart, J.R. The birds of the late Miocene Baynunah Formation, Abu Dhabi Emirate. In *Sands of Time: Late Miocene Fossils from the Baynunah Formation, U.A.E.*; Bibi, F., Kraatz, B., Beech, M., Hill, A., Eds.; Springer: Berlin/Heidelberg, Germany, 2020.

216. Tugarinov, A.Y. Some data on Pliocene ornithofauna of Siberia. *Tr. Paleozoologichaskogo Inst.* **1935**, *4*, 79–85.

217. Deng, T.; Qiu, Z.; Wang, B.; Wang, X.; Hou, S. Late Cenozoic biostratigraphy of the Linxia basin, northwestern China. In *Fossil Mammals of Asia*; Columbia University Press: New York, NY, USA, 2013; pp. 243–273.

218. Hou, L.; Zhou, Z.; Zhang, F.; Wang, Z. A Miocene ostrich fossil from Gansu Province, northwest China. *Chin. Sci. Bull.* **2005**, *50*, 1808–1810. [CrossRef]

219. Kaakinen, A.; Passey, B.H.; Zhang, Z.; Liu, L.; Pesonen, L.J.; Fortelius, M. Stratigraphy and paleoecology of the classical dragon bone localities of Baode County, Shanxi Province. In *Fossil Mammals of Asia: Neogene Biostratigraphy and Chronology*; Wang, X., Flynn, L.J., Fortelius, M., Eds.; Columbia University Press: New York, NY, USA, 2013; pp. 203–217.

220. Lowe, P.R. *Struthious Remains from Northern China and Mongolia: With Descriptions of Struthio wimani, Struthio anderssoni and Struthio mongolicus, Spp. Nov.*; Geological Survey of China: Beijing, China, 1931.

221. Sahni, A.; Kumar, G.; Bajpai, S.; Srinivasan, S. Ultrastructure and taxonomy of ostrich eggshells from Upper Palaeolithic sites of India. *J. Palaeontol. Soc. India* **1989**, *34*, 91–98.
222. Sahni, A.; Kumar, G.; Srinivasan, S.; Bajpai, S. Review of Late Pleistocene ostriches (*Struthio* sp.) of India. *Man Environ.* **1990**, *15*, 41–47.
223. Stern, L.A.; Johnson, G.D.; Chamberlain, C.P. Carbon isotope signature of environmental change found in fossil ratite eggshells from a South Asian Neogene sequence. *Geology* **1994**, *22*, 419–422. [CrossRef]
224. Patnaik, R.; Sahni, A.; Cameron, D.; Pillans, B.; Chatrath, P.; Simons, E.; Williams, M.; Bibi, F. Ostrich-like eggshells from a 10.1 million-yr-old Miocene ape locality, Haritalyangar, Himachal Pradesh, India. *Curr. Sci.* **2009**, *96*, 1485–1495.
225. Davies, W. III.—On some Fossil Bird-Remains from the Siwalik Hills in the British Museum. *Geol. Mag.* **1880**, *7*, 18–27. [CrossRef]
226. Lydekker, R. Indian Tertiary and post-Tertiary Vertebrata. Siwalik and Narbada bunodont Suina. *Mem. Geol. Surv. India Palaeontol. Indica Ser. 10* **1884**, *3*, 35–104.
227. Ginsburg, I.; Heintz, F.; Sen, S. Le gisement pliocène à Mammiferes de Çalta (Ankara, Turquie). *Comptes Rendus De L'académie Des Sci. Paris* **1974**, *278*, 2739–2742.
228. Sen, S. La faune de rangeurs pliocènes de Çalta (Ankara, Turqïe). *Bull. Du Muséum Mational D'histoire Nat. Ser. 3 Sci. De La Terre* **1977**, *61*, 89–171.
229. Janoo, A.; Sen, S. Pliocene vertebrate locality of Çalta, Ankara, Turkey. 2. Aves: Struthionidae. *Geodiversitas* **1998**, *20*, 339–351.
230. Beliaeva, E.I. *Catalogue of Tertiary Fossil Sites of Land Mammals in the USSR*; American Geological Institute: Alexandria, VA, USA, 1962.
231. Sotnikova, M.; Dodonov, A.; Pen'kov, A. Upper Cenozoic bio-magnetic stratigraphy of Central Asian mammalian localities. *Palaeogeogr. Palaeoclimatol. Palaeoecol.* **1997**, *133*, 243–258. [CrossRef]
232. Cai, B.; Zheng, S.; Liddicoat, J.C.; Li, Q. Review of the litho-, bio-, and chronostratigraphy in the Nihewan Basin, Hebei, China. In *Fossil Mammals of Asia*; Wang, X., Flynn, L.J., Fortelius, M., Eds.; Columbia University Press: New York, NY, USA, 2013; pp. 218–242.
233. Buffetaut, E.; Angst, D. A Giant Ostrich from the Lower Pleistocene Nihewan Formation of North China, with a Review of the Fossil Ostriches of China. *Diversity* **2021**, *13*, 47. [CrossRef]
234. Hou, L. Avian fossils of Pleistocene from Zhoukoudian. *Mem. Inst. Vertebr. Palaeontol. Palaeoanthropology Acad. Sin.* **1993**, *19*, 165–297.
235. Vangengeim, E.; Tesakov, A.S. Late Miocene Mammal Localities of Eastern Europe and Western Asia. In *Fossil Mammals of Asia*; Columbia University Press: New York, NY, USA, 2013; pp. 521–537.
236. Kurochkin, E.N.; Lungu, A.N. A new ostrich from the Middle Sarmatian of Moldavia. *Paleontol. Zhurnal* **1970**, *1*, 118–123.
237. Solounias, N.; Rivals, F.; Semprebon, G.M. Dietary interpretation and paleoecology of herbivores from Pikermi and Samos (late Miocene of Greece). *Paleobiology* **2010**, *36*, 113–136. [CrossRef]
238. Bachmayer, F.; Zapfe, H. Reste von *Struthio* aus Pikermi. *Proc. Acad. Athens* **1962**, *37*, 247–253.
239. Michailidis, D.; Roussiakis, S.; Theodorou, G. Palaeoavian remains from the late Miocene localities of Pikermi, Chomateri and Kerassiá: Palaeoecological implications. In Proceedings of the 9th Congress of the Carpathian-Balkan Geological Association, Thessaloniki, Greece, 23–26 September 2010; pp. 23–26.
240. Koufos, G.D.; Kostopoulos, D.S.; Konidaris, G.E. Palaeontology of the upper Miocene vertebrate localiti of Nikiti (Chalkidiki Peninsula, Macedonia, Greece) Foreword. *Geobios* **2016**, *49*, 29–36. [CrossRef]
241. Spassov, N. The Turolian Megafauna of West Bulgaria and the character of the Late Miocene "Pikermian biome". *Boll. Soc. Paleontol. Ital.* **2002**, *41*, 69–82.
242. Boev, Z.; Spassov, N. First record of ostriches (Aves, Struthioniformes, Struthionidae) from the late Miocene of Bulgaria with taxonomic and zoogeographic discussion. *Geodiversitas* **2009**, *31*, 493–507. [CrossRef]
243. Aleksejev, A. *Animaux Fossiles du Village Novo-Elisavetovka*; Zapiski Novorosiyskogo Institute: Odessa, Ukraine, 1915; pp. 273–453.
244. Burchak-Abramovich, N. Fossil ostriches of Caucasus and Southern Ukraine. *Tr. Estestvenoriticheskogo Muzeya Im. G. Zardabi* **1953**, *7*, 1–206.
245. Forsyth Major, C. Sur un gisement d'ossements fossiles dans l'île de Samos, contemporains de l'âge de Pikermi. *Comptes Rendus De L'académie Des Sci. Paris* **1888**, *107*, 1178–1181.
246. Tzankov, T.; Spassov, N.; Stoyanov, K. *Neogene-Quaternary Paleogeography and Geodynamics of the Region of Middle Struma River Valley (South.-Western Bulgaria)*; Publishing House of the South-Western University "N. Rilski": Blagoevgrad, Bulgaria, 2005; p. 199.
247. Spassov, N.; Tzankov, T.; Geraads, D. Late Neogene stratigraphy, biochronology, faunal diversity and environments of South-West Bulgaria (Struma River Valley). *Geodiversitas* **2006**, *28*, 477–498.
248. Theodorou, G.; Athanassiou, A.; Roussiakis, S.; Iliopoulos, G. Preliminary remarks on the late Miocene herbivores of Kerassia (Northern Euboea, Greece). *Deinsea* **2003**, *10*, 519–530.
249. Kampouridis, P.; Michailidis, D.; Kargopoulos, N.; Roussiakis, S.; Theodorou, G. First description of an ostrich from the late Miocene of Kerassia (Euboea, Greece): Remarks on its cervical anatomy. *Hist. Biol.* **2020**, *33*, 1–8. [CrossRef]
250. Burchak-Abramovich, N. New data on the Tertiary ostriches of Southern Ukraine. *Priroda* **1939**, *5*, 94–97.
251. Burchak-Abramovich, N.I.; Vekua, A.K. The Fossil Ostrich from Akchagil Layers of Georgia. *Acta Zool. Crac.* **1971**, *16*, 1–26.
252. Tesakov, A. Biostratigraphy of Middle Pliocene–Eopleistocene of Eastern Europe (based on small mammals). *Trans. Geol. Inst.* **2004**, 1–247.

253. Tesakov, A.S.; Dodonov, A.E.; Titov, V.V.; Trubikhin, V.M. Plio-Pleistocene geological record and small mammal faunas, eastern shore of the Azov Sea, Southern European Russia. *Quat. Int.* **2007**, *160*, 57–69. [CrossRef]

254. Athanassiou, A. A Villafranchian Hipparion-Bearing Mammal Fauna from Sésklo (E. Thessaly, Greece): Implications for the Question of *Hippario-Equus* Sympatry in Europe. *Quaternary* **2018**, *1*, 12. [CrossRef]

255. Ferring, R.; Oms, O.; Agustí, J.; Berna, F.; Nioradze, M.; Shelia, T.; Tappen, M.; Vekua, A.; Zhvania, D.; Lordkipanidze, D. Earliest human occupations at Dmanisi (Georgian Caucasus) dated to 1.85–1.78 Ma. *Proc. Natl. Acad. Sci. USA* **2011**, *108*, 10432–10436. [CrossRef]

256. Burchak-Abramovich, N.; Vekua, A. The fossil ostrich *Struthio dmanisensis* sp. n. from the Lower Pleistocene of eastern Georgia. *Acta Zool. Crac.* **1990**, *33*, 121–132.

257. Lopatin, A.V.; Vislobokova, I.A.; Lavrov, A.V.; Startsev, D.B.; Gimranov, D.O.; Zelenkov, N.V.; Maschenko, E.N.; Sotnikova, M.V.; Tarasenko, K.K.; Titov, V.V. The Taurida Cave, a New Locality of Early Pleistocene Vertebrates in Crimea. *Dokl. Biol. Sci.* **2019**, *485*, 40–43. [CrossRef]

258. Zelenkov, N.V.; Lavrov, A.V.; Startsev, D.B.; Vislobokova, I.A.; Lopatin, A.V. A giant early Pleistocene bird from eastern Europe: Unexpected component of terrestrial faunas at the time of early Homo arrival. *J. Vertebr. Paleontol.* **2019**, *39*, e1605521. [CrossRef]

259. Mayhew, D.F. Revision of the fossil vole assemblage (Mammalia, Rodentia, Arvicolidae) from Pleistocene deposits at Kisláng, Hungary. *Palaeontology* **2012**, *55*, 11–29. [CrossRef]

260. Kretzoi, M. Bericht über die Calabrische (Villafranchische) Fauna von Kisláng, Kom. Feyér. *Jahresber. Der Ung. Geol. Anst.* **1954**, *1953*, 212–264.

261. Rich, P.V. Significance of the Tertiary avifaunas from Africa (with emphasis on a mid to late Miocene avifauna from southern Tunisia). *Ann. Geol. Surv. Egypt* **1974**, *4*, 167–210.

262. Wang, S. Rediscussion in the taxonomic assignment of "*Struthio linxiaensis*" Hou, et al., 2005. *Acta Paleotologica Sin.* **2008**, *47*, 362–368.

263. Milne-Edwards, A. *Recherches Anatomiques et Paléontologiques pour Servir à l'histoire des Oiseaux Fossiles de la France, Tome Second*; Victor Masson: Paris, France, 1869–1871; Volume 1, p. 632.

264. Mlíkovsky, J.; Chenzychenova, F.; Filippov, A. Quaternary birds of the Baikal region, East Siberia. *Acta Soc. Zool. Bohem.* **1997**, *61*, 151–156.

265. Stimpson, C.M.; Lister, A.; Parton, A.; Clark-Balzan, L.; Breeze, P.S.; Drake, N.A.; Groucutt, H.S.; Jennings, R.; Scerri, E.M.L.; White, T.S.; et al. Middle Pleistocene vertebrate fossils from the Nefud Desert, Saudi Arabia: Implications for biogeography and palaeoecology. *Quat. Sci. Rev.* **2016**, *143*, 13–36. [CrossRef]

266. Field, D.J.; Lynner, C.; Brown, C.; Darroch, S.A. Skeletal correlates for body mass estimation in modern and fossil flying birds. *PLoS ONE* **2013**, *8*, e82000. [CrossRef] [PubMed]

267. Buffetaut, E.; Angst, D. How large was the giant ostrich of China. *EVOLUÇÃO Rev. De Geistória E Pré-História* **2017**, *2*, 6–8.

268. Folch, A.; del Hoyo, J.; Christie, D.A.; Collar, N.; Jutglar, F.; Garcia, E.F.J. Lesser Rhea (*Rhea pennata*), version 1.0. *Birds World* **2020**. [CrossRef]

269. Kirwan, G.M.; Korthals, A.; Hodes, C.E. Greater Rhea (*Rhea americana*), version 2.0. *Birds World* **2021**. [CrossRef]

270. Delsuc, F.; Superina, M.; Ferraris, G.; Tilak, M.K.; Douzery, E. Molecular evidence for hybridisation between the two living species of South American ratites: Potential conservation implications. *Conserv. Genet.* **2007**, *8*, 503–507. [CrossRef]

271. Miranda-Ribeiro, A. Notas ornithologicas, tinamidae. *Rev. Do Mus. Paul.* **1938**, *23*, 667–788.

272. Bertelli, S.; Chiappe, L.M.; Mayr, G. Phylogenetic interrelationships of living and extinct Tinamidae, volant palaeognathous birds from the New World. *Zool. J. Linn. Soc.* **2014**, *172*, 145–184. [CrossRef]

273. Rayner, J.M. Form and function in avian flight. In *Current Ornithology*; Springer: New York, NY, USA, 1988; pp. 1–66.

274. Widrig, K.E.; Watanabe, J.; Bhullar, B.S.; Field, D.J. Three-dimensional atlas of pectoral musculoskeletal anatomy in the extant tinamou *Nothoprocta pentlandii* (Palaeognathae: Tinamidae). In Proceedings of the The Society of Vertebrate Paleontology 80th Annual Meeting, Virtual, 12–16 October 2020.

275. Alvarenga, H.M. Uma ave ratitae do paleoceno brasileiro: Bacia calcaria de itaboraí, estado do Rio de Janeiro, Brasil. *Bol. Do Mus. Nacional. Nova Ser. Geol.* **1983**, *41*, 1–8.

276. Woodburne, M.O.; Goin, F.J.; Raigemborn, M.S.; Heizler, M.; Gelfo, J.N.; Oliveira, E.V. Revised timing of the South American early Paleogene land mammal ages. *J. South. Am. Earth Sci.* **2014**, *54*, 109–119. [CrossRef]

277. Agnolín, F.L. Unexpected diversity of ratites (Aves, Palaeognathae) in the early Cenozoic of South America: Palaeobiogeographical implications. *Alcheringa: An. Australas. J. Palaeontol.* **2017**, *41*, 101–111. [CrossRef]

278. Alvarenga, H. *Diogenornis fragilis* Alvarenga, 1985, restudied: A South American ratite closely related to Casuariidae. In Proceedings of the 25th International Ornithological Congress, Campos do Jordão, Brazil, 22–28 August 2010.

279. Tambussi, C. The fossil Rheiformes from Argentina. *Cour. Forsch. Senckenberg* **1995**, *181*, 121–129.

280. Pascual, R.; Ortiz-Jaureguizar, E. The Gondwanan and South American episodes: Two major and unrelated moments in the history of the South American mammals. *J. Mamm. Evol.* **2007**, *14*, 75–137. [CrossRef]

281. Raigemborn, M.S.; Krause, J.M.; Bellosi, E.; Matheos, S.D. Redefinición estratigráfica del grupo Río Chico (Paleógeno Inferior), en el norte de la cuenca del golfo San Jorge, Chubut. *Rev. De La Asoc. Geológica Argent.* **2010**, *67*, 239–256.

282. Krause, J.; Bellosi, E. Paleosols from the Koluel Kaike Formation (Lower-Middle Eocene) in the South-central Chubut, Argentina. A preliminary analysis. In Proceedings of the Actas del IV Congreso Latinoamericano de Sedimentología, La Plata, Argentina, 11–24 November 2006; pp. 125–136.

283. Paredes, J.M.; Colombo, F.; Foix, N.; Allard, J.O.; Nillni, A.; Allo, M. Basaltic Explosive Volcanism in a tuff-dominated intraplate setting, Sarmiento formation (Middle Eocene-lower Miocene), Patagonia, Argentina. *Lat. Am. J. Sedimentol. Basin Anal.* **2008**, *15*, 77–92.

284. Kramarz, A.; Garrido, A.; Ribeiro, A.; Ortiz, R. Nuevos registros de vertebrados fósiles de la Formación Chichinales, Mioceno Temprano de la provincia de Río Negro, Argentina. *Ameghiniana* **2004**, *41*, 53R.

285. Agnolín, F.; Chafrat, P. New fossil bird remains from the Chichinales Formation (Early Miocene) of northern Patagonia, Argentina. *Ann. De Paleontol.* **2015**, *101*, 87–94. [CrossRef]

286. Marshall, L.G.; Patterson, B. *Geology and Geochronology of the Mammal-Bearing Tertiary of the Valle de Santa María and Río Corral Quemado, Catamarca Province, Argentina*; Field Museum of Natural History: Chicago, IL, USA, 1981.

287. Fleagle, J.G.; Perkins, M.E.; Heizler, M.T.; Nash, B.; Bown, T.M.; Tauber, A.A.; Dozo, M.T.; Tejedor, M.F.; Vizcaíno, S.F.; Kay, R.F. Absolute and relative ages of fossil localities in the Santa Cruz and Pinturas Formations. In *Early Miocene Paleobiology in Patagonia: High-Latitude Paleocommunities of the Santa Cruz Formation*; Vizcaíno, S.F., Kay, R.F., Bargo, M.S., Eds.; Cambridge University Press: Cambridge, UK, 2012; pp. 41–58.

288. Blisniuk, P.M.; Stern, L.A.; Chamberlain, C.P.; Idleman, B.; Zeitler, P.K. Climatic and ecologic changes during Miocene surface uplift in the Southern Patagonian Andes. *Earth Planet. Sci. Lett.* **2005**, *230*, 125–142. [CrossRef]

289. Perkins, M.E.; Fleagle, J.G.; Heizler, M.T.; Nash, B.; Bown, T.; Tauber, A.A.; Dozo, M.T. Tephrochronology of the Miocene Santa Cruz and Pinturas Formations, Argentina. In *Early Miocene Paleobiology in Patagonia: High-Latitude Paleocommunities of the Santa Cruz Formation*; Vizcaíno, S.F., Kay, R.F., Bargo, M.S., Eds.; Cambridge University Press: Cambridge, UK, 2012; pp. 23–40.

290. Cuitiño, J.I.; Fernicola, J.C.; Kohn, M.J.; Trayler, R.; Naipauer, M.; Bargo, M.S.; Kay, R.F.; Vizcaíno, S.F. U-Pb geochronology of the Santa Cruz Formation (early Miocene) at the Río Bote and Río Santa Cruz (southernmost Patagonia, Argentina): Implications for the correlation of fossil vertebrate localities. *J. South. Am. Earth Sci.* **2016**, *70*, 198–210. [CrossRef]

291. Ameghino, F. Enumeración de las aves fósiles de la República Argentina. *Rev. Argent. De Hist. Nat.* **1891**, *1*, 441–453.

292. Buffetaut, E. Tertiary ground birds from Patagonia (Argentina) in the Tournouër collection of the Muséum National d'Histoire Naturelle, Paris. *Bull. De La Soc. Geol. De Fr.* **2014**, *185*, 207–214. [CrossRef]

293. Diederle, J.M.; Noriega, J. New Records Of Birds In The Santa Cruz Formation (Early-Middle Miocene) At The Río Santa Cruz Valley, Patagonia, Argentina. *Publicación Electrónica De La Asoc. Paleontológica Argent.* **2020**, *61*, 55–61. [CrossRef]

294. Forasiepi, A.M.; Martinelli, A.G.; de la Fuente, M.; Dieguez, S.; Bond, M. Paleontology and stratigraphy of the Aisol Formation (Neogene), San Rafael, Mendoza. *Cenozoic Geol. Cent. Andes Argentina. SCS Publ. Salta* **2011**, 135–154.

295. Agnolín, F.L.; Noriega, J.I. Una Nueva Especie de Ñandú (Aves: Rheidae) del Mioceno Tardío de la Mesopotamia Argentina. *Ameghiniana* **2012**, *49*, 236–246. [CrossRef]

296. Ganduglia, P. Observaciones geológicas en la región de Ingeniero Jacobacci provincia de Río Negro (con énfasis en el Cretácico y Terciario). Bachelor's Dissertation, Universidad Nacional de Buenos Aires, Buenos Aires, Argentina, 1977.

297. Cione, A.L.; Azpelicueta, M.d.l.M.; Bond, M.; Carlini, A.A.; Casciotta, J.R.; Cozzuol, M.A.; de la Fuente, M.; Gasparini, Z.; Goin, F.J.; Noriega, J. Miocene vertebrates from Entre Ríos province, eastern Argentina. *El Neógeno De Argentina. Ser. Correlación Geológica* **2000**, *14*, 191–237.

298. Cerdeño, E.; Montalvo, C. Los Mesotheriinae (Mesotheriidae, Notoungulata) del Mioceno superior de La Pampa, Argentina. *Rev. Española De Paleontol.* **2001**, *16*, 63–75. [CrossRef]

299. Verzi, D.; Deschamps, C.; Montalvo, C. Bioestratigrafía y biocronología del Mioceno tardío de Argentina central. *Ameghiniana* **2004**, *41*, 21R.

300. Cenizo, M.M.; Tambussi, C.; Montalvo, P.C.I. Late Miocene continental birds from the Cerro Azul Formation in the Pampean region (central-southern Argentina). *Alcheringa: An. Australas. J. Palaeontol.* **2012**, *36*, 47–68. [CrossRef]

301. Bossi, G.E.; Muruaga, C.M. Estratigrafía e inversión tectónica del 'rift' neógeno en el Campo del Arenal, Catamarca, NO Argentina. *Andean Geol.* **2009**, *36*, 311–340.

302. Reguero, M.A.; Candela, A.M. Late Cenozoic mammals from the northwest of Argentina. *Cenozoic Geol. Cent. Andes Argent.* **2011**, *458*, 411–426.

303. Noriega, J.I.; Jordan, E.A.; Vezzosi, R.I.; Areta, J.I. A new species of *Opisthodactylus* Ameghino, 1891 (Aves, Rheidae), from the late Miocene of northwestern Argentina, with implications for the paleobiogeography and phylogeny of rheas. *J. Vertebr. Paleontol.* **2017**, *37*, e1278005. [CrossRef]

304. Deschamps, C.M.; Vucetich, M.G.; Verzi, D.H.; Olivares, A.I. Biostratigraphy and correlation of the Monte Hermoso Formation (early Pliocene, Argentina): The evidence from caviomorph rodents. *J. South. Am. Earth Sci.* **2012**, *35*, 1–9. [CrossRef]

305. Tomassini, R.L.; Montalvo, C.I.; Deschamps, C.M.; Manera, T. Biostratigraphy and biochronology of the Monte Hermoso Formation (early Pliocene) at its type locality, Buenos Aires Province, Argentina. *J. South. Am. Earth Sci.* **2013**, *48*, 31–42. [CrossRef]

306. Rovereto, G. Los estratos araucanos y sus fósiles. *An. Del Mus. Nac. De Hist. Nat. De Buenos Aires* **1914**, *25*, 1–247.

307. Palazzesi, L.; Barreda, V. Fossil pollen records reveal a late rise of open-habitat ecosystems in Patagonia. *Nat. Commun.* **2012**, *1294*, 1–5. [CrossRef]

308. Tambussi, C.P.; Degrange, F.J. *South. American and Antarctic Continental Cenozoic Birds: Paleobiogeographic Affinities and Disparities*; Springer: Dordrecht, The Netherlands, 2013; p. 113.

309. Picasso, M.B.J.; Mosto, M.C. New insights about *Hinasuri nehuensis* (Aves, Rheidae, Palaeognathae) from the early Pliocene of Argentina. *Alcheringa: An. Australas. J. Palaeontol.* **2016**, *40*, 244–250. [CrossRef]

310. Ameghino, C.; Rusconi, C. Nueva subespecie de avestruz fósil del Plioceno de Buenos Aires. *An. De La Soc. Científica Argent.* **1932**, *114*, 38–42.

311. Moreno, F.P.; Mercerat, A. Catálogo de los pájaros fósiles de la República Argentina conservados en el Museo de La Plata. *An. Del Mus. De La Plata Paleontol.* **1891**, *1*, 7–71.

312. Picasso, M.B.J.; Mosto, C. The new taxonomic status of *Rhea anchorenensis* (Ameghino and Rusconi, 1932) (Aves, Palaeognathae) from the Pleistocene of Argentina. *Ann. De Paleontol.* **2016**, *102*, 237–241. [CrossRef]

313. Picasso, M.B.J. Diversity of extinct Rheidae (Aves, Palaeognathae): Historical controversies and the new taxonomic status of *Rhea pampeana* Moreno and Mercerat 1891 from the Pleistocene of Argentina. *Hist. Biol.* **2016**, *28*, 1101–1107. [CrossRef]

314. Chiappe, L. Fossil birds from the Miocene Pinturas Formation of southern Argentina. *J. Vertebr. Paleontol.* **1991**, *11*, 21A–22A.

315. Bertelli, S.; Chiappe, L.M. *Earliest Tinamous (Aves: Palaeognathae) from the Miocene of Argentina and their Phylogenetic Position*; Natural History Museum of Los Angeles County: Los Angeles, CA, USA, 2005.

316. Chandler, R. A New Species of Tinamou (Aves: Tinamiformes, Tinamidae) From the Early-Middle Miocene of Argentina. *PalArch's J. Vertebr. Palaeontol.* **2012**, *9*, 1–8.

317. Woodburne, M.O.; Goin, F.J.; Bond, M.; Carlini, A.A.; Gelfo, J.N.; López, G.M.; Iglesias, A.; Zimicz, A.N. Paleogene Land Mammal Faunas of South America; a Response to Global Climatic Changes and Indigenous Floral Diversity. *J. Mamm. Evol.* **2014**, *21*, 1–73. [CrossRef]

318. Brodkorb, P. Notes on fossil Tinamous. *Auk* **1961**, *78*, 257. [CrossRef]

319. Tonni, E.P. Los Tinamidos Fosiles Argentinos I. El Genero *Tinamisornis* Rovereto, 1914. *Ameghiniana* **1977**, *14*, 224–232.

320. Tambussi, C.; Tonni, E. Un Tinamidae (Aves: Tinamiformes) del Mioceno tardío de La Pampa (República Argentina) y comentarios sobre los tinámidos fósiles argentinos. *Rev. De La Asoc. Paleontológica Argent.* **1985**, *14*, 4.

321. Tambussi, C.P. Catalogo Critico De Los Tinamidae (Aves: Tinamiformes) Fosiles De La Republica Argentina. *Ameghiniana* **1987**, *24*, 241–244.

322. Tambussi, C.P. *Las Aves del Plioceno Tardío-Pleistoceno Temprano de la Provincia de Buenos Aires*; Universidad Nacional de La Plata: Buenos Aires, Argentina, 1989.

323. Marshall, L.G.; Berta, A.; Hofstetter, R.; Pascual, R.; Reig, O.; Bombin, M.; Mones, A. *Mammals and Stratigraphy: Geochronology of the Continental Mammal-Bearing Quaternary of South America*; Laboratoire de paléontologie des vertébrés de l'Ecole pratique des hautes études: Montpellier, France, 1983.

324. Tambussi, C.; Noriega, J. Summary of the avian fossil record from southern South America. *Muenchner Geowiss. Abh.* **1996**, *30*, 245–264.

325. Gill, F.; Donsker, D.; Rasmussen, P. IOC World Bird List. 2021. Available online: https://www.worldbirdnames.org/new/ (accessed on 30 December 2021).

326. Parker, S.A. The extinct Kangaroo Island Emu, a hitherto-unrecognised species. *Bull. Br. Ornithol. Club* **1984**, *104*, 19–22.

327. Spencer, B. The King Island Emu. *Vic. Nat.* **1906**, *23*, 140.

328. Le Souëf, W.H.D. Dromaeus diemenensis. *Bull. Br. Ornithol. Club* **1907**, *21*, 13.

329. Heupink, T.H.; Huynen, L.; Lambert, D.M. Ancient DNA Suggests Dwarf and 'Giant' Emu Are Conspecific. *PLoS ONE* **2011**, *6*, e18728. [CrossRef]

330. Thomson, V.A.; Mitchell, K.J.; Eberhard, R.; Dortch, J.; Austin, J.J.; Cooper, A. Genetic diversity and drivers of dwarfism in extinct island emu populations. *Biol. Lett.* **2018**, *14*, 20170617. [CrossRef]

331. Worthy, T.H.; Nguyen, J.M.T. An annotated checklist of the fossil birds of Australia. *Trans. R. Soc. South. Aust.* **2020**, *144*, 66–108. [CrossRef]

332. Folch, A.; Christie, D.A.; Garcia, E.F.J. Emu (*Dromaius novaehollandiae*), version 1.0. *Birds World* **2020**. [CrossRef]

333. Boles, W. A new emu (Dromaiinae) from the Late Oligocene Etadunna Formation. *Emu Austral. Ornithol.* **2001**, *101*, 317–321. [CrossRef]

334. Woodburne, M.O.; Macfadden, B.J.; Case, J.A.; Springer, M.S.; Pledge, N.S.; Power, J.D.; Woodburne, J.M.; Springer, K.B. Land mammal biostratigraphy and magnetostratigraphy of the Etadunna Formation (late Oligocene) of South Australia. *J. Vertebr. Paleontol.* **1994**, *13*, 483–515. [CrossRef]

335. Megirian, D.; Prideaux, G.J.; Murray, P.F.; Smit, N. An Australian land mammal age biochronological scheme. *Paleobiology* **2010**, *36*, 658–671. [CrossRef]

336. Patterson, C.; Rich, P. The fossil history of the emus, *Dromaius* (Aves: Dromaiinae). *Rec. South. Aust. Mus.* **1987**, *21*, 85–117.

337. Boles, W. Revision of *Dromaius gidju* Patterson and Rich, 1987 from Riversleigh, northwestern Queensland, Australia, with a reassessment of its generic position. *Pap. Avian Paleontol. Honor. Pierce Brodkorb* **1992**, *36*, 195–208.

338. Archer, M.; Hand, S.; Godthelp, H.; Creaser, P. Correlation of the Cainozoic sediments of the Riversleigh World Heritage fossil property, Queensland, Australia. *Mémoires Et Trav. De L'institut De Montp.* **1997**, 131–152.

339. Travouillon, K.J.; Archer, M.; Hand, S.J.; Godthelp, H. Multivariate analyses of Cenozoic mammalian faunas from Riversleigh, northwestern Queensland. *Alcheringa: An. Australas. J. Palaeontol.* **2006**, *30*, 323–349. [CrossRef]

340. Boles, W. Hindlimb proportions and locomotion of *Emuarius gidju* (Patterson & Rich, 1987) (Aves: Casuariidae). *Mem. Qld. Mus.* **1997**, *41*, 235–240.

341. Worthy, T.H.; Hand, S.J.; Archer, M. Phylogenetic relationships of the Australian Oligo-Miocene ratite *Emuarius gidju* Casuariidae. *Integr. Zool* **2014**, *9*, 148–166. [CrossRef]

342. Woodburne, M.; Tedford, R.; Archer, M.; Turnbull, W.; Plane, M. Biochronology of the continental mammal record of Australia and New Guinea. *Spec. Publ. South. Aust. Dep. Mines Energy* **1985**, *5*, 347–363.

343. Rich, P.V. The Dromornithidae, an extinct family of large ground birds endemic to Australia. *Bur. Mineral. Resour. Geol. Geophys. (Aust.) Bull.* **1979**, *184*, 1–196.

344. Rich, P.; Van Tets, J. Fossil birds of Australia and New Guinea: Their biogeographic, phylogenetic and biostratigraphic input. In *The Fossil Vertebrate Record of Australasia*; Rich, P.V., Thompson, E.M., Eds.; Monash University Offset Printing Unit: Melbourne, Australia, 1982; pp. 235–385.

345. Woodburne, M.O. *The Alcoota Fauna, Central Australia: An Integrated Palaeontological and Geological Study*; Australian Government Public Service: Canberra, Australia, 1967; Volume 87.

346. Stirton, R.A.; Tedford, R.H.; Woodburne, M.O. Australian Tertiary deposits containing terrestrial mammals. *Univ. Calif. Publ. Geol. Sci.* **1968**, *77*, 1–30.

347. Woods, J.T. Fossiliferous fluviatile and cave deposits. *J. Geol. Soc. Aust.* **1960**, *7*, 393–403.

348. Stirton, R.; Tedford, R.; Miller, A. Cenozoic stratigraphy and vertebrate paleontology of the Tirari Desert, South Australia. *Rec. South. Aust. Mus.* **1961**, *14*, 19–61.

349. Miller, A.H. Fossil ratite birds of the late Tertiary of South Australia. *Rec. South. Aust. Mus.* **1963**, *14*, 413–420.

350. Hoch, E.; Holm, P. New K/Ar age determinations of the Awe fauna gangue, Papua New Guinea: Consequences for Papuaustralian late Cenozoic biostratigraphy. *Mod. Geol.* **1986**, *10*, 181–195.

351. Plane, M. *Stratigraphy and vertebrate fauna of the Otibanda formation, New Guinea*; Bureau of Mineral Resources, Geology and Geophysics: Canberra, Australia, 1967; p. 64.

352. Miller, A.H. The history and significance of the fossil *Casuarius lydekkeri*. *Rec. Aust. Mus.* **1962**, *25*, 235–238. [CrossRef]

353. Rothschild, L.W.R.B. On the Former and Present Distribution of the So Called Ratitae Or Ostrich-like Birds with Certain Deductions and a Description of a New Form by CW Andrew. In Proceedings of the Fifth International Ornithological Congress, Berlin, Germany, 4 June 1910; pp. 144–169.

354. Williams, P.; McDougall, I.; Powell, J. Aspects of the quaternary geology of the Tari-Koroba area, Papua. *J. Geol. Soc. Aust.* **1972**, *18*, 333–347. [CrossRef]

355. Rich, P.V.; Plane, M.; Schroeder, N. A pygmy cassowary (*Casuarius lydekkeri*) from late Pleistocene bog deposits at Pureni, Papua New Guinea. *BMR J. Aust. Geol. Geophys.* **1988**, *10*, 377–389.

356. Vickers-Rich, P.; Rich, T. *Wildlife of Gondwana*; Indiana University Press: Bloomington, IN, USA, 1993.

357. Martin, H.A. Cenozoic climatic change and the development of the arid vegetation in Australia. *J. Arid Environ.* **2006**, *66*, 533–563. [CrossRef]

358. Yates, A.M.; Worthy, T.H. A diminutive species of emu (Casuariidae: Dromaiinae) from the late Miocene of the Northern Territory, Australia. *J. Vertebr. Paleontol.* **2019**, *39*, e1665057. [CrossRef]

359. Campbell, K.E.; Marcus, L. The relationship of hindlimb bone dimensions to body weight in birds. *Nat. Hist. Mus. Los Angeles Cty. Sci. Ser.* **1992**, *36*, 395–412.

360. Naish, D.; Perron, R. Structure and function of the cassowary's casque and its implications for cassowary history, biology and evolution. *Hist. Biol.* **2014**, *28*, 507–518. [CrossRef]

361. Calder, W.A. The Kiwi. *Sci. Am.* **1978**, *239*, 132–143. [CrossRef]

362. Kinsky, F.C. The consistent presence of paired ovaries in the Kiwi (*Apteryx*) with some discussion of this condition in other birds. *J. Für Ornithol.* **1971**, *112*, 334–357. [CrossRef]

363. Folch, A.; Jutglar, F.; Garcia, E.F.J. Little Spotted Kiwi (*Apteryx owenii*), version 1.0. *Birds World* **2020**. [CrossRef]

364. Weir, J.T.; Haddrath, O.; Robertson, H.A.; Colbourne, R.M.; Baker, A.J. Explosive ice age diversification of kiwi. *Proc. Natl. Acad. Sci. USA* **2016**, *113*, E5580–E5587. [CrossRef] [PubMed]

365. Huynen, L.; Suzuki, T.; Ogura, T.; Watanabe, Y.; Millar, C.D.; Hofreiter, M.; Smith, C.; Mirmoeini, S.; Lambert, D.M. Reconstruction and in vivo analysis of the extinct tbx5 gene from ancient wingless moa (Aves: Dinornithiformes). *BMC Evol. Biol.* **2014**, *14*, 75. [CrossRef]

366. Bunce, M.; Worthy, T.; Ford, T.; Hoppitt, W.; Willerslev, E.; Drummond, A.; Cooper, A. Extreme reversed sexual size dimorphism in the extinct New Zealand moa *Dinornis*. *Nature* **2003**, *425*, 172–175. [CrossRef]

367. Perry, G.L.W.; Wheeler, A.B.; Wood, J.R.; Wilmshurst, J.M. A high-precision chronology for the rapid extinction of New Zealand moa (Aves, Dinornithiformes). *Quat. Sci. Rev.* **2014**, *105*, 126–135. [CrossRef]

368. Greenwood, R.; Atkinson, I. Evolution of divaricating plants in New Zealand in relation to moa browsing. *Proc. New Zealand Ecol. Soc.* **1977**, *24*, 21–33.

369. Bond, W.; Silander, J. Springs and wire plants: Anachronistic defences against Madagascar's extinct elephant birds. *Proc. R. Soc. B* **2007**, *274*, 1985–1992. [CrossRef]

370. Wood, J.R.; Richardson, S.J.; McGlone, M.S.; Wilmshurst, J.M. The diets of moa (Aves: Dinornithiformes). *New Zealand J. Ecol.* **2020**, *44*, 1–21. [CrossRef]

371. Allentoft, M.; Rawlence, N.J. Moa's Ark or volant ghosts of Gondwana? Insights from nineteen years of ancient DNA research on the extinct moa (Aves: Dinornithiformes) of New Zealand. *Ann. Anat.* **2012**, *194*, 36–51. [CrossRef] [PubMed]
372. Tennyson, A.J.D. The origin and history of New Zealand's terrestrial vertebrates. *New Zealand J. Ecol.* **2010**, *34*, 6–27.
373. Suggate, R.P.; Stevens, G.R.; Te Punga, M.T. *The Geology of New Zealand*; EC Keating, Government Printer: Wellington, New Zealand, 1978; Volume 2, p. 819.
374. Landis, C.; Campbell, H.; Begg, J.; Mildenhall, D.; Paterson, A.; Trewick, S. The Waipounamu Erosion Surface: Questioning the antiquity of the New Zealand land surface and terrestrial fauna and flora. *Geol. Mag.* **2008**, *145*, 173–197. [CrossRef]
375. Waters, J.M.; Craw, D. Goodbye Gondwana? New Zealand biogeography, geology, and the problem of circularity. *Syst. Biol.* **2006**, *55*, 351–356. [CrossRef]
376. Trewick, S.A.; Paterson, A.M.; Campbell, H.J. Hello New Zealand. *J. Biogeogr.* **2007**, *34*, 1–6. [CrossRef]
377. Cooper, A.; Cooper, R. The Oligocene bottleneck and New Zealand biota: Genetic record of a past environmental crisis. *Proc. R. Soc. B* **1995**, *261*, 293–302. [CrossRef]
378. Carr, L.; McLenachan, P.; Waddell, P.; Gemmell, N.; Penny, D. Analyses of the mitochondrial genome of *Leiopelma hochstetteri* argues against the full drowning of New Zealand. *J. Biogeogr.* **2015**, *42*, 1066–1076. [CrossRef]
379. Edgecombe, G.D.; Giribet, G. A New Zealand species of the trans-Tasman centipede order Craterostigmomorpha (Arthropoda: Chilopoda) corroborated by molecular evidence. *Invertebr. Syst.* **2008**, *22*, 1–15. [CrossRef]
380. Boyer, S.L.; Giribet, G. Welcome back New Zealand: Regional biogeography and Gondwanan origin of three endemic genera of mite harvestmen (Arachnida, Opiliones, Cyphophthalmi). *J. Biogeogr.* **2009**, *36*, 1084–1099. [CrossRef]
381. Buckley, T.R.; Lord, N.; Ramón-Laca, A.; Allwood, J.; Leschen, R. Multiple lineages of hyper-diverse Zopheridae beetles survived the New Zealand Oligocene Drowning. *J. Biogeogr.* **2020**, *47*, 927–940. [CrossRef]
382. Wallis, G.P.; Jorge, F. Going under down under? Lineage ages argue for extensive survival of the Oligocene marine transgression on Zealandia. *Mol. Ecol.* **2018**, *27*, 4368–4396. [CrossRef] [PubMed]
383. Sharma, P.P.; Wheeler, W.C. Revenant clades in historical biogeography: The geology of New Zealand predisposes endemic clades to root age shifts. *J. Biogeogr.* **2013**, *40*, 1609–1618. [CrossRef]
384. Strogen, D.P.; Bland, K.J.; Nicol, A.; King, P.R. Paleogeography of the Taranaki Basin region during the latest Eocene–Early Miocene and implications for the 'total drowning' of Zealandia. *N. Z. J. Geol. Geophys.* **2014**, *57*, 110–127. [CrossRef]
385. Mildenhall, D.C.; Pocknall, D.T. Miocene-Pleistocene spores and pollen from Central Otago, South Island, New Zealand. *New Zealand Geol. Surv. Paleontol. Bull.* **1989**, 1–128.
386. Pole, M.; Douglas, B. A quantitative palynostratigraphy of the Miocene Manuherikia Group, New Zealand. *J. R. Soc. New Zealand* **1998**, *28*, 405–420. [CrossRef]
387. Worthy, T.H.; Worthy, J.P.; Tennyson, A.; Salisbury, S.; Hand, S.; Scofield, R. Miocene fossils show that kiwi (*Apteryx*, Apterygidae) are probably not phyletic dwarves. In *Paleornithological Research 2013—Proceedings of the 8th International Meeting of the Society of Avian Paleontology and Evolution*; Natural History Museum: Vienna, Austria, 2013; pp. 63–80.
388. Calder, W.A. *Size, Function, and Life History*; Harvard University Press: Cambridge, MA, USA, 1984; p. 448.
389. Gould, S.J. Of kiwi eggs and the Liberty Bell. *Nat. Hist.* **1986**, *95*, 20–29.
390. Ducatez, S.; Field, D.J. Disentangling the avian altricial-precocial spectrum: Quantitative assessment of developmental mode, phylogenetic signal, and dimensionality. *Evolution* **2021**, *75*, 2717–2735. [CrossRef]
391. Worthy, T.H.; Holdaway, R.N. *The Lost World of the Moa: Prehistoric Life of New Zealand*; Indiana University Press: Bloomington, IN, USA, 2002; p. 718.
392. Tennyson, A.J.D.; Tomotani, B.M. A new fossil species of kiwi (Aves: Apterygidae) from the mid-Pleistocene of New Zealand. *Hist. Biol.* **2021**, 1–9. [CrossRef]
393. Worthy, T. A mid-Pleistocene rail from New Zealand. *Alcheringa* **1997**, *21*, 71–78. [CrossRef]
394. Worthy, T.H.; Tennyson, A.J.D.; Jones, C.; McNamara, J.A.; Douglas, B.J. Miocene waterfowl and other birds from central Otago, New Zealand. *J. Syst. Palaeontol.* **2007**, *5*, 1–39. [CrossRef]
395. Tennyson, A.; Worthy, T.; Jones, C.M.; Scofield, R.; Hand, S. Moa's Ark: Miocene fossils reveal the great antiquity of moa (Aves: Dinornithiformes) in Zealandia. In Proceedings of the VII International Meeting of the Society of Avian Paleontology and Evolution, ed. W.E. Boles and T.H. Worth. *Rec. Aust. Mus.* **2010**, *62*, 105–114. [CrossRef]
396. Tennyson, A.J.D.; Martinson, P. *Extinct birds of New Zealand*; Te Papa Press: Wellington, New Zealand, 2006.
397. Worthy, T.H.; Edwards, A.R.; Millener, P.R. The fossil record of moas (Aves: Dinornithiformes) older than the Otira (last) Glaciation. *J. R. Soc. New Zealand* **1991**, *21*, 101–118. [CrossRef]
398. Forbes, H. On avian remains found under a lava-flow near Timaru. *Trans. Proc. New Zealand Inst.* **1891**, *23*, 366–372.
399. Mathews, W.H.; Curtis, G. Date of the Pliocene-Pleistocene boundary in New Zealand. *Nature* **1966**, *212*, 979–980. [CrossRef]
400. Beu, A.; Edwards, A. New Zealand Pleistocene and late Pliocene glacio-eustatic cycles. *Palaeogeogr. Palaeoclimatol. Palaeoecol.* **1984**, *46*, 119–142. [CrossRef]
401. Oliver, W.R.B. *The Moas of New Zealand and Australia*; Dominion Museum: Wellington, New Zealand, 1949.
402. Marshall, P. Occurence of Fossil Moa-bones in the Lower Wanganui Strata. *Trans. Proc. New Zealand Inst.* **1919**, *51*, 250–253.
403. Coates, G. *The Rise and Fall of the Southern Alps*; Canterbury University Press: Christchurch, New Zealand, 2002; p. 80.
404. Dewar, R.E. Extinctions in Madagascar: The loss of the subfossil fauna. In *Quaternary Extinctions: A Prehistoric Revolution*; Martin, P.S., Klein, R.G., Eds.; The University of Arizona Press: Tuscon, CA, USA, 1984; pp. 574–593.

405. Samonds, K.E.; Godfrey, L.R.; Ali, J.R.; Goodman, S.M.; Vences, M.; Sutherland, M.R.; Irwin, M.T.; Krause, D.W. Spatial and temporal arrival patterns of Madagascar's vertebrate fauna explained by distance, ocean currents, and ancestor type. *Proc. Natl. Acad. Sci. USA* **2012**, *109*, 5352–5357. [CrossRef]
406. Samonds, K.E. Late Pleistocene bat fossils from Anjohibe Cave, northwestern Madagascar. *Acta Chiropterologica* **2007**, *9*, 39–65. [CrossRef]
407. Angst, D.; Buffetaut, É. *Palaeobiology of Giant Flightless Birds*; Elsevier: London, UK, 2017; p. 281.
408. Clarke, S.; Miller, G.; Fogel, M.; Chivas, A.; Murray-Wallace, C. The amino acid and stable isotope biogeochemistry of elephant bird (*Aepyornis*) eggshells from southern Madagascar. *Quat. Sci. Rev.* **2006**, *25*, 2343–2356. [CrossRef]
409. Torres, C.R.; Clarke, J.A. Nocturnal giants: Evolution of the sensory ecology in elephant birds and other palaeognaths inferred from digital brain reconstructions. *Proc. R. Soc. B* **2018**, *285*, 20181540. [CrossRef] [PubMed]
410. Dewar, R.E.; Radimilahy, C.; Wright, H.T.; Jacobs, Z.; Kelly, G.O.; Berna, F. Stone tools and foraging in northern Madagascar challenge Holocene extinction models. *Proc. Natl. Acad. Sci. USA* **2013**, *110*, 12583–12588. [CrossRef] [PubMed]
411. Hansford, J.; Wright, P.C.; Rasoamiaramanana, A.; Pérez, V.R.; Godfrey, L.R.; Errickson, D.; Thompson, T.; Turvey, S.T. Early Holocene human presence in Madagascar evidenced by exploitation of avian megafauna. *Sci. Adv.* **2018**, *4*, eaat6925. [CrossRef] [PubMed]
412. Anderson, A.; Clark, G.; Haberle, S.; Higham, T.; Nowak-Kemp, M.; Prendergast, A.; Radimilahy, C.; Rakotozafy, L.M.; Ramilisonina; Schwenninger, J.L.; et al. New evidence of megafaunal bone damage indicates late colonization of Madagascar. *PLoS ONE* **2018**, *13*, e0204368. [CrossRef] [PubMed]
413. Douglass, K.; Hixon, S.; Wright, H.T.; Godfrey, L.R.; Crowley, B.E.; Manjakahery, B.; Rasolondrainy, T.; Crossland, Z.; Radimilahy, C. A critical review of radiocarbon dates clarifies the human settlement of Madagascar. *Quat. Sci. Rev.* **2019**, *221*, 105878. [CrossRef]
414. Hixon, S.W.; Douglass, K.G.; Crowley, B.E.; Rakotozafy, L.M.A.; Clark, G.; Anderson, A.; Haberle, S.; Ranaivoarisoa, J.F.; Buckley, M.; Fidiarisoa, S. Late Holocene spread of pastoralism coincides with endemic megafaunal extinction on Madagascar. *Proc. R. Soc. B* **2021**, *288*, 20211204. [CrossRef]
415. Goodman, S.M.; Patterson, B.D. *Natural Change and Human Impact in Madagascar*; Smithsonian Institution Press: Washington, DC, USA, 1997.
416. Crowley, B.E. A refined chronology of prehistoric Madagascar and the demise of the megafauna. *Quat. Sci. Rev.* **2010**, *29*, 2591–2603. [CrossRef]
417. Buffetaut, E. Elephant Birds Under the Sun King? Etienne de Flacourt and the Vouron patra. *Bol. Do Cent. Port. De Geo-História E Pré-História* **2018**, *1*, 1.
418. Barrett, P. Antarctic palaeoenvironment through Cenozoic times-a review. *Terra Antarct.* **1996**, *3*, 103–119.
419. Pross, J.; Contreras, L.; Bijl, P.K.; Greenwood, D.R.; Bohaty, S.M.; Schouten, S.; Bendle, J.A.; Röhl, U.; Tauxe, L.; Raine, J.I.; et al. Persistent near-tropical warmth on the Antarctic continent during the early Eocene epoch. *Nature* **2012**, *488*, 73–77. [CrossRef]
420. Contreras, L.; Pross, J.; Bijl, P.K.; Koutsodendris, A.; Raine, J.; van de Schootbrugge, I.; Bas Brinkhuis, H. Early to Middle Eocene vegetation dynamics at the Wilkes Land Margin (Antarctica). *Rev. Palaeobot. Palynol.* **2013**, *197*, 119–142. [CrossRef]
421. Poole, I.; Hunt, R.J.; Cantrill, D. A Fossil Wood Flora from King George Island: Ecological Implications for an Antarctic Eocene Vegetation. *Ann. Bot.* **2001**, *88*, 33–54. [CrossRef]
422. Tambussi, C.; Noriega, J.; Gbzdzicki, A.; Tatur, A.; Reguero, M.; Vizcaíno, S. Ratite bird from the Paleogene La Meseta formation, Seymour Island, Antarctica. *Pol. Polar Res.* **1994**, *15*, 15–20.
423. Field, D.J. Preliminary paleoecological insights from the Pliocene avifauna of Kanapoi, Kenya: Implications for the ecology of *Australopithecus anamensis*. *J. Hum. Evol.* **2017**, *140*, 1–10. [CrossRef]
424. Case, J.A.; Woodburne, M.O.; Chaney, D.S. A gigantic phororhacoid(?) bird from Antarctica. *J. Paleontol.* **1987**, *61*, 1280–1284. [CrossRef]
425. Alvarenga, H.M.; Höfling, E. Systematic revision of the Phorusrhacidae (Aves: Ralliformes). *Papéis Avulsos De Zool.* **2003**, *43*, 55–91. [CrossRef]
426. Chávez, M. Fossil birds of Chile and Antarctic Peninsula. *Arq. Do Mus. Nac. Rio De Jan.* **2007**, *65*, 551–572.
427. Tambussi, C.; Acosta Hospitaleche, C. Antarctic birds (Neornithes) during the Cretaceous-Eocene times. *Rev. De La Asoc. Geológica Argent.* **2007**, *62*, 604–617.
428. Alvarenga, H.M.F.; Chiappe, L.M.; Bertelli, S. Phorusrhacids: The Terror Birds. In *Living Dinosaurs: The Evolutionary History of Modern Birds*; Dyke, G., Kaiser, G., Eds.; John Wiley & Sons: Hoboken, NJ, USA, 2011; pp. 187–208.
429. Cenizo, M. Review of the putative Phorusrhacidae from the Cretaceous and Paleogene of Antarctica: New records of ratites and pelagornithid birds. *Pol. Polar Res.* **2012**, *33*, 239–258. [CrossRef]
430. Acosta Hospitaleche, C.; Jadwiszczak, P.; Clarke, J.; Cenizo, M. The fossil record of birds from the James Ross Basin, West Antarctica. *Adv. Polar Sci.* **2019**, *30*, 251–273. [CrossRef]
431. Pujana, R.R.; Santillana, S.N.; Marenssi, S.A. Conifer fossil woods from the La Meseta Formation (Eocene of Western Antarctica): Evidence of Podocarpaceae-dominated forests. *Rev. Palaeobot. Palynol.* **2014**, *200*, 122–137. [CrossRef]
432. Amenábar, C.R.; Montes, M.; Nozal, F.; Santillana, S. Dinoflagellate cysts of the La Meseta Formation (middle to late Eocene), Antarctic Peninsula: Implications for biostratigraphy, palaeoceanography and palaeoenvironment. *Geol. Mag.* **2020**, *157*, 351–366. [CrossRef]
433. Tambussi, C.; Noriega, J.; Santillana, S.; Marenssi, S. Falconid bird from the Middle Eocene La Meseta Formation, Seymour Island. West Antarctica. *J. Vertebr. Paleontol.* **1995**, *15*, 55A.

434. Cenizo, M.; Noriega, J.I.; Reguero, M.A. A stem falconid bird from the Lower Eocene of Antarctica and the early southern radiation of the falcons. *J. Ornithol.* **2016**, *157*, 885–894. [CrossRef]

435. Goin, F.J.; Reguero, M.A.; Pascual, R.; von Koenigswald, W.; Woodburne, M.O.; Case, J.A.; Marenssi, S.A.; Vieytes, C.; Vizcaíno, S.F. First gondwanatherian mammal from Antarctica. *Geol. Soc. Lond. Spec. Publ.* **2006**, *258*, 135–144. [CrossRef]

436. Gelfo, J.; Goin, F.; Bauzá, N.; Reguero, M. The fossil record of Antarctic land mammals: Commented review and hypotheses for future research. *Adv. Polar Sci.* **2019**, *30*, 274–292. [CrossRef]

437. Woodburne, M.O.; Zinsmeister, W.J. The First Land Mammal from Antarctica and Its Biogeographic Implications. *J. Paleontol.* **1984**, *58*, 913–948.

438. Case, J.A.; Woodburne, M.O.; Chaney, D.S. A new genus of polydolopid marsupial from Antarctica. *Geol. Soc. Am. Mem.* **1988**, *169*, 505–522.

439. Goin, F.J.; Carlini, A.A. An early Tertiary microbiotheriid marsupial from Antarctica. *J. Vertebr. Paleontol.* **1995**, *15*, 205–207. [CrossRef]

440. Goin, F.J.; Case, J.A.; Woodburne, M.O.; Vizcaíno, S.F.; Reguero, M.A. New Discoveries of "Opposum-Like" Marsupials from Antarctica (Seymour Island, Medial Eocene). *J. Mamm. Evol.* **1999**, *6*, 335–365. [CrossRef]

441. Bond, M.; Reguero, M.; Vizcaíno, S.; Marenssi, S. A new 'South American ungulate' (Mammalia: Litopterna) from the Eocene of the Antarctic Peninsula. *Geol. Soc. Lond. Spec. Publ.* **2006**, *258*, 163–176. [CrossRef]

442. Bond, M.; Pascual, R.; Reguero, M.; Santillana, S.; Marenssi, S. Los primeros ungulados extinguidos sudamericanos de la Antártida. *Ameghiniana* **1990**, *16*, 240.

443. Bond, M.; Kramarz, A.; Macphee, R.D.E.; Reguero, M. A New Astrapothere (Mammalia, Meridiungulata) from La Meseta Formation, Seymour (Marambio) Island, and a Reassessment of Previous Records of Antarctic Astrapotheres. *Am. Mus. Novit.* **2011**, *2011*, 1–16. [CrossRef]

444. Hooker, J.J. An additional record of a placental mammal (Order Astrapotheria) from the Eocene of West Antarctica. *Antarct. Sci.* **1992**, *4*, 107–108. [CrossRef]

445. Gelfo, J.N.; López, G.M.; Santillana, S.N. Eocene ungulate mammals from West Antarctica: Implications from their fossil record and a new species. *Antarct. Sci.* **2017**, *29*, 445–455. [CrossRef]

446. Vizcaíno, S.F.; Reguero, M.A.; Goin, F.J.; Tambussi, C.P.; Noriega, J.I. Community structure of Eocene terrestrial vertebrates from Antarctic Peninsula. *Publicación Electrónica De La Asoc. Paleontológica Argent.* **1998**, *5*.

447. Scher, H.D.; Martin, E.E. Timing and climatic consequences of the opening of Drake Passage. *Science* **2006**, *312*, 428–430. [CrossRef]

448. Bijl, P.K.; Bendle, J.A.P.; Bohaty, S.M.; Pross, J.; Schouten, S.; Tauxe, L.; Stickley, C.E.; McKay, R.M.; Röhl, U.; Olney, M.; et al. Eocene cooling linked to early flow across the Tasmanian Gateway. *Proc. Natl. Acad. Sci. USA* **2013**, *110*, 9645–9650. [CrossRef]

449. Wang, Z.; Zhang, J.; Xu, X.; Witt, C.; Deng, Y.; Chenc, G.; Meng, G.; Feng, S.; Xu, L.; Szekely, T. Phylogeny and Sex Chromosome Evolution of Palaeognathae. *J. Genet. Genom.* **2021**. [CrossRef]

450. Kimball, R.T.; Wang, N.; Heimer-McGinn, V.; Ferguson, C.; Braun, E.L. Identifying localized biases in large datasets: A case study using the avian tree of life. *Mol. Phylogenetics Evol.* **2013**, *69*, 1021–1032. [CrossRef]

451. Field, D.J.; Berv, J.; Hsiang, A.; Lanfear, R.; Landis, M.; Dornburg, A. Timing the Extant Avian Radiation: The Rise of Modern Birds, and the Importance of Modeling Molecular Rate Variation. *Bull. Am. Mus. Nat. History* **2020**, *440*. [CrossRef]

452. Berv, J.S.; Field, D.J. Genomic Signature of an Avian Lilliput Effect across the K-Pg Extinction. *Syst. Biol.* **2017**, *67*, 1–13. [CrossRef] [PubMed]

453. Beaulieu, J.M.; O'Meara, B.C.; Crane, P.; Donoghue, M.J. Heterogeneous Rates of Molecular Evolution and Diversification Could Explain the Triassic Age Estimate for Angiosperms. *Syst. Biol.* **2015**, *64*, 869–878. [CrossRef] [PubMed]

454. Mayr, G.; De Pietri, V.L.; Scofield, R.P.; Worthy, T.H. On the taxonomic composition and phylogenetic affinities of the recently proposed clade Vegaviidae Agnolín et al., 2017—Neornithine birds from the Upper Cretaceous of the Southern Hemisphere. *Cretac. Res.* **2018**, *86*, 178–185. [CrossRef]

455. Field, D.J.; Hsiang, A.Y. A North American stem turaco, and the complex biogeographic history of modern birds. *BMC Evol. Biol.* **2018**, *18*. [CrossRef]

456. Mayr, G. Old World Fossil Record of Modern-Type Hummingbirds. *Science* **2004**, *304*, 861–864. [CrossRef]

457. Mayr, G.; Alvarenga, H.; Mourer-Chauviré, C. Out of Africa: Fossils shed light on the origin of the hoatzin, an iconic Neotropic bird. *Naturwissenschaften* **2011**, *98*, 961–966. [CrossRef]

458. Olson, S.L. Evolution of the rails of the South Atlantic islands (Aves: Rallidae). *Smithson. Contrib. Zool.* **1973**, *152*, 1–153. [CrossRef]

459. Wright, N.A.; Steadman, D.W.; Witt, C.C. Predictable evolution toward flightlessness in volant island birds. *Proc. Natl. Acad. Sci. USA* **2016**, *113*, 4765–4770. [CrossRef]

460. McNab, B.K. Energy Conservation and the Evolution of Flightlessness in Birds. *Am. Nat.* **1994**, *144*, 628–642. [CrossRef]

461. Paxinos, E.E.; James, H.F.; Olson, S.L.; Sorenson, M.D.; Jackson, J.; Fleischer, R.C. mtDNA from fossils reveals a radiation of Hawaiian geese recently derived from the Canada goose (*Branta canadensis*). *Proc. Natl. Acad. Sci. USA* **2002**, *99*, 1399–1404. [CrossRef] [PubMed]

462. Black, K.; Archer, M.; Hand, S.; Godthelp, H. The Rise of Australian Marsupials: A Synopsis of Biostratigraphic, Phylogenetic, Palaeoecologic and Palaeobiogeographic Understanding. In *Earth and Life: Global Biodiversity, Extinction Intervals and Biogeographic Perturbations Through Time*; Talent, J., Mulder, E.F.J.d., Derbyshire, E., Eds.; International Year of Planet Earth; Springer: Berlin/Heidelberg, Germany, 2012; pp. 983–1078.

463. Lyson, T.; Miller, I.; Bercovici, A.; Weissenburger, K.; Fuentes, A.J.; Clyde, W.; Hagadorn, J.W.; Butrim, M.J.; Johnson, K.R.; Fleming, R.; et al. Exceptional continental record of biotic recovery after the Cretaceous Paleogene mass extinction. *Science* **2019**, *366*, 977–983. [CrossRef] [PubMed]

464. Van Valkenburgh, B. Major patterns in the history of carnivorous mammals. *Annu. Rev. Earth Planet. Sci.* **1999**, *27*, 463–493. [CrossRef]

465. Figueirido, B.; Palmqvist, P.; Pérez-Claros, J.A.; Janis, C.M. Sixty-six million years along the road of mammalian ecomorphological specialization. *Proc. Natl. Acad. Sci. USA* **2019**, *116*, 12698–12703. [CrossRef] [PubMed]

466. Maxwell, E.E.; Larsson, H.C.E. Osteology and myology of the wing of the Emu (*Dromaius novaehollandiae*), and its bearing on the evolution of vestigial structures. *J. Morphol.* **2007**, *268*, 423–441. [CrossRef] [PubMed]

467. Hudson, G.; Schreiweis, D.; Wang, S.C.; Lancaster, D. A numerical study of the wing and leg muscles of tinamous (Tinamidae). *Northwest. Sci.* **1972**, *46*, 207–255.

468. Suzuki, D.; Chiba, K.; VanBuren, C.S.; Ohashi, T. *The Appendicular Anatomy of the Elegant Crested Tinamou (Eudromia elegans)*; Kitakyushu Museum of Natural History and Human History: Kitakyushu, Japan, 2014.

469. Witmer, L.M. The extant phylogenetic bracket and the importance of reconstructing soft tissues in fossils. *Funct. Morphol. Vertebr. Paleontol.* **1995**, *1*, 19–33.

470. Hosner, P.A.; Tobias, J.A.; Braun, E.L.; Kimball, R.T. How do seemingly non-vagile clades accomplish trans-marine dispersal? Trait and dispersal evolution in the landfowl (Aves: Galliformes). *Proc. R. Soc. B* **2017**, *284*. [CrossRef]

471. Chiappe, L.M.; Walker, C.A. Skeletal morphology and systematics of the Cretaceous Euenantiornithes (Ornithothoraces: Enantiornithes). In *Mesozoic Birds: Above the Heads of Dinosaurs*; Chiappe, L.M., Witmer, L.M., Eds.; University of California Press: Berkeley, CA, USA, 2002; pp. 240–267.

472. O'Connor, J.K.; Chiappe, L.M.; Bell, A. Pre-modern birds: Avian divergences in the Mesozoic. In *Living Dinosaurs: The Evolutionary History of Modern Birds*; Dyke, G., Kaiser, G., Eds.; Wiley-Blackwell: Hoboken, NJ, USA, 2011; pp. 39–114.

473. Chiappe, L.M.; Qingjin, M. *Birds of Stone: Chinese Avian Fossils from the Age of Dinosaurs*; JHU Press: Baltimore, MD, USA, 2016.

474. Tschudy, R.H.; Pillmore, C.L.; Orth, C.J.; Gilmore, J.S.; Knight, J.D. Disruption of the Terrestrial Plant Ecosystem at the Cretaceous-Tertiary Boundary, Western Interior. *Science* **1984**, *225*, 1030–1032. [CrossRef]

475. Vajda, V.; Raine, J.I.; Hollis, C.J. Indication of Global Deforestation at the Cretaceous-Tertiary Boundary by New Zealand Fern Spike. *Science* **2001**, *294*, 1700–1702. [CrossRef]

476. Vajda, V.; Bercovici, A. The global vegetation pattern across the Cretaceous–Paleogene mass extinction interval: A template for other extinction events. *Glob. Planet. Chang.* **2014**, *122*, 29–49. [CrossRef]

477. Klein, C.G.; Pisani, D.; Field, D.J.; Lakin, R.; Wills, M.A.; Longrich, N.R. Evolution and dispersal of snakes across the Cretaceous-Paleogene mass extinction. *Nat. Commun.* **2021**, *12*, 5335. [CrossRef] [PubMed]

478. Hughes, J.J.; Berv, J.S.; Chester, S.G.B.; Sargis, E.J.; Field, D.J. Ecological selectivity and the evolution of mammalian substrate preference across the K-Pg boundary. *Ecol. Evol.* **2021**, *11*, 14540–14554. [CrossRef] [PubMed]

479. Anders, E.; Wolbach, W.S.; Gilmour, I. Major wildfires at the Cretaceous-Tertiary boundary. In *Global Biomass Burning*; Levine, J.S., Ed.; The MIT Press: Cambridge, MA, USA, 1991; pp. 485–492.

480. Robertson, D.S.; Lewis, W.M.; Sheehan, P.M.; Toon, O.B. K-Pg extinction: Reevaluation of the heat-fire hypothesis. *J. Geophys. Res. Biogeosci.* **2013**, *118*, 329–336. [CrossRef]

481. Clyde, W.C.; Ramezani, J.; Johnson, K.R.; Bowring, S.A.; Jones, M.M. Direct high-precision U–Pb geochronology of the end-Cretaceous extinction and calibration of Paleocene astronomical timescales. *Earth Planet. Sci. Lett.* **2016**, *452*, 272–280. [CrossRef]

482. Johnson, K.R.; Ellis, B. A Tropical Rainforest in Colorado 1.4 Million Years After the Cretaceous-Tertiary Boundary. *Science* **2002**, *296*, 2379–2383. [CrossRef]

483. Bell, A.; Chiappe, L.M. A species-level phylogeny of the Cretaceous Hesperornithiformes (Aves: Ornithuromorpha): Implications for body size evolution amongst the earliest diving birds. *J. Syst. Palaeontol.* **2014**, *14*, 239–251. [CrossRef]

484. Dumont, M.; Tafforeau, P.; Bertin, T.; Bhullar, B.; Field, D.; Schulp, A.; Strilisky, B.; Thivichon-Prince, B.; Viriot, L.; Louchart, A. Synchrotron imaging of dentition provides insights into the biology of *Hesperornis* and *Ichthyornis*, the last toothed birds. *BMC Evol. Biol.* **2016**, *16*, 178. [CrossRef]

485. Field, D.J.; Hanson, M.; Burnham, D.; Wilson, L.E.; Super, K.; Ehret, D.; Ebersole, J.A.; Bhullar, B.A.S. Complete Ichthyornis skull illuminates mosaic assembly of the avian head. *Nature* **2018**, *557*, 96–100. [CrossRef]

486. Alegret, L.; Thomas, E.; Lohmann, K.C. End-Cretaceous marine mass extinction not caused by productivity collapse. *Proc. Natl. Acad. Sci. USA* **2012**, *109*, 728–732. [CrossRef]

487. Gottschaldt, K. Structure and function of avian somatosensory receptors. In *Form and Function in Birds*; King, A.S., McLelland, J., Eds.; Academic Press: London, UK, 1985; Volume 3, pp. 375–461.

488. Cunningham, S.; Castro, I.; Alley, M. A new prey detection mechanism for kiwi (*Apteryx* spp.) suggests convergent evolution between paleognathous and neognathous birds. *J. Anat.* **2007**, *211*, 493–502. [CrossRef] [PubMed]

489. Crole, M.; Soley, J. Bony Pits in the Ostrich (*Struthio camelus*) and Emu (*Dromaius novaehollandiae*) Bill Tip. *Anat. Rec.* **2017**, *300*, 1705–1715. [CrossRef] [PubMed]

490. Zelenitsky, D.K.; Therrien, F.; Ridgely, R.C.; McGee, A.R.; Witmer, L.M. Evolution of olfaction in non-avian theropod dinosaurs and birds. *Proc. R. Soc. B* **2011**, *278*, 3625–3634. [CrossRef] [PubMed]

diversity

Article

New Comparative Data on the Long Bone Microstructure of Large Extant and Extinct Flightless Birds

Aurore Canoville [1,2], Anusuya Chinsamy [1,*] and Delphine Angst [1]

[1] Department of Biological Sciences, University of Cape Town, Private Bag X3, Rondebosch 7701, South Africa; canoville.aurore08@gmail.com (A.C.); angst.delphine@gmail.com (D.A.)

[2] Stiftung Schloss Friedenstein, Schloss Friedenstein, Schlossplatz 1, 99867 Gotha, Germany

* Correspondence: achinsamyturan@gmail.com or anusuya.chinsamy-turan@uct.ac.za

Abstract: Here, we investigate whether bone microanatomy can be used to infer the locomotion mode (cursorial vs. *graviportal*) of large terrestrial birds. We also reexamine, or describe for the first time, the bone histology of several large extant and extinct flightless birds to (i) document the histovariability between skeletal elements of the hindlimb; (ii) improve our knowledge of the histological diversity of large flightless birds; (iii) and reassess previous hypotheses pertaining to the growth strategies of modern palaeognaths. Our results show that large extinct terrestrial birds, inferred as *graviportal* based on hindlimb proportions, also have thicker diaphyseal cortices and/or more bony trabeculae in the medullary region than cursorial birds. We also report for the first time the occurrence of growth marks (not associated with an outer circumferential layer-OCL) in the cortices of several extant ratites. These observations support earlier hypotheses that flexible growth patterns can be present in birds when selection pressures for rapid growth within a single year are absent. We also document the occurrence of an OCL in several skeletally mature ratites. Here, the high incidence of pathologies among the modern species is attributed to the fact that these individuals were probably long-lived zoo specimens.

Keywords: terrestrial birds; flightless birds; Palaeognathae; bone histology; microanatomy; growth marks; avian pathologies

Citation: Canoville, A.; Chinsamy, A.; Angst, D. New Comparative Data on the Long Bone Microstructure of Large Extant and Extinct Flightless Birds. *Diversity* **2022**, *14*, 298. https://doi.org/10.3390/d14040298

Academic Editor: Luc Legal

Received: 31 January 2022
Accepted: 19 March 2022
Published: 15 April 2022

Publisher's Note: MDPI stays neutral with regard to jurisdictional claims in published maps and institutional affiliations.

1. Introduction

Large flightless birds are represented today by members of the palaeognaths, which include ostriches, emus, rheas, and cassowaries. Despite having different habitats, body sizes, and hindlimb proportions, all these birds are considered cursorial, although there is a continuum between their locomotor styles [1]. Cursoriality refers to the capability of a terrestrial animal to have sustained running [2–4]. Studies have shown that the degree of cursoriality is correlated to the relative proportions of the hindlimb elements [5–8], and, more specifically the length-width ratio (i.e., Lg/Dm) of the tarsometatarsus [9]. Thus, ostriches, with their long and slender hindlimbs, are the largest but also the fastest-running living ratites [1], while cassowaries (such as *Casuarius casuarius*) that are more heavily built with the shortest and widest tarsometatarsus among extant cursorial birds [9], are also the slowest runners among their relatives [1]. Birds having a slow-walking type of locomotion (also often referred to as having "graviportal" locomotion; [9,10]) are represented by large terrestrial extinct forms [9]. These birds are hypothesized to have been capable of running, but only at a slow pace and for short durations [11–13]. Among the extinct terrestrial birds, several groups inferred as "graviportal" (e.g., [9]) are known from the Cenozoic fossil record, such as the Gastornithidae and Dromornithidae (Neognathae; [9,14–18]), and two recently extinct ratite clades [19]: the New-Zealand moas (Dinornithiformes, Palaeognathae) and the Madagascan elephant birds (Aepyornithiformes, Palaeognathae).

However, it is often difficult to infer the locomotion mode of avian fossil specimens since, more often than not, the osteological material is fragmentary. For example, Storer's [5]

method to deduce the locomotory style of extinct birds ideally requires linear measurements of three complete elements of the same hindlimb for a single individual. The model proposed by Angst et al. [9] also relies on the measurement of complete and undistorted tarsometatarsi, i.e., the ratio Lg/Dm of the tarsometatarsus is <12 for birds inferred as "graviportal" and hence, slow walking. However, this element is often incomplete or even missing in the paleontological collections for many giant fossil terrestrial birds. Thus, it is often challenging to decipher the locomotion of extinct species using the current available methods. For example, in the case of *Gargantuavis philoinos* from the Late Cretaceous of Europe [10], only the femur is known from its hindlimb, and its locomotion has therefore been estimated based on the morphology of its synsacrum [10]. Inferences of the locomotor habits of *Gargantuavis* are therefore not directly comparable with the results drawn for other birds. Another example concerns *Remiornis*, a large terrestrial bird from the Upper Palaeocene of France [20], for which only a partial tarsometatarsus, a coracoid, and a thoracic vertebra are known. Thus, for these common cases of incomplete skeletal remains of large terrestrial birds, it would be useful to have a different estimation method of locomotory habits that does not require complete hindlimb elements.

In a comparative and standardized framework, the long limb bone mid-diaphyseal microanatomy and cross-sectional geometry are known to provide insights into the ecology and/or locomotor activity (and biomechanical loading) of tetrapods (e.g., [21]). Hence, several studies reported that cursorial terrestrial animals generally have tubular limb bones, with open medullary cavities and thicker bone walls than their flying relatives [21–24]. Moreover, a recent study by Houssaye et al. [25] showed that graviportal tetrapods generally exhibit thicker compact cortices and more trabeculae in the medullary region of the diaphysis than their cursorial counterparts. However, to date, no comparative study has been carried out to investigate the relationship between locomotion patterns and hindlimb bone microanatomy in large flightless birds.

Furthermore, although isolated studies of the limb bone microstructure of ratites and other Cenozoic terrestrial birds exist, their results have never been considered within a comparative framework. A notable exception is a study by Legendre et al. [26], where the authors sampled several large terrestrial birds, including all extant ratite genera, as well as Dinornithidae and Aepyornithidae. However, in the latter study, the researchers mostly focused on quantitative histological parameters related to the vascularization pattern and osteocyte lacunae size and density and did not discuss microanatomical differences between these animals.

Most Neornithes (modern birds including Palaeognathae and Neognathae), are considered to have a fast and uninterrupted growth strategy (with deposition of uninterrupted, well-vascularized fibrolamellar bone tissue (e.g., [27]). Among these birds, skeletal maturity is achieved in less than a year, as opposed to the discontinuous and prolonged growth of non-avian dinosaurs and most basal, non-ornithurine birds [27–34]. With the exception of a few modern species (such as the kiwi; see [35]), growth marks are thus usually lacking in the cortices of birds, although some closely spaced lines of arrested growth (LAGs) can sometimes be observed in the avian outer circumferential layer (abbreviated OCL [36,37]; and equivalent to the fundamental external system observed in tetrapods in general [30,38–40]). Thus, most extant palaeognaths (basal among modern birds), such as tinamous, cassowaries, emus, rheas, and ostriches, have been described as lacking growth marks in their limb bone cortices because they are considered to reach skeletal maturity within a single year (e.g., [26,28,30,35,41–45]). In addition, previous studies (e.g., [28,29,46]) pointed out that the OCL was poorly developed to absent in extant ratites, such as *Struthio* and *Casuarius*. It appears that kiwis (Apterygidae), the smallest extant ratites endemic to New Zealand, are the exception in presenting LAGs in their long bone cortices [35], which led to the deduction that these unique birds experience protracted growth and delayed sexual maturity. LAGs have also been observed in the cortex of extinct large ratites, such as some Dinornithidae [26,30,44,45], and Aepyornithidae [18,33,47]. Multiple growth marks have also been documented in large dromornithids such as *Dromornis* [15] and *Genyor-*

nis [48], suggesting a similar protracted growth for all these giant extinct terrestrial birds. Finally, such growth marks have been reported in *Gargantuavis* [32], although its avian phylogenetic affinities are still debated (e.g., [49–51]).

Protracted growth strategies are also evident in the extinct solitaire of Rodrigues *Pezophaps solitarius* [52]. Such extended growth dynamics have been interpreted as a convergent adaptive strategy to unusual environmental factors and insular life where there are relaxed pressures for rapid growth [33,35,45,53,54]. Finally, in extant neognaths, a single LAG has been observed in the cortex of a parrot metatarsal [46] and in the limb bones of the king penguin *Aptenodytes patagonicus* [55]; the latter having a well-studied, discontinuous growth during the austral winter, with a fasting period associated with a pause of skeletal growth [56–58]). Finally, a LAG has also been observed in a tibia of the giant extinct Eocene bird *Gastornis* (*Diatryma* is now considered as synonymous of *Gastornis* [30,44]).

In this preliminary comparative study, we investigate whether bone microanatomy could be used to infer the locomotion mode (i.e., cursorial vs. "graviportal") of large terrestrial birds as a complement or alternative to previously proposed methods (when the fossil material is fragmentary). We also describe the bone histology of some extant and extinct large terrestrial birds, in order to (i) understand the histovariability between the different skeletal elements of the hindlimb (femur, tibiotarsus, tarsometatarsus); (ii) improve our knowledge of terrestrial bird bone microstructural diversity; (iii) and reassess previous hypotheses pertaining to the growth strategies of some palaeognaths.

2. Materials and Methods

2.1. Biological Sample

To assess the inter-elements as well as inter-specific histovariability in large terrestrial birds, we sampled 47 hindlimb bones (femora, tibiotarsi and tarsometatarsi) of at least seven extant and extinct bird genera (Table 1).

Our sample encompassed the diversity of the largest extant terrestrial birds represented by the ratites (Palaeognathae) and included the common ostrich *Struthio camelus*, the greater rhea *Rhea americana*, the emu *Dromaius novaehollandiae*, and the Southern cassowary *Casuarius casuarius* (Table 1). Our (sub)fossil material comprised three different groups of Cenozoic giant terrestrial birds: (i) limb bones from the recently extinct elephant birds from Madagascar, Aepyornithidae (Palaeognathae); (ii) limb bones from the giant moas (Dinornithidae, Palaeognathae) from New Zealand; (iii) and bone fragments from *Gastornis* (Neognathae), a giant flightless bird from the Paleocene and Eocene of Europe, North Africa, and China [14,59].

All the bones sampled in this study were derived from specimens housed in the collections of paleontology and comparative anatomy of the Muséum National d'Histoire Naturelle (MNHN, Paris, France). The aepyornithid material was undiagnosed when we originally sampled it, however, Hansford and Turvey [60] subsequently undertook a comprehensive taxonomic assessment of all the aepyornithid specimens. In our earlier study of the aepyornithid bone histology [33], James Hansford assisted with the taxonomic identification of our aepyornithid material. Thus, our sampled aepyornithid material (Table 1) comprised of two femora (Ae-fm 2 & 3) that have been identified as belonging to *Aepyornis maximus*, while one femur (Ae-fm-1), four tibiotarsi (Ae-tb-1, Ae-tb-5, Ae-tb-6 and Ae-tb-8) and two tarsometatarsi (Ae-tm-1 & 2) are considered to belong to *Vorombe titan*. Four tarsometatarsi remained undiagnosed because of insufficient data (see [33] for more details). For the (sub)fossil material of Dinornithidae, identification at the species level was not possible. Provenance of this material was also not recorded in the catalogues of the MNHN, and it is likely that some extant specimens were captive zoo individuals.

Table 1. Large terrestrial bird material sampled in this study for histological analysis and simple linear measurements recorded from the limb bones. Abbreviations: BC, bone core taken at mid-diaphysis; Cm, circumference at a minimal diameter of the shaft (which could be used to estimate the body mass of birds, see [9]); CT, complete cross-section taken at mid-diaphysis; Dm, minimal diameter of the shaft measured in dorsal or ventral view; l, left; Lg., maximal length of the limb bone; pe, proximal extremity; r, right.

Taxon	Sample Number	Bone	MNHN Collection Number	Sample Type	Lg (mm)	Dm (mm)	Lg/Dm	Cm (mm)
EXTANT PALAEOGNATHS								
Casuarius casuarius	Cc-fm-1	Femur (l)	1923-879 SS-S.B.S2.M1.C1	CT	224	-	-	87
Casuarius casuarius	Cc-fm-2	Femur (r)	1923-879 SS-S.B.S2.M1.C1	CT	-	-	-	-
Casuarius casuarius	Cc-tb-1	Tibiotarsus (r)	1923-877 SS-S.B.S2.M1.C1	CT	389	-	-	78
Casuarius casuarius	Cc-tb-2	Tibiotarsus (l)	1923-877 SS-S.B.S2.M1.C1	CT	386	-	-	78
Casuarius casuarius	Cc-tm-1	Tarsometatarsus (l)	1923-878 SS-S.B.S2.M1.C1	CT	295	22.8	12.94	76
Casuarius casuarius	Cc-tm-2	Tarsometatarsus (l)	1923-878	CT	290	23.2	12.5	74
Dromaius novaehollandiae	Dn-fm-1	Femur (r)	1923-900 SS-S.B.S2.M1.C23	CT	221 *	-	-	95 *
Dromaius novaehollandiae	Dn-fm-2	Femur (l)	1923-897 SS-S.B.S2.M1.C23	CT	234	-	-	99
Dromaius novaehollandiae	Dn-tb-1	Tibiotarsus (l)	1923-899 SS-S.B.S2.M1.C23	CT	456	-	-	91
Dromaius novaehollandiae	Dn-tb-2	Tibiotarsus (l)	1923-899 SS-S.B.S2.M1.C23	CT	454	-	-	87
Dromaius novaehollandiae	Dn-tm-1	Tarsometatarsus (l)	1923-898 SS-S.B.S2.M1.C23	CT	387	17.92	21.60	76
Dromaius novaehollandiae	Dn-tm-2	Tarsometatarsus (r)	1923-898 SS-S.B.S2.M1.C23	CT	400	19.28	20.75	84
Rhea americana	Ra-fm-1	Femur (l)	1920-116	CT	-	-	-	-
Rhea americana	Ra-fm-2	Femur (l)	-	CT	186	-	-	60
Rhea americana	Ra-tb-1	Tibiotarsus (r)	-	CT	286	-	-	55
Rhea americana	Ra-tb-2	Tibiotarsus (l)	1920-116	CT	354	-	-	73
Rhea americana	Ra-tm-1	Tarsometatarsus (l)	1920-116	CT	316	15.64	20.20	58
Rhea americana	Ra-tm-2	Tarsometatarsus (l)	-	CT	272	13.15	20.68	48
Struthio camelus	Sc-fm-1	Femur (l)	A4751 SS-S.B.S2.M1.C21	CT	301 *	-	-	140 *
Struthio camelus	Sc-fm-2	Femur (l)	-	CT	327	-	-	140.5
Struthio camelus	Sc-tb-1	Tibiotarsus (l)	-	CT	597	-	-	109
Struthio camelus	Sc-tb-2	Tibiotarsus (r)	A-5-100 SS-S.B.S2.M1.C21	CT	470	-	-	97
Struthio camelus	Sc-tm-1	Tarsometatarsus (r)	-	CT	514	32.95	15.60	97
Struthio camelus	Sc-tm-2	Tarsometatarsus (l)	-	CT	460	28.90	15.92	85
EXTINCT PALAEOGNATHS								-
Vorombe titan	Ae-fm-1	Femur (r)	1937-62	BC	360	-	-	-
Aepyornis maximus	Ae-fm-2	Femur (l)	MAD367 1906-17	CT	380	-	-	244
Aepyornis maximus	Ae-fm-3	Femur (r)	MAD 365	CT	350	-	-	244
Vorombe titan	Ae-tb-1	Tibiotarsus	1910-12	CT	-	-	-	190
Aepyornithidae indet.	Ae-tb-2	Tibiotarsus pe (l)	-	CT	-	-	-	-
Aepyornithidae indet.	Ae-tb-3	Tibiotarsus pe (r)	-	CT	-	-	-	-
Aepyornithidae indet.	Ae-tb-4	Tibiotarsus (l)	1937-62	BC	612	-	-	-
Vorombe titan	Ae-tb-5	Tibiotarsus (r)	1937-62	BC	645	-	-	-
Vorombe titan	Ae-tb-6	Tibiotarsus (l)	1937-62	BC	670	-	-	-
Aepyornithidae indet.	Ae-tb-7	Tibiotarsus (r)	1937-62	BC	615	-	-	-

Table 1. *Cont.*

Taxon	Sample Number	Bone	MNHN Collection Number	Sample Type	Lg (mm)	Dm (mm)	Lg/Dm	Cm (mm)
Vorombe titan	Ae-tb-8	Tibiotarsus (l)	1937-62	BC	760	-	-	-
Aepyornithidae indet.	Ae-tm-1	Tarsometatarsus (l)	1906-16 Mr Belo	CT	430	81.8	5.26	190
Aepyornithidae indet.	Ae-tm-2	Tarsometatarsus (l)	1937-62	BC	481	88.03	5.46	-
Dinornis sp.	Di-fm-1	Femur (l)	-	CT	315	-	-	182
Dinornis sp.	Di-tb-1	Tibiotarsus (l)	-	CT	-	-	-	160 **
Dinornithidae indet.	Di-tb-3	Tibiotarsus (l)	-	CT	-	-	-	-
Dinornithidae indet.	Di-tb-4	Tibiotarsus (?)	-	CT	-	-	-	-
Dinornis sp.	Di-tm-1	Tarsometatarsus (l)	-	CT	235	42	5.60	107
EXTINCT NEOGNATHS								
Gastornis sp.	Ga-fm-1	Fem/tibiotarus (?)	Mu12734	Frag.	-	-	-	-
Gastornis sp.	Ga-fm-2	Femur (r)	BR12419	BC	-	-	-	-
Gastornis sp.	Ga-fm-3	Femur (r)	BR12420	Frag.	-	-	-	-
Gastornis sp.	Ga-tb-1	Tibiotarsus (?)	-	Frag.	-	-	-	-
Gastornis sp.	Ga-tb-2	Tibiotarsus (?)	CRL2457	BC	-	-	-	-

* Estimated lengths from the counterpart element. ** Not at the diaphyseal level. (?) side of skeletal element unknown.

2.2. Thin-Section Preparation and Histological Descriptions

When possible, we sampled the mid-diaphyseal complete cross-sections of the main hindlimb bones involved in terrestrial locomotion: femur, tibiotarsus and tarsometatarsus (Table 1). In some cases, the (sub)fossil material was either too fragmentary or rare to acquire complete mid-diaphyseal cross-sections. We thus sampled bone cores at the mid-diaphyseal level using the technique described in Stein and Sander [61]. This sampling method leads to the least possible damage to the overall anatomy of skeletal elements and is frequently used in paleohistological studies (e.g., [62–64]).

Thin sections were prepared at the MNHN (Paris, France) and at the Biological Sciences Department of the University of Cape Town (Cape Town, South Africa) following standard procedures described in Chinsamy and Raath [65] and Padian and Lamm [39]. The histological descriptions follow the terminology used in Francillon-Vieillot et al. [66] and Chinsamy-Turan [31].

Note that since the histology of the aepyornithid material has already been comprehensively described by our team [33], we mostly report on its global compactness in this manuscript.

2.3. Simple Linear Measurements and Global Compactness

Each bone was measured and photographed before thin sectioning or core drilling. For each skeletal element, we measured, when possible, its maximal length (Lg) between the proximal and distal epiphyses. For comparison with the study of Angst et al. [9], we also calculated the ratio between the maximal length of the tarsometatarsus and the minimum shaft diameter (Dm) in dorsal or ventral view (Table 1).

Microanatomical investigations were carried out on a sub-sample of extant and extinct bird specimens (21 skeletal elements), when complete cross-sections were available (as shown in Figures 1–10), and the skeletal elements did not show signs of bone pathology drastically affecting the overall cortical thickness (Table 2). The complete cross-sections were transformed into binary images to obtain their global compactness values (Comp.) using Bone Profiler [67]. In some cases (especially sub-fossil specimens), the obtained global compactness values are underestimated, since some fragile bone trabeculae occupying the medullary region broke during the sampling process. We also recorded the presence or absence of bone trabeculae in the mid-diaphyseal region of each sampled element, since this has been linked to a graviportal adaptation in some large amniotes [25].

Table 2. Bone microanatomical data (on a subsample). The global compactness values were obtained in Bone Profiler [67]. Abbreviations: Comp., global compactness of the mid-diaphyseal cross-section; fm, femur; Lg, maximal length of the limb bone; MD, maximal diameter of the section; tb, tibiotarsus; tm, tarsometatarsus.

Species	Locomotion Type	Sampling Number	MD (mm)	Comp.	Bone Trabeculae
Casuarius casuarius	Cursorial [1]	Cc-fm-1	29.0	0.395	NO
Dromaius novaehollandiae	Cursorial [1]	Dn-fm-2	32.7	0.389	YES
		Dn-tb-2	28.5	0.500	NO
		Dn-tm-1	-	0.655	YES
		Dn-tm-2	-	0.560	NO
Rhea americana	Cursorial [1]	Ra-fm-1	-	0.656	NO
		Ra-fm-2	22.6	0.404	NO
		Ra-tb-1	17.0	0.459	NO
		Ra-tb-2	21.6	0.644	NO
		Ra-tm-1	20.3	0.826	NO
		Ra-tm-2	16.5	0.618	NO
Struthio camelus	Cursorial [1]	Sc-fm-1	46.3	0.392	NO
		Sc-fm-2	55.1	0.289	YES (broken)
		Sc-tb-1	37.2	0.593	NO
		Sc-tm-2	28.6	0.702	NO
Aepyornithidae	Graviportal [2,3]	Ae-fm-2	96.5	0.512	YES
		Ae-fm-3	82.4	0.386	YES
		Ae-tb-1	71.6	0.706	YES (broken)
		Ae-tm-1	76.7	0.785	YES (broken)
Dinornithidae	Graviportal [2]	Di-fm-1	55.8	0.741	NO
		Di-tb-3	68.5	0.733	YES (broken)

[1] [1]; [2] [9]; [3] [13].

Figure 1. Long bone histology of the Southern cassowary *Casuarius casuarius.* (**A**) Mid-diaphyseal cross-section of the femur Cc-fm-1. (**B**) Mid-diaphyseal cross-section of the femur Cc-fm-2. Note the presence of pathological bone tissues on the endosteal and periosteal (red box) margins. (**C,D**) Close-ups of the cortex of Cc-fm-1. Most of the cortex is formed of laminar or reticular fibrolamellar bone tissue. Three to four growth marks are visible in the cortex (arrows). An outer circumferential layer (OCL) with several closely spaced LAGs (arrowheads) is present in the outermost cortex. (**E**) Close-up of the cortex of Cc-fm-1, showing a highly remodeled zone of the cortex, with numerous secondary osteons. (**F**) Close-up of the periosteal surface of Cc-fm-2 showing a layer of periosteal pathological bone tissue (per. PB) overlaying the laminar cortical tissue.

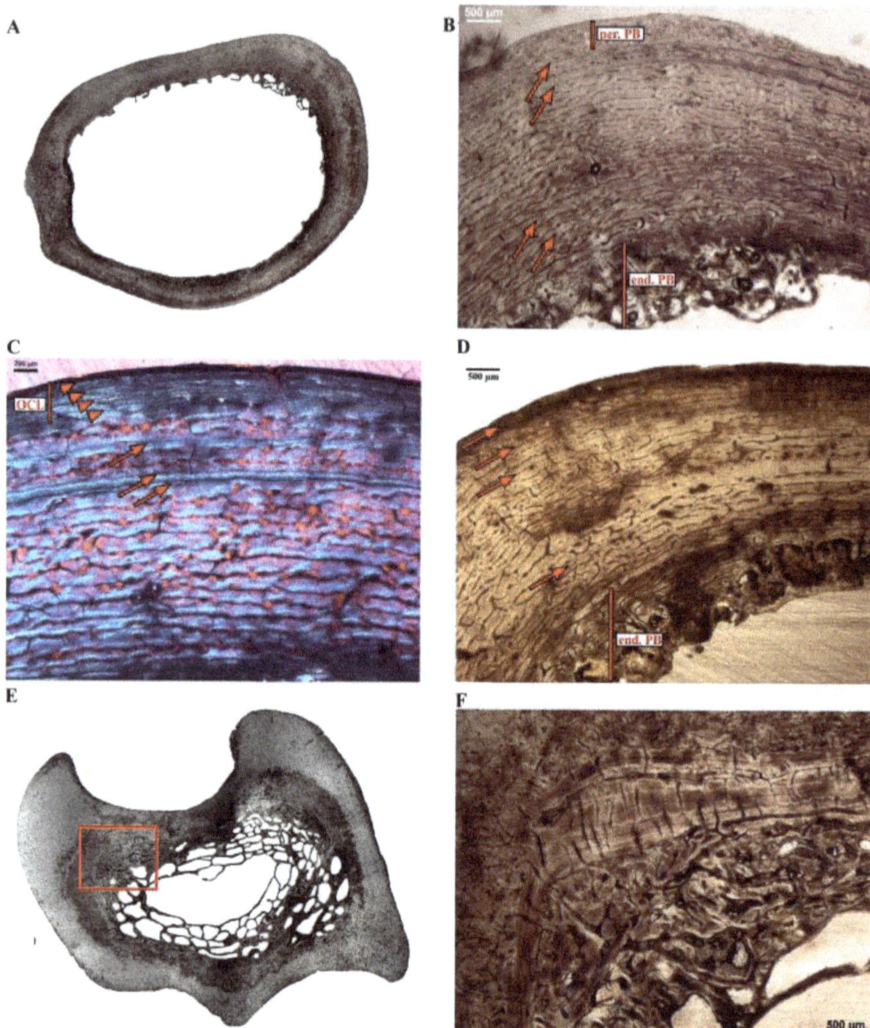

Figure 2. Long bone histology of the Southern cassowary *Casuarius casuarius* (continued). (**A**) Mid-diaphyseal cross-section of the tibiotarsus Cc-tb-2. (**B**) Close-up of a region of the cortex of Cc-tb-2, showing pathological bone tissues on both the periosteal (per. PB) and endosteal (end. PB) margins. Note that LAGs (arrows) are visible throughout the cortical thickness. (**C**) Close-up of a region of the cortex of Cc-tb-2 showing that the primary cortex is composed of a well-vascularized laminar to reticular fibro-lamellar bone (in polarized light). Three LAGs (arrows) are visible in the outer-third of the cortex, attesting of an interrupted growth. An OCL containing four closely spaced LAGs (arrowheads) is visible in the outermost cortex. (**D**) Close-up of a region of the cortex of Cc-tb-1, showing endosteal pathological bone tissue (end. PB), as well as at least four LAGs (arrows) interrupting the reticular fibrolamellar bone tissue of the cortex. (**E**) Mid-diaphyseal cross-section of the tarsometatarsus Cc-tm-1. (**F**) close-up of the framed region in (**E**), showing endosteal pathological bone tissue.

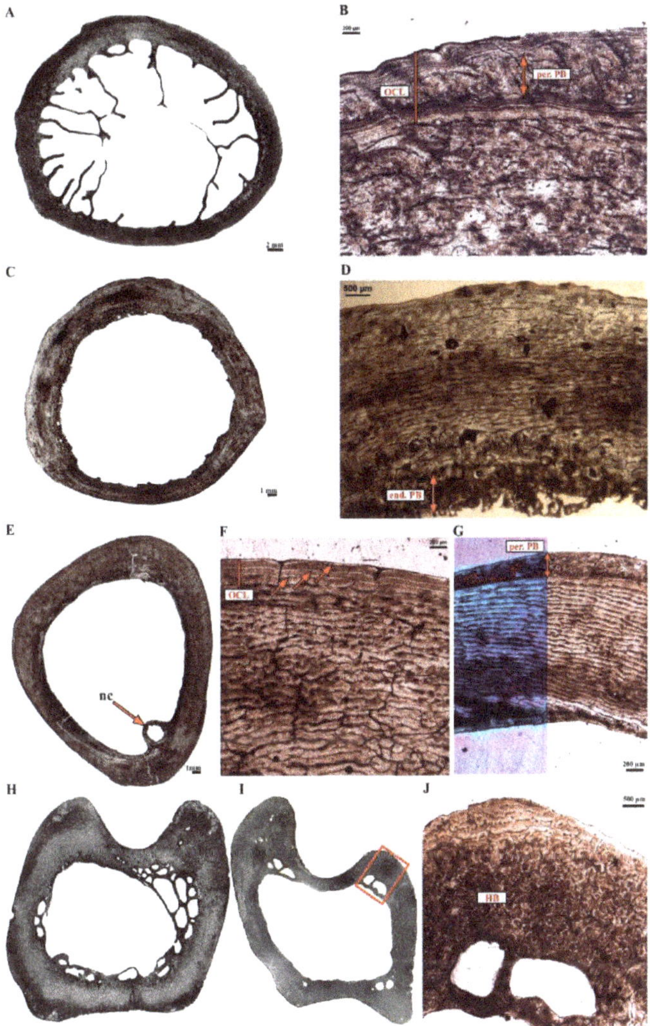

Figure 3. Long bone histology of the common emu *Dromaius novaehollandiae*. (**A**) Mid-diaphyseal cross-section of the femur Dn-fm-2. (**B**) Close-up of the cortex of Dn-fm-2. Note the numerous closely spaced LAGs forming an OCL in the outermost cortex. A layer of vascularized and partly remodeled pathological tissue (per. PB) is also intercalated between some of the LAGs of the OCL in some parts of the cross-section. (**C**) Mid-diaphyseal cross-section of the tibiotarsus Dn-tb-1. (**D**) Close-up of the cortex of Dn-tb-1 showing the radially organized endosteal pathological bone (end. PB), as well as its uneven periosteal surface. (**E**) Mid-diaphyseal cross-section of the tibiotarsus Dn-tb-2 with the nutrient canal (nc) visible. (**F**) Close-up of the cortex of Dn-tb-2. Most of the cortex is composed of a highly vascularized reticular fibrolamellar bone tissue. An OCL with several closely spaced LAGs (arrows) is visible in the outermost cortex. (**G**) Close-up of the cortex of Dn-tb-2 with a layer of vascularized and remodeled pathological tissue (per. PB) intercalated within the OCL. (**H**) Mid-diaphyseal cross-section of the tarsometatarsus Dn-tm-1. (**I**) Mid-diaphyseal cross-section of the tarsometatarsus Dn-tm-2. (**J**) Close-up of the framed region of the cortex of the tarsometatarsus Dn-tm-2 in (**I**) that consists mostly of dense Haversian bone (HB).

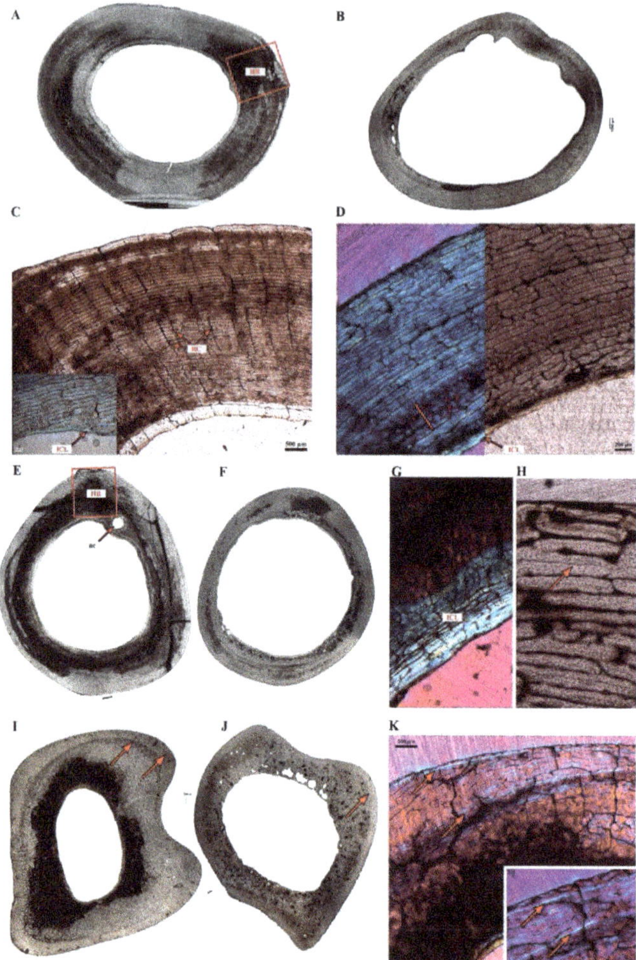

Figure 4. Long bone histology of the greater rhea *Rhea americana*. (**A**) Mid-diaphyseal cross-section of the femur Ra-fm-1. (**B**) Mid-diaphyseal cross-section of the femur Ra-fm-2. (**C**) Close-up of the framed region of the cortex of Ra-fm-1 in (A). Most of the cortex is formed of a well-vascularized reticular to laminar fibrolamellar bone tissue with radial anastomoses (RC). Insert showing the thin birefringent ICL. (**D**) Close-up of the cortex of Ra-fm-2. The cortex is mostly formed of a reticular to poorly characterized laminar bone tissue with cyclical modulations in the bone vascularization pattern. No decrease in vascular density is observed in the outer cortex of this element. Note the presence of a thin ICL along the medullary margin. (**E**) Mid-diaphyseal cross-section of the tibiotarsus Ra-tb-2. Note that the nutrient canal (nc) is visible. (**F**) Mid-diaphyseal cross-section of the tibiotarsus Ra-tb-1. (**G**) Close-up of the deep cortex and the well-vascularized layer of endosteal bone (ICL) lining the medullary region in Ra-tb-2. (**H**) Close-up of outer cortex in Ra-tb-1. A faint LAG (arrow) interrupts the well-vascularized fibrolamellar bone tissue close to the periosteal surface. Vascular canals are visible at the bone surface and no clear decrease in bone vascularization is observed in the outer cortex of this element. (**I**) Mid-diaphyseal cross-section of the tarsometatarsus Ra-tm-1. (**J**) Mid-diaphyseal cross-section of the tarsometatarsus Ra-tm-2. (**K**) Close-up of the cortex of Ra-tm-1. The cortex is stratified and presents a first LAG (arrow) marking the transition from a partly remodeled plexifom tissue to a more reticular fibrolamellar bone. A second LAG (arrow) under the periosteal surface.

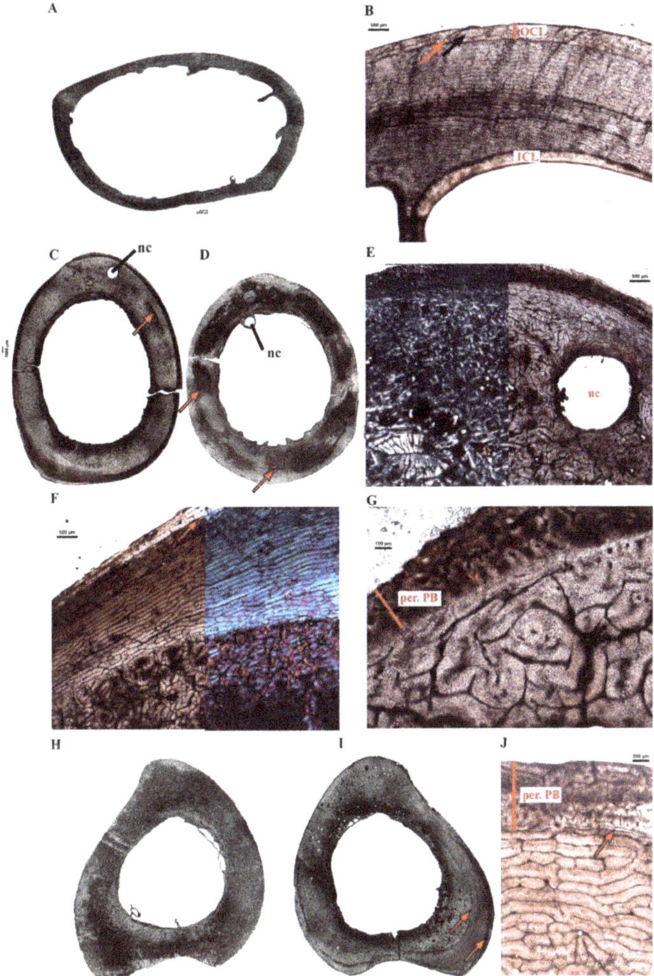

Figure 5. Long bone histology of the common ostrich *Struthio camelus*. (**A**) Mid-diaphyseal cross-section of the femur Sc-fm-2. (**B**) Close-up of the cortex of Sc-fm-2. A thin and discontinuous layer of endosteal lamellar bone forming the onset of the ICL lines the medullary cavity. Most of the cortex is formed of a well-vascularized fibrolamellar bone with reticular to laminar organization. Few radial anastomoses are visible throughout the cortex. Two LAGs (arrows) and a decrease in bone vascularization towards the periosteal surface mark the onset of the OCL. (**C**) Mid-diaphyseal cross-section of the tibiotarsus Sc-tb-1. Note the presence of the nutrient canal (nc) in the section. (**D**) Mid-diaphyseal cross-section of the tibiotarsus Sc-tb-2. Note the presence of the nutrient canal (nc) in the section. (**E,F**) Close-ups of the bone wall of Sc-tb-1, with a thick layer of remodeled coarse cancellous bone forming half to two-thirds of the cortex and a layer of non-remodeled fibrolamellar bone with a reticular to laminar organization. A LAG (arrow), close to the bone surface, marks a change in bone vascularization. (**G**) Close-up of the cortex of Sc-tb-1. In some parts of the section, the outermost cortex consists of a thin bone layer reminiscent of pathological bone (per. PB) containing numerous bundles of Sharpey's fibers (Sf). (**H**) Mid-diaphyseal cross-section of the tarsometatarsus Sc-tm-2. (**I**) Mid-diaphyseal cross-section of the tarsometatarsus Sc-tm-1. (**J**) Close-up of the outer cortex of Sc-tm-1. A LAG marks the transition between a "normal" fibrolamellar bone tissue and a well-vascularized bone layer that is most likely pathological (per. PB).

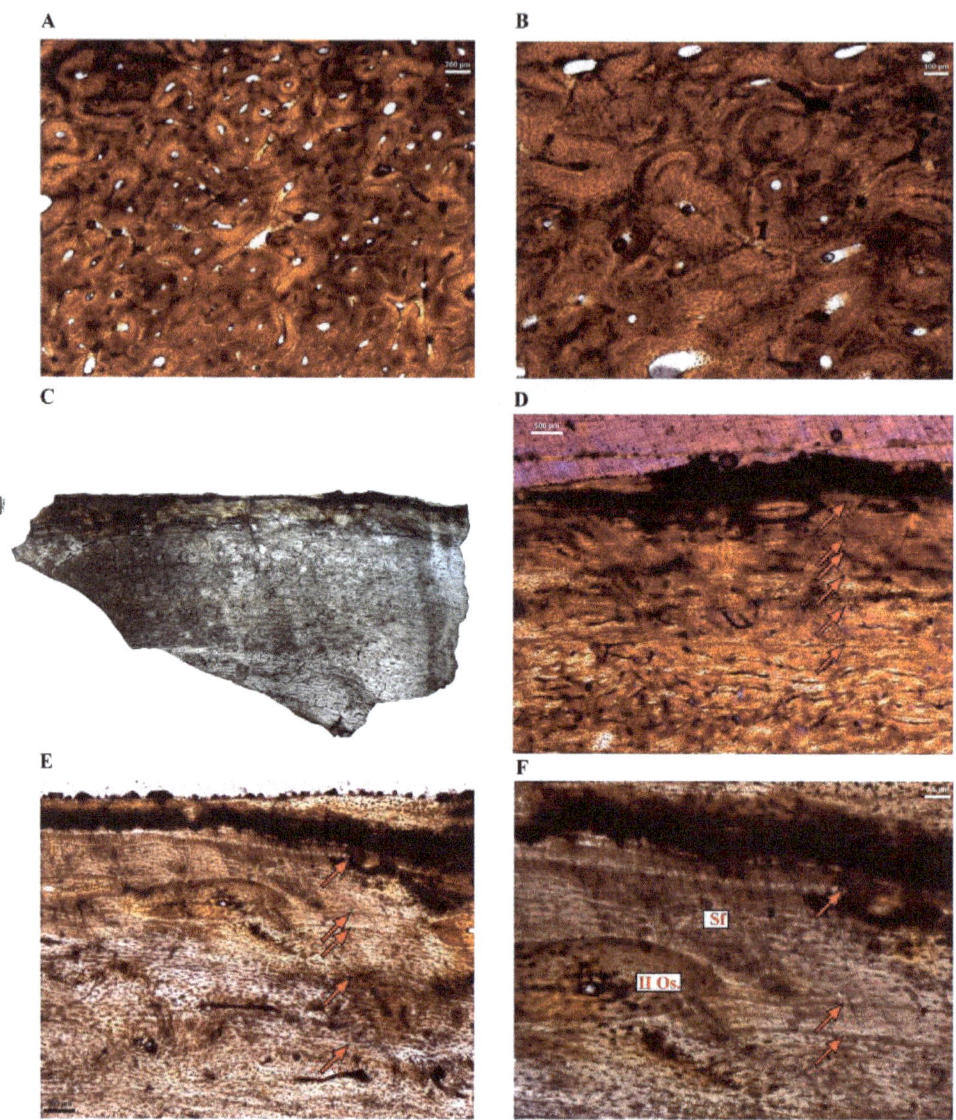

Figure 6. Femoral histology of the extinct terrestrial bird *Gastornis*. All pictures but D (polarized light) were taken in direct light. (**A,B**) Close-ups of the cortex in the femur Ga-fm-1. Haversian bone constitutes most of the cortex. The secondary osteons exhibit different orientations and sizes. (**C**) Part of cortical bone of the diaphysis of femur Ga-fm-2. The deepest cortex and the periosteal surface are missing. (**D–F**) Close-ups of the bone cortex of Ga-fm-2, primarily formed of a well-vascularized fibrolamellar bone tissue with a reticular to pseudo-laminar organization. At least six LAGs (arrows) were identified in this fragment. Secondary osteons (II Os.), as well as bundles of Sharpey's fibers (Sf) are scattered throughout the cortical bone fragment.

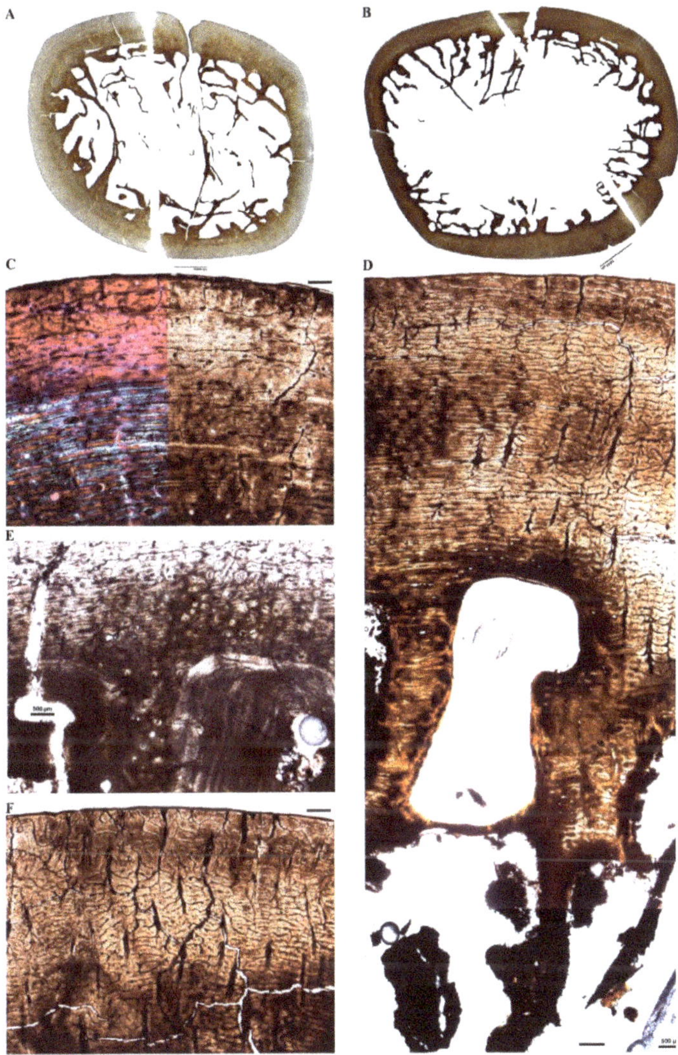

Figure 7. Femoral histology of Aepyornithidae. (**A**) Mid-diaphyseal cross-section of the femur Ae-fm-2. (**B**) Mid-diaphyseal cross-section of the femur Ae-fm-3. (**C**) Close-up of the cortex of Ae-fm-2 (left: in polarized light; right: in direct light). Most of the cortex is formed of a highly vascularized fibrolamellar bone tissue with a reticular or a laminar organization. Towards the periphery, a clear transition between the well-vascularized fibrolamellar bone tissue and a less vascularized bone interrupted by several LAGs is visible. (**D**) Close-up of the cortex of Ae-fm-3. Numerous bone trabeculae occupy part of the medullary cavity. They result from the deep and prolonged erosion of the deep cortical bone during growth. Most of the cortex is formed of a highly vascularized fibrolamellar bone tissue with a reticular or a laminar organization. Numerous radial anastomoses are also present. (**E**) Close-up of the perimedullary region in the cortex of Ae-fm-2. Haversian substitution is extensive in the deep cortex. Large erosion spaces, as well as the bone trabeculae, have been lined by sequential layers of endosteal lamellar bone. (**F**) Close-up of the outer cortex in Ae-fm-3, where a change in vascular orientation and density is observed.

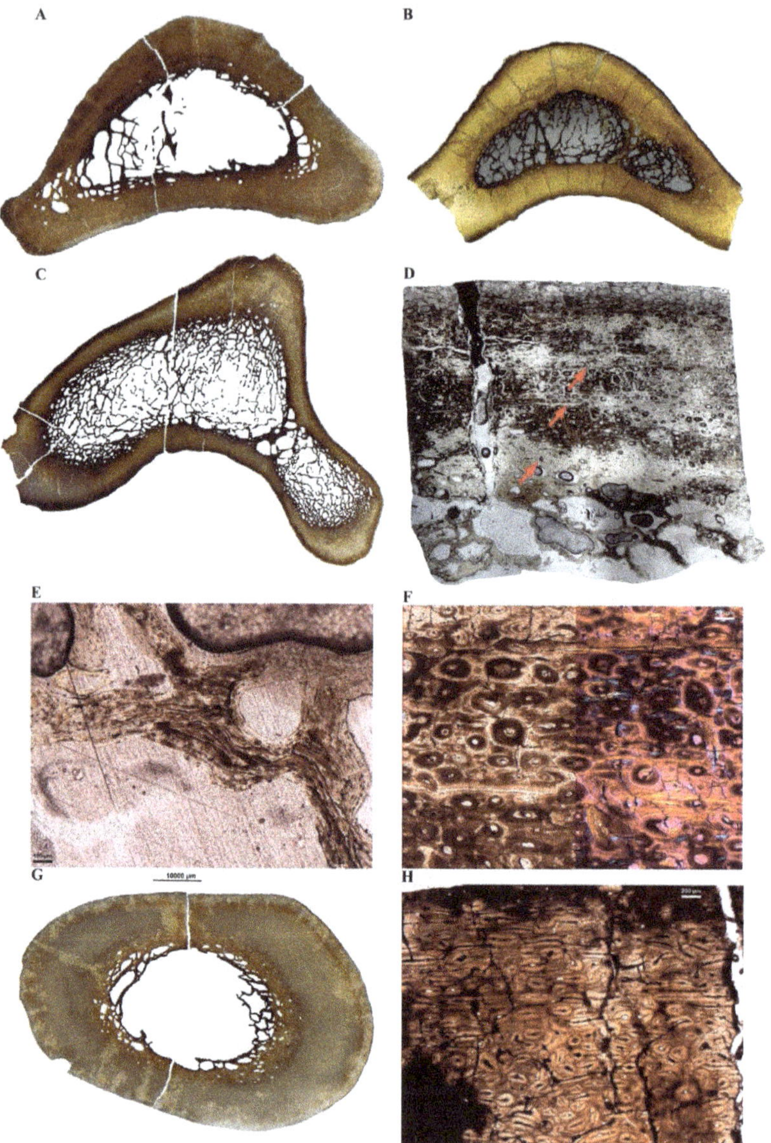

Figure 8. Long bone histology of Aepyornithidae. (**A**) Mid-diaphyseal cross-section of the tibiotarsus Ae-tb-1. (**B**) Proximal cross-section of the tibiotarsus Ae-tb-2. (**C**) Proximal cross-section of the tibiotarsus Ae-tb-3. (**D**) Close-up of the cortex of the tibiotarsus Ae-tb-5. Although Haversian bone is extensive, at least three LAGs (arrows) are visible in the cortex. (**E**) Close-up of the perimedullary region and the bone trabeculae in Ae-tb-5. Several events of desorption–redeposition at the surface of the bone trabeculae, resulted into an accumulation of crosscutting layers of endosteal lamellar bone. (**F**) Close-up of the cortex of the tibiotarsus Ae-tb-5 (direct light to the left; polarized light to the right). (**G**) Mid-diaphyseal cross-section of the tarsometatarsus Ae-tm-1. (**H**) Close-up of the heavily remodeled cortex of Ae-tm-1. Secondary osteons vary in size and orientation.

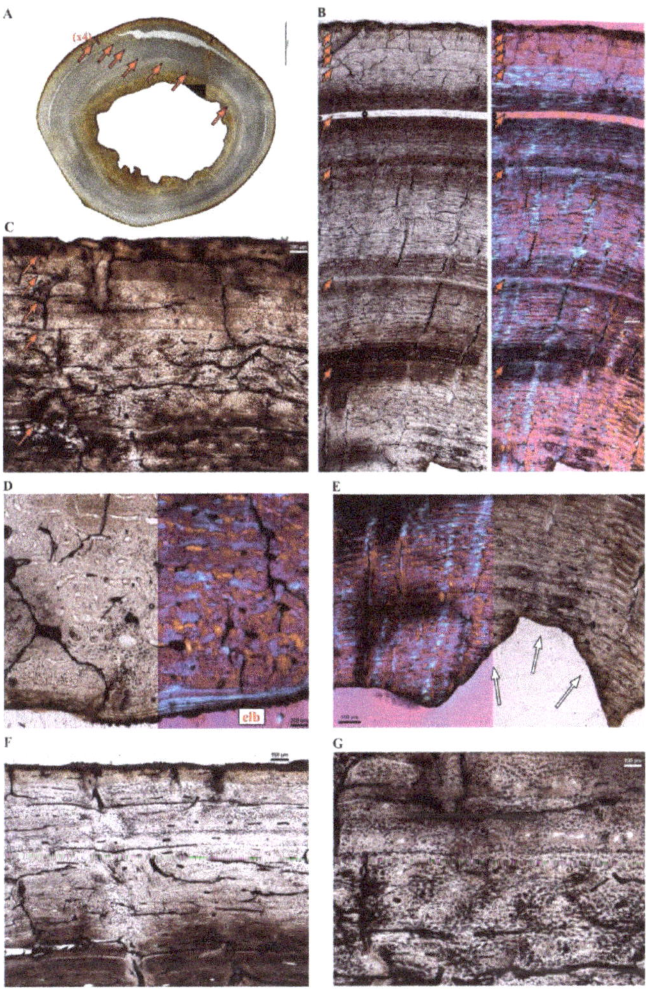

Figure 9. Femoral histology of the giant moa *Dinornis*. (**A**) Mid-diaphyseal cross-section of the femur Di-fm-1. At low magnification, at least seven growth marks (arrows) are visible throughout the cortical thickness. (**B**) Close-up of the cortex of Di-fm-1 [in direct (left) and polarized (right) light]. Most of the cortex consists of fibrolamellar bone tissue (mostly reticular in the deep cortex and more laminar in the mid-cortex; with numerous radial anastomoses). A very well preserved growth record is visible throughout the cortex. Here, five annuli associated to LAGs are visible from the deep cortex to two-third of the bone wall. (**C**) Close-up of the outer cortex of Di-fm-1. A LAG marks the transition to a progressive decrease in bone vascularization and a change in bone matrix (with mostly longitudinal primary osteons in the outermost cortex) and is followed by four closely and regularly spaced LAGs marking the onset of an OCL. (**D,E**) Close-ups of the deep cortex in Di-fm-1. The fibrolamellar bone tissue has a preferential reticular organization in the deepest cortex, with primary osteons that preserve large lumens (**D**). The endosteal margin is irregular because deep resorption lacunae (see arrows) have been formed in the deep cortex (**E**). In some parts, a thin layer of avascular endosteal lamellar bone marking the ICL borders the medullary cavity (**D**). (**F,G**) Close-ups of the cortex of Di-fm-1, showing the numerous bundles of short Sharpey's fibers (Sf).

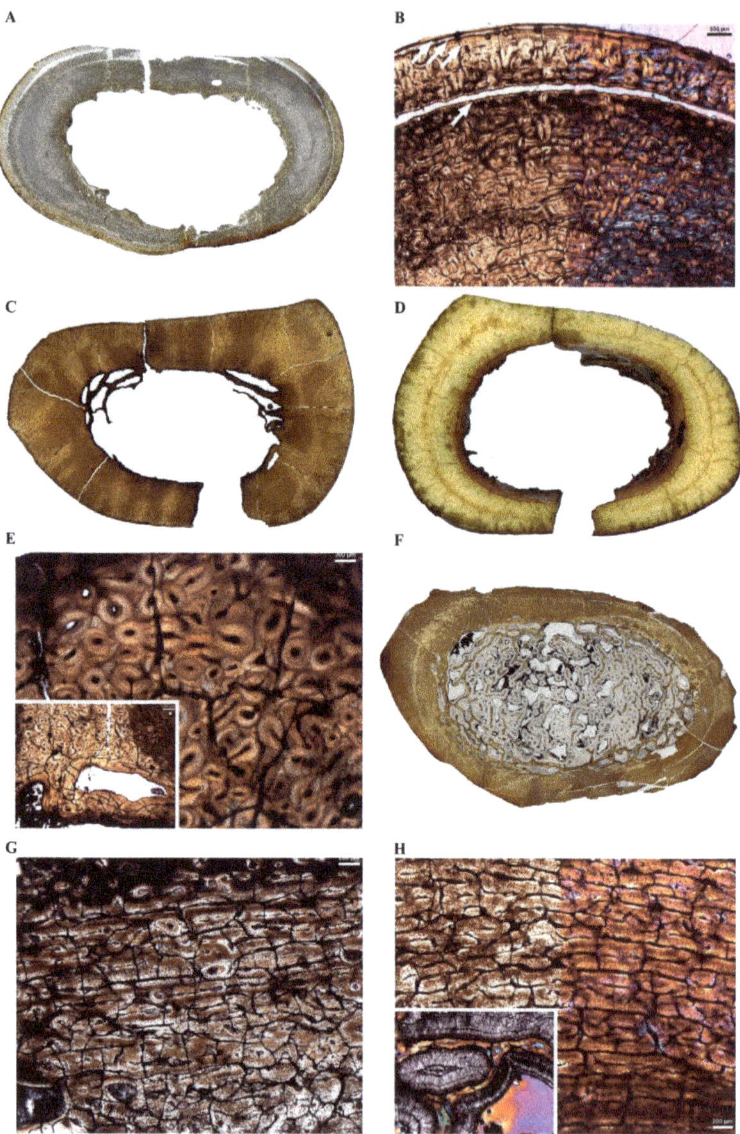

Figure 10. Long bone histology of Dinornithidae. (**A**) Mid-diaphyseal cross-section of the tibiotarsus Di-tb-1. (**B**) Close-up of the highly remodeled cortex of Di-tb-1 (normal light to the left; polarized light to the right). The fracture observed in the outer-third of the cortex follows a LAG (white arrow). At least 3 closely spaced LAGs are also visible in the outermost cortex, marking the onset of an OCL (white arrowheads). (**C**) Mid-diaphyseal cross-section of the tibiotarsus Di-tb-3. (**D**) Mid-diaphyseal cross-section of the tibiotarsus Di-tb-4. (**E**) Close-up of the cortex of Di-tb-3. Most of the cortex is formed of dense Haversian bone tissue. In the deep cortex, resorption cavities are lined up with endosteal lamellar bone (insert). (**F**) Mid-diaphyseal cross-section of the tarsometatarsus Di-tm-2. (**G**,**H**) Close-ups of the partly remodeled cortice of Di-tm-2 (normal light to the left; polarized light to the right). LAGs are visible in the cortex.

3. Results

3.1. Tarsometatarsi Proportions and Type of Locomotion

As indicated in Table 1, both Aepyornithid tarsometatarsi have Lg/Dm ratios of 5.26 and 5.46, and the ratio for the *Dinornis* tarsometatarsus equals 5.60. All extant cursorial birds sampled have a Lg/Dm of the tarsometatarsus >12; with values ranging from 12.5 for Cc-tm-2 of *Casuarius casuarius* to 21.6 for the Dn-tm-1 of *Dromaius novaehollandiae* (Table 1). It is worth noting that both sampled tarsometatarsi of *Casuarius casuarius* have the lowest ratios among cursorial birds.

3.2. Microanatomical Observations

Our results show that the global compactness of a given long bone varies between individuals of the same species. In the rhea, when a given element was sampled from two different-sized individuals (when available, Lg was used as a proxy for size; Table 1), the largest (and also oldest, based on bone histology; see below) specimen shows thicker bone walls and thus higher compactness values than the smallest one (e.g., specimens Ra-tb-1 and Ra-tb-2; Figure 3; Table 2). The reverse pattern was observed for the tarsometatarsi of the emu and the femora of the ostrich (e.g., specimens Sc-fm-1 and Sc-fm-2; Tables 1 and 2).

However, even though our sample size is small, our results suggest that there is a proximo-distal increase of bone compactness along the hindlimb of extant ratites (Table 2). Thus, when the average compactness values are considered, the femur is always less compact than the tibiotarsus, and the tibiotarsus less compact than the tarsometatarsus, in each extant ratite sampled (Table 2). This is not exactly the case in the sub-fossil material sampled. The average compactness of the femur (0.449) is lower than the compactness of the sampled tibiotarsus (0.785) and tarsometatarsus (0.706) in the Aepyornithidae, although the tibiotarsus has a slightly higher compactness value than the tarsometatarsus (Table 2). The same pattern is observed between the sampled femur (Comp. = 0.741) and tibiotarsus (Comp. = 0.733) of Dinornithidae. Overall, based on this preliminary study of the link between locomotion and bone microanatomy, it seems that graviportal birds (Aepyornithidae and Dinornithidae) have higher average compactness values than their extant cursorial relatives (Table 2). Finally, the compactness value of the femur (0.741) of the sampled Dinornithidae is higher than for any other individual considered in this study.

Regarding the presence of bone trabeculae in the mid-diaphyseal region, only the emu *Dromaius novaehollandiae* shows well-defined trabeculae in its femur and, to some extent, in its tarsometatarsus (Figure 3A,H). One femur of *Struthio camelus* presents structures in the perimedullary region that could correspond to small, non-pathological (probably broken) trabeculae (Figure 5A). In our sub-fossil sample, the femur of *Dinornis* (Di-fm-1) does not show any bone trabeculae in the medullary region (Figure 9A). Most sampled bones of Aepyornithidae show a network of thin bone trabeculae in the medullary region (Figures 7 and 8; some sections show little or no trabeculae, but this is an artifact, i.e., trabeculae broke during the sampling process and where observed during sampling).

3.3. Histological Descriptions

3.3.1. Extant Ratites

Casuarius casuarius (Southern cassowary–Figures 1 and 2)—Two femora of the Southern cassowary were sampled at the mid-diaphyseal level. Whereas femur Cc-fm-1 shows normal cortical tissues (Figure 1A), the contralateral element Cc-fm-2 presents unusual pathological tissues reminiscent of avian osteopetrosis [68] on both the endosteal margin and the periosteal surface (Figure 1B,F). The bone histology of Cc-fm-2 is thus abnormal, and this bone was not further considered for microanatomical comparisons.

The left femur Cc-fm-1 (Figure 1A) is 224 mm in length, and the maximal diameter of the mid-diaphyseal cross-section is 29 mm. A thin inner circumferential layer (ICL) of endosteally formed lamellar bone bordered the medullary cavity. The deep cortex is partly remodeled with several generations of longitudinal secondary osteons. However, most of the mid-cortex is formed of a primary laminar to the reticular fibrolamellar bone, with only sparse secondary osteons (Figure 1C,D), and there is a tendency towards a decrease in vascularization towards the periphery of the bone (Figure 1C,D), attesting of a reduction of depositional rate. From the mid-cortical region to the periosteal surface, four lines of arrested growth (LAGs) interrupt the well-vascularized fibrolamellar bone tissue (Figure 1C,D). The fourth LAG (closest to the periosteal surface) seems to mark the onset of the OCL (Figure 1D); a layer of avascular parallel fibered to lamellar bone interrupted by up to 6 closely spaced LAGs at the periphery of the bone (Figure 1D). Note that a region of the section is remodeled up to the periosteal surface, suggesting a zone of muscle or ligament insertion (Figure 1E). All these histological features suggest that this bone belonged to a somatically mature individual.

Two tibiotarsi (Cc-tb-1 and CC-tb-2) of relatively the same size and belonging to the same individual were sampled. Both skeletal elements present small amounts of endosteal and periosteal pathological tissues in parts of the sections (Figure 2A,B,D). In both elements, the cortex is mostly primary and composed of a well-vascularized laminar or reticular (depending on the region of the sections) fibro-lamellar bone (Figure 2B–D). In both elements, only a restricted region of the cortex is remodeled up to the surface, suggesting a zone of muscle insertion. Additionally, they both show several LAGs are visible in the fibro-lamellar cortex, attesting to an interrupted growth through ontogeny (Figure 2C,D). As in the femur Cc-fm-1, an OCL, with closely spaced LAGs (Figure 2C, arrowheads), is also visible in the outermost cortex of both elements.

Periosteally cc-tb-2 shows a distinctive change in the bone tissue just under the periosteal surface. The latter appears to be more richly vascularized than the tissue just preceding this deposition, and the vascular canals tend to be longitudinally organized, as opposed to the more laminar organization evident in the earlier-formed bone tissue (Figure 2B). These findings suggest that the most recently formed periosteal bone tissue and the endosteal tissue are pathological [68].

Two similar-sized tarsometatarsi belonging to different individuals were sampled (Table 1) and present similar histology. In both elements, the primary bone tissue shows extensive remodeling and in some parts, the cortex-dense Haversian bone tissue extends right up to the bone surface. In both elements, an unusual, thick layer of endosteal bone has been deposited along the perimedullary margin (Figure 2E,F).

Dromaius novaehollandiae (common emu–Figure 3)—Two different-sized femora that show similar bone histology (Dn-fm-1 and Dn-fm-2) were sampled for this species (Table 1). Remnants of slender bone trabeculae extend into the medullary cavity (Figure 3A), which is partially bordered by a thin layer of endosteal lamellar bone. In both elements, most of the cortex is composed of highly vascularized reticular fibrolamellar bone, with some patches of Haversian tissue. Both elements present a slow-down of the deposition rate towards the periphery, which signals the presence of an OCL composed of parallel-fibered to lamellar bone tissue interrupted by several and closely spaced LAGs (Figure 3B). In a localized region of the outermost cortex of Dn-fm-2, a band of vascularized and partly remodeled tissue is intercalated within the OCL, suggesting a localized increase in bone rate deposition when the animal had already reached skeletal maturity (Figure 3C). Numerous Sharpey's fibers running obliquely to the bone surface are visible throughout the primary bone in both sections (Figure 3B).

Two similar-sized tibiotarsi (Table 1) belonging to different individuals were sampled and exhibited different histologies. Dn-tb-1 presents an unusual radially organized bone tissue on its endosteal margin (Figure 3C,D), an uneven periosteal surface, as well as resorption cavities throughout the cortex (Figure 3C,D). These features affect its overall cortical thickness and this skeletal element was therefore not considered for microanatomical comparisons. Dn-tb-2 shows a more homogeneous cortical thickness (Figure 3E) and overall cortical histology (Figure 3F,G). Its cross-section was taken close to the initial center of ossification and thus encloses the nutrient canal (Figure 3E). The medullary cavity, free of bone trabeculae, is bordered by an ICL consisting of endosteally formed lamellar bone (Figure 3G). Apart from one highly remodeled, but the restricted region, most of the cortex is composed of a highly vascularized laminar to reticular fibrolamellar bone tissue (Figure 3F,G). Around most of the periosteal surface, three LAGs intercepting a thin layer of poorly vascularized parallel-fibered bone mark the onset of an OCL (Figure 3F). Finally, as observed in the femur Dn-fm-2, an extra layer of vascularized and partly remodeled bone tissue is visible in localized regions of the periosteal surface. This histology of this bone layer contrasts with the underlying primary cortical tissue and is intercalated between the first LAG of the OCL and the outermost periosteal layer (Figure 3G).

Two similar-sized tarsometatarsi were sampled for this species (Table 1) and show slightly different cortical thicknesses (Figure 3H,I). With a relatively thicker bone wall, Dn-tm-1 (Figure 3H) presents a higher number of resorption spaces and bone trabeculae in the perimedullary region than Dn-tm-2 (Figure 3I). In both elements, a thin layer of endosteal lamellar bone tissue is visible along the medullary margin and most of the cortex consists of a dense Haversian tissue that extends right up to the periosteal surface; however, some regions are less remodeled and a decrease in vascularization is visible in the outer cortex (Figure 3J).

Rhea americana (greater rhea–Figure 4)—Two different-sized femora (Table 1) were sampled and present different bone microstructures. The larger femur, Ra-fm-1 (Figure 4A), presents a much thicker bone wall than the smaller one (Figure 4B). A narrow layer of lamellar bone forms the ICL lines most of the medullary cavity in both elements (Figure 4C insert, 4D). In Ra-fm-1, a decrease in bone vascular density is visible towards the bone periphery and a thin layer of poorly vascularized parallel-fibered bone marks the onset of an OCL at the periosteal surface (Figure 4C). These observations suggest that growth in diameter had slowed down already when this animal died. On the contrary, Ra-fm-2 exhibits a very thin layer of endosteally formed lamellar bone (initiation of the ICL, Figure 4D), attesting that the expansion of the medullary cavity had just ceased. Also, no clear decrease in vascular density (that would indicate a decrease in bone deposition rate) is observed close to the periphery of this bone (Figure 4D). In both femora, most of the cortical bone is primary (except a zone of muscle attachment with Haversian tissue in Ra-fm-1; Figure 4A) and formed of a highly vascularized fibrolamellar bone. The deep cortex of Ra-fm-1 is composed of a reticular to poorly organized laminar bone that turns into a well-defined laminar bone tissue towards the periosteal margin (Figure 4C). Numerous radial anastomoses are also visible throughout the section (Figure 4C). In Ra-fm-2, the cortex is mostly formed of a reticular to poorly characterized laminar bone tissue (Figure 4D). No LAGs are visible in the cortex of both elements.

Two different-sized tibiotarsi were sampled (Table 1). The larger tibiotarsus Ra-tb-2 belongs to the same individual as femur Ra-fm-1; the same applies to Ra-tb-1 and Ra-fm-2 (Table 1). As for the femora, the tibiotarsi histologies are congruent with different ontogenetic stages. The larger tibiotarsus Ra-tb-2 possesses a relatively thicker bone wall (Figure 4E) than the smallest tibiotarsus Ra-tb-1 (Figure 4F). Ra-tb-2 also shows a well-developed layer of endosteal bone (ICL) that appears anisotropic in polarized light and contrasts with the rest of the more isotropic cortical bone (Figure 4G). Interestingly, this ICL is well vascularized (Figure 4G). The rest of the cortex is mostly formed of a highly vascularized fibrolamellar bone (the organization of the vascular canals is variable within the section); with few patches of Haversian bone tissue, especially in one region of muscle insertion (Figure 4E). A faint annulus, but no clear LAG, was observed in the mid-cortex of this bone. Bone vascular density decreases towards the periphery of this femur, and a distinct annulus marks the transition from well-vascularized fibrolamellar bone tissue to a poorly vascularized parallel-fibered bone layer marking the onset of an OCL. Ra-tb-1 presents a thinner ICL than Ra-tb-2. There is also less Haversian substitution in this specimen as compared to Ra-tb-2, and when present, it tends to be limited to the perimedullary region. As for its associated femur Ra-fm-2, several resorption spaces are present in the perimedullary region of Ra-tb-1 (Figure 4F). Finally, at least one clear LAG is visible in the outer-third of the cortex (Figure 4H). However, the bone deposited after the LAG is a highly vascularized fibrolamellar bone and vascular canals are still piercing the bone surface (Figure 4H). These observations confirm that this individual was still growing at the time of death.

Two different-sized tarsometatarsi, belonging to the same above-mentioned individuals, were sampled. The histology of both tarsometatarsi is congruent with the histology of the other sampled limb bones; however, Haversian substitution is more advanced and makes up a significant portion of the cortex in both elements (Figure 4I,J). Again, the largest tarsometatarsus Ra-tm-1 presents a relatively thicker cortex (Figure 4I) than the smaller one (Figure 4J). As in the tibiotarsus Ra-tb-2, the ICL is well vascularized. The cortex is stratified and presents a first LAG that marks the transition from a well-vascularized (and partly remodeled) plexiform to a more reticular (depending on the part of the section) bone tissue (Figure 4I,K). A second LAG occurs much closer to the bone surface (Figure 4K). In the smaller tarsometatarsus Ra-tm-2, two-thirds of the cortex comprises compacted coarse cancellous bone with still numerous resorption cavities. A narrow ICL, made of a layer of lamellar bone tissue, lines the medullary cavity. Finally, there is no obvious decrease in vascular bone density towards the bone periphery and numerous vascular canals pierce the periosteal surface. Again, these features confirm that this skeletal element was still growing in diameter at the time of the individual's death. Finally, it is important to note that the skeletal elements Ra-fm-1, Ra-tb-2 and Ra-tm-1 all belonged to the same individual (specimen 1920-116) and present congruent bone microstructures. The skeletal elements Ra-fm-2, Ra-tb-1 and Ra-tm-2 also present very similar bone microstructures (e.g., thinner bone walls than the larger specimen 1920-116, resorption cavities in the deep cortex, etc.) and belonged to a single individual.

Struthio camelus (common ostrich–Figure 5)—Two different-sized femora were sampled (Table 1) and present similar bone microstructure, although, the largest femur Sc-fm-2 (Figure 5A) has a thinner bone wall than the smaller femur Sc-fm-1, because of extensive resorption along the endosteal margin. While the smaller femur shows no bone trabeculae in the medullary region, the largest one has few short trabeculae (Figure 5A). Both elements show a discontinuous and thin ICL bordering the medullary cavity (Figure 5B). This endosteal bone layer is made of lamellar bone tissue that contains numerous osteocyte lacunae. Most of the cortex is primary and formed of a well-vascularized fibrolamellar bone with a reticular to laminar organization (depending on the region of the section), and few short radial anastomoses (Figure 5B). Aside from a narrow zone of Haversian tissue corresponding to an attachment site, only few isolated secondary osteons are visible in rest of the cortices. In both sections, there is a decrease in vascularization close to the bone surface, accompanied by a change in the refringence of the bone matrix which comprises one to two LAGs, which marks the onset of the OCL (Figure 5B).

Two tibiotarsi, belonging to 2 different individuals were sampled (Figure 5C,D). The smaller one, Sc-tb-2 (Figure 5D) is associated with the smaller femur Sc-fm-1. Both tibiotarsi have been sampled close to the neutral region of growth of the bone, as attests to the presence of the nutrient canal (Figure 5C–E). The larger tibiotarsus Sc-tb-1 presents unusual histological features. Its cortex is stratified (Figure 5E,F) into different types of bone tissue; i.e., from the medullary margin to the periosteal surface: (i) a thin lamellar bone tissue with some radially oriented vascular canals (ICL), (ii) a thick layer of compacted coarse cancellous bone with numerous small resorption cavities forming half to two-third of the cortex (depending on the region of the section); (iii) a layer of primary fibrolamellar bone with a reticular to laminar organization of the vascular canals; (iv) a clear LAG, close to the bone surface, marking a change in bone vascularization under the periosteal surface (Figure 5F). In some parts of the section, the outermost cortex consists of a thin layer reminiscent of pathological bone and shows numerous bundles of Sharpey's fibers (Figure 5G). The smaller tibiotarsus Sc-tb-2 is less remodeled and is mostly composed of a well-vascularized fibrolamellar bone with a reticular to laminar organization (Figure 5D). At least one clear LAG is visible in the mid-cortex and thus attests that the growth in diameter of this bone has been discontinuous (Figure 5D).

Two different-sized tarsometatarsi were sampled (Table 1) and showed distinct bone microstructures. The larger one, Sc-tm-1 (Figure 5I), presents a very similar bone microstructure to the larger tibiotarsus Sc-tb-1 (see above; Figure 5C) and most probably belongs to the same individual. In the outermost cortex, a clear LAG marks the transition between a "normal" fibrolamellar bone tissue and an outermost region of predominantly plexiform bone tissue that is likely pathological (Figure 5I,J). The primary bone tissue in the cortex appears to be reticular fibrolamellar bone. The endosteal region is highly resorptive, and in the perimedullary region, large erosion cavities are present. The smaller tarsometatarsus, Sc-tm-2 (Figure 5H), presents a clear ICL made of a poorly vascularized lamellar bone tissue. Most of the cortex is formed of a dense Haversian bone. A narrow layer of lamellar bone is visible in some parts of the outermost cortex that have not been completely remodeled. The histological features of Sc-tm-2 show that this bone belonged to an adult individual.

3.3.2. Extinct Terrestrial Birds

Gastornis sp. (Figure 6)—Five different limb bones, representing femora and tibiotarsi, have been sampled for this genus (Table 1). We were unfortunately allowed to only sample superficial fragments of the bone walls or take bone cores. Moreover, the preservation of the bone microstructure was poor in most of the specimens. The information gathered on *Gastornis* bone histology is thus limited and incomplete. Most elements show that the cortex underwent secondary reconstruction, and that LAGs were deposited. In Ga-fm-1, most of the cortex appears to be remodeled. Several generations of secondary osteons are visible, although dense Haversian proportions are not attained. The secondary osteons exhibit different orientations and sizes, and several preserve large lumens (Figure 6A,B). The endosteal margin is resorptive, but in some places, a narrow band of lamellar tissue forms the ICL. The bone core sampled from Ga-fm-2 consists of the periosteal surface of bone, and part of the cortical bone, although the deepest part of the cortex and the endosteal region are missing (Figure 6C). However, this bone fragment reveals that the mid-cortex of *Gastornis* femur had well-vascularized fibrolamellar bone with a reticular organization (Figure 6D). Six closely spaced LAGs were identified in this bone fragment (relatively close to the bone surface); the deepest one marks the transition from a well vascularized fibromellar bone to a less vascularized lamellar bone tissue that forms the OCL in which numerous Sharpey's fibres are visible (Figure 6D,E). Secondary osteons, as well as bundles of Sharpey's fibers, are scattered throughout this cortical fragment (Figure 6E). The bone wall of Ga-Fm-3B is slightly fractured, and the periosteal surface of the bone is not well preserved. Here too, the primary cortical bone is interrupted by several LAGs. The periosteal and endosteal part of the *Gastornis* tibiotarsus (Ga-tb-2) are diagenetically altered, but it is evident that the cortex is comprised of fibrolamellar bone with predominantly longitudinally oriented vascular canals. Deeper in the cortex, many secondary osteons are visible. A few growth marks are visible in the mid-cortical regions. The bone tissue of tibiotarsus Ga-tb-1B is better preserved. However, in this bone overlying the OCL, which comprises four to five LAGs, a richly vascularized periosteal reactive bone tissue is present (Figure 6F).

Aepyornithidae (the elephant birds, Figures 7 and 8)—Three femora, eight tibiotarsi and two tarsometatarsi were sampled from various Aepyornithidae specimens (Table 1). Since a comprehensive description of the bone tissues of these specimens is provided in Chinsamy et al. (2020), here, our focus is on the microanatomical structure of the bones.

Complete cross-sections from two different-sized femora (Ae-fm-2 and Ae-fm-3) belonging to two individuals of *A. maximus* (Figure 7) were obtained for histological analyses. Although the cortical thickness is variable within a section, the longer femur, Ae-fm-2 (Figure 7A) has a relatively thicker cortex and higher bone compactness than the smaller one (Figure 7B; Tables 1 and 2). Both sections contain numerous and slender bone trabeculae occupying part of the medullary cavity (Figure 7A,B). The cores of these trabeculae result from the deep and prolonged erosion of the deep cortical bone during growth. Thus, they are either formed of well-vascularized fibrolamellar bone tissue (similar to the mid-cortex; Figure 7D) or Haversian bone tissue (resulting from the remodeling of the deep cortex before being eroded and transformed into trabecular bone; Figure 7E). In the deep cortex of Ae-fm-2, Haversian substitution is more extensive than in Ae-fm-3. Also, in Ae-fm-2, the large erosion spaces present in the perimedullary region, as well as the bony trabeculae have been lined with sequential layers of endosteal lamellar bone (Figure 7E). In Ae-fm-3, this deposition of endosteal lamellar bone is much more limited (Figure 7D). In both sections, most of the cortex is formed of a highly vascularized fibrolamellar bone with a preferential reticular or laminar organization and with numerous radial anostomoses (Figure 7C,D,F). In the largest femur Ae-fm-2, a clear transition between a well-vascularized fibrolamellar bone tissue and a less vascularized bone interrupted by several LAGs is visible in the outercortex and testify of a slowdown of the growth in this individual (Figure 7C). On the contrary, only a change in vascular orientation and density is observed at the periphery of Ae-fm-3 (Figure 7F). All these histological features indicate that the largest femur belonged to an older, more mature individual than the smaller one (that was probably juvenile or sub-adult).

Complete cross-sections were made from 3 different tibiotarsi referred respectively to *Vorombe titan* (Ae-tb-1; Figure 8A) and Aepyornithidae indet. (Ae-tb-2, and Ae-tb-3; Figure 8C,D). These sections were sampled at different levels along the shafts of these skeletal elements, which explains part of the differences observed in cross-sectional geometry and cortical thickness. All three sections are heavily remodeled and the cortex mostly formed of a dense Haversian bone tissue. The three sections present numerous and slender bone trabeculae, which occupy most of the medullary cavity. The small amount of bone trabeculae in the mid-diaphyseal cross-section of Ae-tb-1 is an artefact (Figure 8A) since numerous, fragile trabeculae were broken during the sampling of the specimen. The process underlying the formation of these bone trabeculae is the same as in the femur, i.e., during growth, there is an intense and prolonged resorption of the deep cortex. The remains of the un-resorbed deep cortical bone become integrated into the core of the trabeculae. Several events of resorption–redeposition occurred along the surface of the trabeculae, resulting in an accumulation of cross-cutting layers of endosteal lamellar bone (Figure 8E). The bone core of Ae-tb-5 (Figure 8D) reveals that Haversian substitution was also extensive in this bone. Although Haversian bone is present up to the bone surface (with secondary osteons that vary in size and orientation; Figure 8D), several LAGs (at least 3; Figure 8D,F) are still visible in the cortex, and suggest that the growth in diameter of this bone was discontinuous.

Two tarsometatarsi were sampled and presented completely remodeled cortices, with secondary osteons varying in size and orientation (Figure 8H). Again, the diaphyseal cross-section of Ae-tm-1 (Figure 8G) presents a thick compact cortex and slender bone trabeculae occupying the medullary cavity (although most of the trabeculae have been broken during the processing of the bone for thin sectioning).

Dinornithidae (the moas, Figures 9 and 10)—One femur, three tibiotarsi and one tarsometatarsus were sampled for this clade (Table 1). Three elements were identified as belonging to the genus *Dinornis*.

The mid-diaphyseal cross-section of the *Dinornis* femur Di-fm-1, presents a thick bone wall and a medullary cavity free of bone trabeculae (Figure 9A). The endosteal margin is irregular because of extensive resorption of the deep cortex (Figure 9E). However, once the expansion of the medullary region was completed, a deposition of thin layers of endosteally formed lamellar bone occurred (Figure 9D). Most of the cortex consists of primary bone tissue, and a well-preserved growth record is visible throughout the cortex (Figure 9A–C). At least five narrow annuli coupled with LAGs (irregularly spaced) are visible from the deep cortex to two-thirds of the bone wall, and they alternate with a highly vascularized fibrolamellar bone tissue (Figure 9B). This fibrolamellar bone tissue has a reticular organization in the deepest part of the cortex, with primary osteons that preserve large lumens (Figure 9D), but becomes laminar in the mid-cortex. Numerous radial anastomoses are visible throughout the cortex (Figure 9B,E). In the external third of the cortex, a clear LAG marks the transition to a progressive decrease in bone vascularization.

The external layer of the cortex is poorly vascularized (with longitudinal primary osteons) and contains four closely and regularly spaced LAGs. Although the appositionial growth of this bone drastically decreased, it was still growing slowly in diameter as attests the outermost layer of bone deposited, which is still vascularized (Figure 9C). Finally, numerous bundles of short Sharpey fibers are visible throughout the cortex (Figure 9F,G).

All three tibiotarsi exhibit different states of preservation. Trabeculae in the medullary cavity are present in a part of the sections (Di-tb-3 and Di-tb-4, respectively Figure 10C,D), but broke during bone sampling. All sections present thick and highly remodeled cortices (Figure 10B,E). However, despite the intense remodeling, some LAGs are still visible in most elements (e.g., Figure 10B), which indicate that growth in diameter of these bones was discontinuous.

The sampled tarsometatarsus Di-tm-2 shows numerous and small bone trabeculae in the medullary cavity (Figure 10F). Unlike the tibiotarsi, the trabecular network has been well preserved during sampling of this element, because the medullary cavity was filled with diagenetic minerals. The cortex is partly remodeled (Figure 10G,H) and at least one LAG is visible in the mid-cortex.

4. Discussion

4.1. Limb Bone Proportions

A previous study by Angst and colleagues [9] has shown that the locomotion type of large terrestrial birds can be deciphered from simple linear measurements of the tarsometatarsus. According to these researchers, all extant cursorial ratites (c.f. [1]) have a tarsometatarsal length-width ratio >12, while all extinct giant terrestrial birds sampled had a length-width ratio <12, which was thus interpreted as indicative of a graviportal–slow walking locomotion (an inference also based on morphological correlates; [9]). These results further show that large extant ratites have relatively long and slender tarsometatarsi, as expected for cursorial animals [1,6–8], whereas giant extinct terrestrial birds, such as Aepyornithidae, some members of the Dinornithidae and Gastornithidae, have relatively shorter, stouter tarsometatarsi, which is an adaptation for biomechanical loading. In our current study, we found that the Southern cassowary *Casuarius casuarius*, which is considered to be the slowest of all living ratites [1], had the lowest tarsometatarsal length-width ratio for the extant species, which agrees with the earlier findings reported by Angst et al. [9].

4.2. Long Bone Microanatomy and Locomotor Patterns in Large Terrestrial Birds

4.2.1. Interskeletal Element Variability

When a skeletal element was sampled from different-sized individuals of the same species (and thus, most likely, individuals at different ontogenetic stages), we observed variability in terms of the compactness of the mid-diaphysis, with larger specimens generally presenting thicker bone walls. This observation reflects the changes in bone microanatomy and global compactness that accompany bone growth through ontogeny (e.g., [8,43]). Furthermore, our preliminary results show that, in general, among extant terrestrial birds, there is a proximo-distal gradient of bone compactness in the hindlimb with a progressive increase in bone compactness from the femur to the tarsometatarsus. However, in the sampled graviportal extinct birds, this pattern was not necessarily observed, although the femur was still the least compact of the limb bones in Aepyornithidae.

4.2.2. Differences between Cursorial Ratites and Extinct Terrestrial Birds

Although our sample size is small and would need to be increased in future studies, extinct terrestrial birds, such as the Aepyornithidae and some Dinornithidae, inferred as graviportal by previous authors using limb bone proportions [9,13], have a tendency to show overall higher compactness values in their hindlimb bones than most extant ratites sampled in this study (see Table 2). These higher global compactness values are the result of relatively thicker compact cortices in the limb bones of the sampled subfossil birds (such as, the *Dinornis* sp. femur Di-fm-1; Figure 9A) and/or the presence of a well-developed network of bone trabeculae along the shaft, especially in the hindlimb bones of Aepyornithidae (e.g., [33]) that is usually poorly developed (such as *Casuarius* and *Dromais*, Figures 1–3) or absent (such as in *Rhea*, Figure 4) in the mid-shaft of extant cursorial ratites (see also the ratite femoral cross-sections in Foote [41]: plate 5), and among modern birds in general (e.g., see bird femoral cross-sections in Foote [41]: plates 5–7). These microanatomical features are congruent with a graviportal mode of locomotion in the sampled Aepyornithidae and Dinornithidae [25]. A large body size correlates with a large body mass, which has consequences for the biomechanical adaptations of the hindlimb bones. Thus, our bone compactness results, together with the histological findings showing thick bone walls often with extensive development of bone trabeculae in the medullary cavity, concur with the expectations of a slow graviportal type of locomotion.

4.3. New Data on the Long Bone Histology of Extant Ratites

4.3.1. Interskeletal Element Variability

As previously documented in the hindlimb bones of the ostrich [28,46], but also in Aepyornithidae [33,47] and Dinornithidae [45], we found that Haversian substitution is limited in the femur but can be extensive in the tarsometatarsus of all extant ratites sampled (ostrich, emu, rhea and cassowary). The tibiotarsus presents an intermediate level of Haversian substitution. This pattern has also been described in *Apteryx* [35]. Table S1 (in Supplementary Materials) summarizes the bone histological observations made for all the taxa studied.

4.3.2. Growth Marks in the Cortex of Extant Ratites

Most modern birds (Neornithes) are considered to have a rapid, uninterrupted rate of bone deposition. They reach skeletal maturity in less than a year and, except for a few closely spaced LAGs that can be present in the OCL at the adult stage [36], they generally do not show any growth marks in the cortices of their limb bones [27,44,54,69]. A few exceptions to this pattern of rapid, uninterrupted growth among the Neornithes are known among island birds such as, *Apteryx* [35,70], and several large flightless extinct birds such as the New Zealand moa [45], the aepyornithids [33,47], as well as the dromornithids [15,48]. These findings are consistent with the hypothesis outlined by Starck and Chinsamy [54] and Chinsamy-Turan [31] that in the absence of selection pressures for rapid growth, birds will adopt a more flexible growth strategy of slower, episodic growth.

In the current study, we document that the limb bones of several large extant ratites present annuli or even LAGs that interrupt events of rapid bone deposition in their cortices. For example, the femur Cc-fm-1 and tibiotarsus Cc-tb-2 of *Casuarius casuarius* present several LAGs within the well-vascularized mid-cortex, i.e., clearly not part of the OCL (Figure 1D,G,H), and several skeletal elements of the greater rhea *Rhea americana* also exhibited growth marks in their cortices (Figure 3). Thus, growth marks appear to be more common than previously reported in ratites.

When Bourdon et al. [35] first observed the presence of discontinuous and periodic growth in the kiwi, they proposed that growth marks may have been absent in the last common ancestor of ratites and may therefore represent an apomorphy of the clade Apterygidae-Dinornithiformes; However, their hypothesis is poorly supported because the inter-relationships within Palaeognathae are still debated, and the clade Apterygidae-Dinornithiformes is not recognized in most recent ratite phylogenies [19,71,72]. Moreover, we also documented LAGs interrupting fibrolamellar bone in the cortices of the rhea and cassowary. We propose the alternative hypothesis that such flexible growth strategies may represent the plesiomorphic condition of Neornithes (and thus palaeognaths) and their extinct theropod ancestors [31,73].

Since growth marks have not been observed in modern ratites that were previously studied (e.g., a growth series of *Struthio* [28]), but appeared commonly in several of our birds, we further raise the possibility that perhaps living in France, outside of their natural environments, these birds would have been subjected to captivity, periodic cold spells and even snow, which may have resulted in physiological stress and the formation of LAGs. This would agree with our hypothesis that flexible growth strategies are inherent among the ratites.

4.3.3. Comments on OCL Formation and Ontogenetic Status in Birds

In the large comparative study of the OCL in extant birds, Ponton et al. [37] hypothesized that the poor development to complete absence of this structure in ratites noted by previous authors [28,29,46] might be related to the large size of these bird species. However, more recently Woodward et al. [74] reported the presence of an OCL in a specimen of *Struthio camelus*, and indeed we observed that several extant and extinct ratite specimens (rheas, cassowaries, ostriches, emus, and the giant moa) presented distinctive OCL, often accompanied by several closely spaced LAGs (e.g., Figures 1D, 2C, and 9C). In addition, Bourdon et al. [35] and Heck et al. [70] have described the presence of an OCL in *Apteryx*. Thus, it is apparent that OCLs are quite common in ratites, and we propose that perhaps the ratites examined in previous studies [28,29,46] were not from skeletally mature individuals.

Our observations suggest that all *Casuarius* skeletal elements studied were from somatically mature individuals since they show a distinctive OCL in the outermost cortex. Of the two *Rhea* femora studied, Ra-Fm-1 has an OCL indicating that it was from a fully-grown individual, whereas Ra-fm-2 appears to have been actively growing at the time of death and it was most likely a subadult or juvenile. The differences in bone microstructure observed between the two femora are congruent with the difference in the size of these elements (Table 1), which also suggests that they represent different ontogenetic stages.

4.4. New Data on the Long Bone Histology of Extinct Large Terrestrial Birds
4.4.1. Aepyornithidae

The hindlimb bone histology of Aepyornithidae has been previously described [26,33,47,53]. Of these, the Chinsamy et al. [33] study comprehensively sampled three aepyornitid taxa [60], i.e., *Aepyornis maximus*, *Aepyornis hildebrandti* and *Vorombe titan*, as well as some taxonomically unidentifiable juvenile Aepyornithiformes. All these studies showed that, like their small relative-the kiwi-these large ratites, also experienced protracted, episodic growth. Chinsamy et al. [33] proposed that the periodic interruptions in the bones of the aepyornithids were caused by seasonally variable growth rates mediated by environmental conditions.

4.4.2. Dinornithidae

The limb bone histology of the moas has been examined by previous researchers [26, 30,44,45,75]. In agreement with these earlier studies, we found that the cortical bone of the moas, when not completely remodeled (as in most tibiotarsi sampled) is made of a highly vascularized, fibrolamellar bone, regularly interrupted by LAGs, suggesting a discontinuous and prolonged growth strategy for this family [45]. Up to seven widely spaced growth cycles were observed throughout the cortex of the *Dinornis* femur Di-fm-1 (Figure 9A), followed by a drastic slow-down in growth and the deposition of a poorly vascularized bone tissue (marking the onset of an OCL), interrupted by four closely-spaced LAGs (Figure 9C).

Considering that the innermost growth cycles may have been resorbed with the expansion of the medullary cavity, our observations suggest that it took at least seven years for this individual to reach skeletal maturity and that it was at least 11 years-old when it died. Our observations show again that in the sampled Dinornithidae, the femur is less affected by Haversian substitution than the other limb bone elements. Indeed, all three tibiotarsi and the tarsometatarsus sampled present some degree of Haversian bone tissue (Figure 10), whereas the adult femur (Di-fm-1) does not show any secondary osteons in its cortex (Figure 9). These findings concur with those reported for the aepyornithids [33,47]. This may reflect differential biomechanical loading between these skeletal elements [76].

4.4.3. *Gastornis* sp.

The histology of a single tibia of *Diatryma* (now recognized as synonymous with *Gastornis*; [57]) has been briefly described in Ricqlès et al. [30]. These authors observed the presence of one LAG, in the outer cortex of this individual, although not associated with the OCL. In the present study, we sampled cortical fragments and bone cores from 5 skeletal elements (femora and tibiotarsi) belonging to Gastornithidae. We observed that, contrary to most other extinct large terrestrial birds studied (e.g., [33,48]), all femora presented at least partially remodeled cortices, with extensive Haversian bone tissue. These findings suggest some biomechanical differences between *gastornithids* and other large flightless birds. Furthermore, we counted up to six closely spaced LAGs in the outer cortex of femur Ga-fm-2, suggesting that this animal's growth already slowed down at the time of death and it was skeletally mature.

4.5. Pathologies Evident in Modern Samples

In the current study we observed several histological features that appeared to be the result of bone pathologies: (i) the avian osteopetrosis-like peripherally- and endosteally-formed tissues [66] in a femur of *Casuarius* (Figure 1B) and a tibiotarsus of *Dromaius* (Figure 3C,D); (ii) another tibiotarsus of *Dromaius* with periosteal deposits of a highly vascularized bone tissue that results in an uneven bone wall margin (Figure 3G); (iii) a tibiotarsus and a tarsometatarsus of *Struthio* (Figure 5E–G) with periosteal reactive tissues.

Except for the *Casuarius* and *Dromaius*, which have features reminiscent of osteopeterosis, we are uncertain about the etiology of the pathologies for the other bones. Unfortunately, the MNHN collection records do not give any provenance for the material, but it is more than likely that the modern birds were obtained from zoos. Most of the birds sampled in our study appear to be mature individuals, which agrees with the possibility that they were zoo animals (which generally tend to have long lives since they are in a protected environment without any threats, regular food, etc.), and perhaps they are therefore more prone to diseases of old age.

5. Conclusions

Our preliminary microanatomical observations tend to show that extinct terrestrial birds, inferred as graviportal based on limb proportions, have thicker cortices and/or more bony trabeculae in their hindlimb bone diaphyses, than extant cursorial groups. However, we note that our sample size is small and further investigations are warranted to ascertain the relationship between locomotion mode and the inner architecture of hindlimb bones in terrestrial birds.

In this study, we also documented for the first time the presence of growth marks (not associated with an OCL) in the cortices of several extant ratites. These observations support the hypotheses of Starck and Chinsamy [54] and Chinsamy-Turan [31] that episodic and therefore protracted growth can be present in birds living in an environments where the selection pressure for rapid growth within a single year is absent. Thus, although most modern birds grow rapidly and reach maturity within a few weeks to months, like *Apteryx* [35,70], several other extant ratites (current study), and several other large flightless birds [33,45,48] are also capable of growing in an interrupted manner and can take several years to reach skeletal maturity.

This study also documents for the first time, the presence of a distinctive OCL in several skeletally mature ratites, and raises the possibility that earlier studies may have examined immature individuals.

The high incidence of pathologies among the modern bird specimens studied here is attributed to the fact that these birds were probably long-lived, zoo specimens that may have been more susceptible to diseases.

Supplementary Materials: The following supporting information can be downloaded at: https://www.mdpi.com/article/10.3390/d14040298/s1, Table S1: Summary of histological features seen in different species studied. ICL, inner circumferential layer; OCL, outer circumferential layer; CCCB, compacted coarse cancellous bone; FLB, fibrolamellar bone; R-L = reticular to laminar FLB. Note that the absence of information in the table means that those features were not observed/not preserved.

Author Contributions: Conceptualization, A.C. (Aurore Canoville), A.C. (Anusuya Chinsamy) and D.A.; methodology, A.C. (Aurore Canoville) and D.A.; formal analysis, A.C. (Aurore Canoville), A.C. (Anusuya Chinsamy) and D.A.; resources, A.C. (Aurore Canoville), A.C. (Anusuya Chinsamy). and D.A.; writing—original draft preparation, A.C. (Aurore Canoville), A.C. (Anusuya Chinsamy) and D.A.; writing—review and editing, A.C. (Aurore Canoville), A.C. (Anusuya Chinsamy). and D.A. All authors have read and agreed to the published version of the manuscript.

Funding: This research was partly funded by The National Research Foundation (NRF, South Africa) grant number 466596.

Institutional Review Board Statement: Not applicable.

Data Availability Statement: All the data are provided within this manuscript.

Acknowledgments: We are grateful to Christine Lefèvre (The Comparative Anatomy Collections, Museum National d'Histoire Naturelle, Paris, France) and Ronan Allain (The Paleontological Collections, Museum National d'Histoire Naturelle, Paris, France) who facilitated our access to specimens used in this study. We would also like to acknowledge Eric Buffetaut (Centre National de la Recherche Scientifique, Ecole Normale Supérieure, Paris, France) for assistance during the sampling process. Vivian de Buffrénil and Vincent Rommevaux (Museum National d'Histoire Naturelle, Paris, France) are thanked for granting us access to the equipment necessary to treat, sample some of the skeletal elements and process some thin sections. We are also grateful to Silvia Kolomaznik for formatting the references.

Conflicts of Interest: The authors declare no conflict of interest.

References

1. Abourachid, A.; Renous, S. Bipedal locomotion in ratites (Paleognatiform): Examples of cursorial birds. *Ibis* **2000**, *142*, 538–549. [CrossRef]
2. Coombs, W.P. Theoretical aspects of cursorial adaptations in dinosaurs. *Q. Rev. Biol.* **1978**, *53*, 393–418. [CrossRef]
3. Stein, B.R.; Casinos, A. What is a cursorial mammal? *J. Zool.* **1997**, *242*, 185–192. [CrossRef]
4. Carrano, M.T. What, if anything, is a cursor? Categories versus continua for determining locomotor habit in mammals and dinosaurs. *J. Zool.* **1999**, *247*, 29–42. [CrossRef]
5. Storer, R. Adaptive radiation in birds. In *Biology and Comparative Physiology of Birds*; Academic Press: New York, NY, USA, 1960; Volume 1, pp. 15–55.
6. Alexander, R.; Maloiy, G.M.O.; Njau, R.; Jayes, A.S. Mechanics of running of the ostrich (*Struthio camelus*). *J. Zool.* **1979**, *187*, 169–178. [CrossRef]
7. Picasso, M.B.J. Postnatal ontogeny of the locomotor skeleton of a cursorial bird: Greater rhea. *J. Zool.* **2012**, *286*, 303–311. [CrossRef]
8. Gilbert, M.M.; Snively, E.; Cotton, J. The tarsometatarsus of the ostrich *Struthio camelus*: Anatomy, bone densities, and structural mechanics. *PLoS ONE* **2016**, *11*, e0149708. [CrossRef]
9. Angst, D.; Buffetaut, E.; Lecuyer, C.; Amiot, R. A new method for estimating locomotion type in large ground birds. *Palaeontology* **2015**, *59*, 217–223. [CrossRef]
10. Buffetaut, E.; Angst, D. The giant flightless bird *Gargantuavis philoinos* from the Late Cretaceous of southwestern Europe: A review. In *Cretaceous Period: Biotic Diversity and Biogeography*; Khosla, A., Lucas, S.G., Eds.; NMMNH&S Bulletin: Albuquerque, NM, USA, 2016; Volume 71, pp. 45–50.
11. Gregory, W.K. Notes on the principles of quadrupedal locomotion and on the mechanism of the limbs in hoofed animals. *Ann. N. Y. Acad. Sci.* **1912**, *22*, 267–294. [CrossRef]
12. Smith, J.M.; Savage, R.J.G. Some locomotory adaptations in mammals. *Zool. J. Linn. Soc.* **1956**, *42*, 603–622. [CrossRef]
13. Murray, P.F.; Vickers-Rich, P. *Magnificent Mihirungs: The Colossal Flightless Birds of the Australian Dreamtime (Life of the Past)*; Indiana University Press: Bloomington, IN, USA, 2004; 416p.
14. Buffetaut, E.; Angst, D. Stratigraphic distribution of large flightless birds in the Palaeogene of Europe and its palaeobiological and palaeogeographical implications. *Earth Sci. Rev.* **2014**, *138*, 394–408. [CrossRef]
15. Handley, W.D.; Chinsamy, A.; Yates, A.M.; Worthy, T.H. Sexual dimorphism in the late Miocene mihirung Dromornis stirtoni (Aves: Dromornithidae) from the Alcoota Local Fauna of central Australia. *J. Vertebr. Paleontol.* **2016**, *36*, e1180298. [CrossRef]

16. Worthy, T.H.; Handley, W.D.; Archer, M.; Hand, S.J. The extinct flightless mihirungs (Aves, Dromornithidae): Cranial anatomy, a new species, and assessment of Oligo-Miocene lineage diversity. *J. Vertebr. Paleontol.* **2016**, *36*, e1031345. [CrossRef]

17. Worthy, T.H.; Degrange, F.J.; Handley, W.D.; Lee, M.S. The evolution of giant flightless birds and novel phylogenetic relationships for extinct fowl (Aves, Galloanseres). *R. Soc. Open Sci.* **2017**, *4*, 170975. [CrossRef]

18. Chinsamy-Turan, A.; Worthy, T.H.; Handley, W. Growth strategies linked to prevailing environmental conditions in Australian giant flightless mihirung birds (Aves: Dromornithidae). *J. Vertebr. Palaeontol.* **2019**, 79.

19. Mitchell, K.J.; Llamas, B.; Soubrier, J.; Rawlence, N.J.; Worthy, T.H.; Wood, J.; Lee, M.S.Y.; Cooper, A. Ancient DNA reveals elephant birds and kiwi are sister taxa and clarifies ratite bird evolution. *Science* **2014**, *344*, 898–900. [CrossRef]

20. Martin, L.D. The status of the late Paleocene birds *Gastornis* and *Remiornis*. *Nat. Hist. Mus. Los Angeles Cty. Sci. Ser.* **1992**, *36*, 97–108.

21. Canoville, A.; De Buffrénil, V.; Laurin, M. Bone Microanatomy and Lifestyle in Tetrapods. In *Vertebrate Skeletal Histology and Paleohistology*; De Buffrénil, V., De Ricqlès, A.J., Zylberberg, L., Padian, K., Eds.; CRC Press: Boca Raton, FL, USA, 2021; pp. 724–743.

22. Currey, J.D.; Alexander, R. The thickness of the walls of tubular bones. *J. Zool.* **1985**, *206*, 453–468. [CrossRef]

23. Canoville, A.; Laurin, M. Evolution of humeral microanatomy and lifestyle in amniotes, and some comments on palaeobiological inferences. *Biol. J. Linn. Soc.* **2010**, *100*, 384–406. [CrossRef]

24. Quemeneur, S.; De Buffrénil, V.; Laurin, M. Microanatomy of the amniote femur and inference of lifestyle in limbed vertebrates. *Biol. J. Linn. Soc.* **2013**, *109*, 644–655. [CrossRef]

25. Houssaye, A.; Waskow, K.; Hayashi, S.; Cornette, R.; Lee, A.H.; Hutchinson, J.R. Biomechanical evolution of solid bones in large animals: A microanatomical investigation. *Biol. J. Linn. Soc.* **2016**, *117*, 350–371. [CrossRef]

26. Legendre, L.J.; Bourdon, E.; Scofield, R.P.; Tennyson, A.J.; Lamrous, H.; Ricqlès, A.; Cubo, J. Bone histology, phylogeny, and palaeognathous birds (Aves: Palaeognathae). *Biol. J. Linn. Soc.* **2014**, *112*, 688–700. [CrossRef]

27. Chinsamy, A.; Elzanowski, A. Bone histology: Evolution of growth pattern in birds. *Nature* **2001**, *412*, 402–403. [CrossRef] [PubMed]

28. Chinsamy, A. Histological perspectives on growth in the birds *Struthio camelius* and *Sagittarius serpentarius*. *Cour. Forsch.-Inst. Senckenberg* **1995**, *181*, 317–323.

29. Chinsamy, A.; Chiappe, L.M.; Dodson, P. Mesozoic avian bone microstructure: Physiological implications. *Paleobiology* **1995**, *21*, 561–574. [CrossRef]

30. De Ricqlès, A.; Padian, K.; Horner, J.R. The bone histology of basal birds in phylogenetic and ontogenetic perspectives. In *New Perspective on the Origin and Evolution of Birds, Proceedings of the International Symposium in Honor of John, H. Ostrom, New Haven, USA, 13–14 February 1999*; Gauthier, J., Gall, L.F., Eds.; Yale University Press: New Haven, CT, USA, 2001; pp. 411–426.

31. Chinsamy-Turan, A. *The Microstructure of Dinosaur Bone: Deciphering Biology with Fine-Scale Techniques*; Johns Hopkins University Press: Baltimore, MD, USA, 2005; 216p.

32. Chinsamy, A.; Buffetaut, E.; Canoville, A.; Angst, D. Insight into the growth dynamics and systematic affinities of the Late Cretaceous *Gargantuavis* from bone microstructure. *Naturwissenschaften* **2014**, *101*, 447–452. [CrossRef]

33. Chinsamy, A.; Angst, D.; Canoville, A.; Göhlich, U.B. Bone histology yields insights into the biology of the extinct elephant birds (Aepyornithidae) from Madagascar. *Biol. J. Linn. Soc.* **2020**, *130*, 268–295. [CrossRef]

34. Atterholt, J.; Poust, A.W.; Erickson, G.M.; O'Connor, J.K. Intraskeletal osteohistovariability reveals complex growth strategies in a Late Cretaceous enantiornithine. *Front. Earth Sci.* **2021**, *9*, 118. [CrossRef]

35. Bourdon, E.; Castanet, J.; De Ricqlès, A.; Scofield, P.; Tennyson, A.; Lamrous, H.; Cubo, J. Bone growth marks reveal protracted growth in New Zealand kiwi (Aves, Apterygidae). *Biol. Lett.* **2009**, *5*, 639–642. [CrossRef]

36. Van Soest, R.W.M.; Van Utrecht, W.L. The layered structure of bones of birds as a possible indication of age. *Bijdr Dierkd.* **1971**, *41*, 61–66. [CrossRef]

37. Ponton, F.; Elżanowski, A.; Castanet, J.; Chinsamy, A.; De Margerie, E.; De Ricqlès, A.; Cubo, J. Variation of the outer circumferential layer in the limb bones of birds. *Acta Ornithol.* **2004**, *39*, 137–140. [CrossRef]

38. Horner, J.R.; De Ricqlès, A.; Padian, K. Long bone histology of the hadrosaurid dinosaur *Maiasaura peeblesorum*: Growth dynamics and physiology based on an ontogenetic series of skeletal elements. *J. Vertebr. Paleontol.* **2000**, *20*, 115–129. [CrossRef]

39. Padian, K.; Lamm, E.T. *Bone Histology of Fossil Tetrapods: Advancing Methods, Analysis, and Interpretation*; University of California Press: Oakland, CA, USA, 2013; 298p.

40. Angst, D.; Chinsamy, A.; Steel, L.; Hume, J.P. Bone histology sheds new light on the ecology of the dodo (*Raphus cucullatus*, Aves, Columbiformes). *Sci. Rep.* **2017**, *7*, 1–10. [CrossRef] [PubMed]

41. Foote, J.S. *A Contribution to the Comparative Histology of the Femur*; Smithsonian Contributions to Knowledge: Washington, DC, USA, 1916; 242p.

42. Houde, P. Histological evidence for the systematic position of *Hesperornis* (Odontornithes: Hesperornithiformes). *Auk* **1987**, *104*, 125–129. [CrossRef]

43. Castanet, J.; Rogers, K.C.; Cubo, J.; Jacques-Boisard, J. Periosteal bone growth rates in extant ratites (ostriche and emu). Implications for assessing growth in dinosaurs. *Comptes Rendus de l'Académie des Sciences-Series III-Sci. de la Vie* **2000**, *323*, 543–550. [CrossRef]

44. Padian, K.; De Ricqlès, A.; Horner, J.R. Dinosaurian growth rates and bird origins. *Nature* **2001**, *412*, 405–408. [CrossRef]

45. Turvey, S.T.; Green, O.R.; Holdaway, R.N. Cortical growth marks reveal extended juvenile development in New Zealand moa. *Nature* **2005**, *435*, 940–943. [CrossRef]

46. Amprino, R.; Godina, G. La struttura delle ossa nei vertebrati: Ricerche comparative negli anfibi e negli amnioti. *Comment. Pont. Acad. Sei.* **1947**, *11*, 329–464.

47. De Ricqlès, A.; Bourdon, E.; Legendre, L.J.; Cubo, J. Preliminary assessment of bone histology in the extinct elephant bird *Aepyornis* (Aves, Palaeognathae) from Madagascar. *Comptes Rendus Palevol* **2016**, *15*, 205–216. [CrossRef]

48. Chinsamy, A.; Angst, D.; Canoville, A.; Göhlich, U. Bone histology and biology of the giant insular extinct bird, *Aepyornis maximus*. In *Abstract Book of the 9th International Meeting of the Society of Avian Paleontology and Evolution*; Palaeontological Association: Diamante, Argentina, 2016.

49. Chinsamy, A.; Worthy, T.H. Histovariability and Palaeobiological Implications of the Bone Histology of the Dromornithid, *Genyornis newtoni*. *Diversity* **2021**, *13*, 219. [CrossRef]

50. Buffetaut, E.; Angst, D. A femur of the Late Cretaceous giant bird *Gargantuavis* from Cruzy (southern France) and its systematic implications. *Palaeovertebrata* **2019**, *42*, e3. [CrossRef]

51. Buffetaut, E.; Angst, D. *Gargantuavis* is an insular basal ornithurine: A comment on Mayr et al., 2020, 'A well-preserved pelvis from the Maastrichtian of Romania suggests that the enigmatic *Gargantuavis* is neither an ornithurine bird nor an insular endemic'. *Cretac. Res.* **2020**, *112*, 104438. [CrossRef]

52. Mayr, G.; Codrea, V.; Solomon, A.; Bordeianu, M.; Smith, T. A well-preserved pelvis from the Maastrichtian of Romania suggests that the enigmatic *Gargantuavis* is neither an ornithurine bird nor an insular endemic. *Cretac. Res.* **2020**, *106*, 104271. [CrossRef]

53. Steel, L. Bone histology and skeletal pathology of two recently-extinct flightless pigeons: *Raphus cucullatus* and *Pezophaps solitarius*. *J. Vertebr. Paleontol.* **2009**, *29*, 185.

54. Starck, J.M.; Chinsamy, A. Bone microstructure and developmental plasticity in birds and other dinosaurs. *J. Morphol.* **2002**, *254*, 232–246. [CrossRef]

55. Buffetaut, E. The giant bird *Gastornis* in Asia: A revision of *Zhongyuanus xichuanensis* Hou, 1980, from the Early Eocene of China. *Paleontol. J.* **2013**, *47*, 1302–1307. [CrossRef]

56. Canoville, A.; de Buffrénil, V. Ontogenetic development and intraspecific variability of bone microstructure in the king penguin *Aptenodytes patagonicus*: Considerations for paleoecological inferences in Sphenisciformes. *Anat. Rec.* **2016**, *299*, 270.

57. Barrat, A. Quelques aspects de la biologie et de l'écologie du manchot royal (*Aptenodytes patagonicus*) des îles Crozet. *CNFRA* **1976**, *40*, 9–52.

58. Cherel, Y.; Le Maho, Y. Five months of fasting in king penguin chicks: Body mass loss and fuel metabolism. *Am. J. Physiol. Regul. Integr. Comp. Physiol.* **1985**, *249*, 387–392. [CrossRef]

59. Buffetaut, E. New remains of the giant bird *Gastornis* from the Upper Paleocene of the eastern Paris Basin and the relationships between *Gastornis* and *Diatryma*. *Neues Jahrbuch für Geologie und Paläontologie-Monatshefte* **1997**, *3*, 179–190. [CrossRef]

60. Hansford, J.P.; Turvey, S.T. Unexpected diversity within the extinct elephant birds (Aves: Aepyornithidae) and a new identity for the world's largest bird. *R. Soc. Open Sci.* **2018**, *5*, 181295. [CrossRef] [PubMed]

61. Stein, K.; Sander, M. Histological core drilling: A less destructive method for studying bone histology. In *Methods in Fossil Preparation: Proceedings of the First Annual Fossil Preparation and Collections Symposium*; Brown, M.A., Kane, J.F., Parker, W.G., Eds.; Petrified Forest: Petrified Forest National Park, AZ, USA, 2009; pp. 69–80.

62. Redelstorff, R.; Sander, P.M. Long and girdle bone histology of *Stegosaurus*: Implications for growth and life history. *J. Vertebr. Paleontol.* **2009**, *4*, 1087–1099. [CrossRef]

63. Redelstorff, R.; Hübner, T.R.; Chinsamy, A.; Sander, P.M. Bone histology of the stegosaur *Kentrosaurus aethiopicus* (Ornithischia: Thyreophora) from the Upper Jurassic of Tanzania. *Anat. Rec.* **2013**, *296*, 933–952. [CrossRef] [PubMed]

64. Canoville, A.; Chinsamy, A. Bone microstructure of pareiasaurs (Parareptilia) from the Karoo Basin, South Africa: Implications for growth strategies and lifestyle habits. *Anat. Rec.* **2017**, *300*, 1039–1066. [CrossRef] [PubMed]

65. Chinsamy, A.; Raath, M.A. Preparation of fossil bone for histological examination. *Palaeont. Afr.* **1992**, *29*, 39–44.

66. Francillon-Vieillot, H.; De Buffrénil, V.; Castanet, J.; Géraudie, J.; Meunier, F.J.; Sire, J.Y.; Zylberberg, L.; De Ricqlés, A. Skeletal biomineralization: Patterns, processes and evolutionary trends. In *Microstructure and Mineralization of Vertebrate Skeletal Tissues*; Van Nostrand Reinhold: New York, NY, USA, 1990; pp. 471–548.

67. Girondot, M.; Laurin, M. Bone profiler: A tool to quantify, model, and statistically compare bone-section compactness profiles. *J. Vertebr. Paleontol.* **2003**, *23*, 458–461. [CrossRef]

68. Chinsamy, A.; Tumarkin-Deratzian, A. Pathologic bone tissues in a turkey vulture and a nonavian dinosaur: Implications for interpreting endosteal bone and radial fibrolamellar bone in fossil dinosaurs. *Anat. Rec.* **2009**, *292*, 1478–1484. [CrossRef]

69. Watanabe, J. Ontogeny of surface texture of limb bones in modern aquatic birds and applicability of textural ageing. *Anat. Rec.* **2018**, *301*, 1026–1045. [CrossRef]

70. Heck, C.T.; Woodward, H.N. Intraskeletal bone growth patterns in the North Island Brown Kiwi (*Apteryx mantelli*): Growth mark discrepancy and implications for extinct taxa. *J. Anat.* **2021**, *239*, 1075–1095. [CrossRef]

71. Phillips, M.J.; Gibb, G.C.; Crimp, E.A.; Penny, D. Tinamous and moa flock together: Mitochondrial genome sequence analysis reveals independent losses of flight among ratites. *Syst. Biol.* **2010**, *59*, 90–107. [CrossRef]

72. Baker, A.J.; Haddrath, O.; McPherson, J.D.; Cloutier, A. Genomic support for a moa–tinamou clade and adaptive morphological convergence in flightless ratites. *Mol. Biol. Evol.* **2014**, *31*, 1686–1696. [CrossRef] [PubMed]

73. Chinsamy, A.; Marugán-Lobón, J.; Serrano, F.J.; Chiappe, L.M. Life history traits and biology of the basal pygostylian, *Confuciusornis sanctus*. *Anat. Rec.* **2020**, *303*, 949–962. [CrossRef] [PubMed]

74. Woodward, H.N.; Tremaine, K.; Williams, S.A.; Zanno, L.E.; Horner, J.R.; Myhrvold, N. Growing up *Tyrannosaurus rex*: Osteohistology refutes the pygmy "*Nanotyrannus*" and supports ontogenetic niche partitioning in juvenile *Tyrannosaurus*. *Sci. Adv.* **2020**, *6*, eaax6250. [CrossRef] [PubMed]
75. Enlow, D.H.; Brown, S.O. A comparative histological study of fossil and recent bone tissues. Part II. *Tex. J. Sci.* **1957**, *9*, 186–204.
76. McFarlin, S.C.; Terranova, C.J.; Zihlman, A.L.; Enlow, D.H.; Bromage, T.G. Regional variability in secondary remodeling within long bone cortices of catarrhine primates: The influence of bone growth history. *J. Anat.* **2008**, *213*, 308–324. [CrossRef] [PubMed]

diversity

Article

Contrasting Patterns of Sensory Adaptation in Living and Extinct Flightless Birds

Peter Johnston [1,*] and Kieren J. Mitchell [2]

[1] Department of Anatomy and Medical Imaging, University of Auckland, P.O. Box 92019, Auckland 1142, New Zealand
[2] Department of Zoology, University of Otago, 340 Great King Street, Dunedin 9054, New Zealand; kieren.mitchell@otago.ac.nz
* Correspondence: petersjohnston54@gmail.com

Abstract: Avian cranial anatomy is constrained by the competing (or complementary) requirements and costs of various facial, muscular, sensory, and central neural structures. However, these constraints may operate differently in flighted versus flightless birds. We investigated cranial sense organ morphology in four lineages of flightless birds: kiwi (*Apteryx*), the Kakapo (*Strigops habroptilus*), and the extinct moa (Dinornithiformes) from New Zealand; and the extinct elephant birds from Madagascar (Aepyornithidae). Scleral ring and eye measurements suggest that the Upland Moa (*Megalapteryx didinus*) was diurnal, while measurements for the Kakapo are consistent with nocturnality. Kiwi are olfactory specialists, though here we postulate that retronasal olfaction is the dominant olfactory route in this lineage. We suggest that the Upland Moa and aepyornithids were also olfactory specialists; the former additionally displaying prominent bill tip sensory organs implicated in mechanoreception. Finally, the relative size of the endosseous cochlear duct revealed that the Upland Moa had a well-developed hearing sensitivity range, while the sensitivity of the kiwi, Kakapo, and aepyornithids was diminished. Together, our results reveal contrasting sensory strategies among extant and extinct flightless birds. More detailed characterisation of sensory capacities and cranial anatomy in extant birds may refine our ability to make accurate inferences about the sensory capacities of fossil taxa.

Keywords: moa; *Aepyornis*; kiwi; kakapo; olfaction; vision; hearing

Citation: Johnston, P.; Mitchell, K.J. Contrasting Patterns of Sensory Adaptation in Living and Extinct Flightless Birds. *Diversity* **2021**, *13*, 538. https://doi.org/10.3390/d13110538

Academic Editors: Eric Buffetaut and Delphine Angst

Received: 1 October 2021
Accepted: 22 October 2021
Published: 26 October 2021

Publisher's Note: MDPI stays neutral with regard to jurisdictional claims in published maps and institutional affiliations.

1. Introduction

The concept of the complementary use of avian senses—and trade-offs among them—provides a framework for the inference of sensory function and sensory ecology in both living and extinct taxa [1]. Indeed, there has been increasing interest recently in 'avian palaeoneurology'—the inference of sensory and other function from the shape of the brain as a whole and from brain components related to specific functions [2], particularly for fossil taxa (e.g., [3,4]). However, the power of these inferences is limited by our incomplete understanding of the complex links between avian cranial anatomy and function. Of particular importance for improving our understanding are flightless birds, many of which appear to have evolved unusual combinations of sensory capacities. Detailed study of the sensory structures of flightless birds—both living and extinct—may therefore help to illuminate the full diversity of possible avian sensory patterns, and the trade-offs involved in their evolution.

Flightlessness has evolved multiple times in the ancestors of a number of bird species from New Zealand (NZ), including a large, flightless, nocturnal parrot—the Kakapo (*Strigops habroptilus*). In addition, two lineages of flightless palaeognathous birds evolved in New Zealand—the extant kiwi (*Apteryx*) and the recently extinct moa (Dinornithiformes). Molecular phylogenetic work has revealed that kiwi and moa are only distant relatives [5–8], with the kiwi being the sister group of the extinct elephant birds (Aepyornithidae) from

Madagascar, and moa the sister group of tinamous from South America (Figure 1A). Based on the phylogeny and distribution of these lineages, it is argued that the ancestors of the kiwi, moa, and elephant birds all independently evolved flightlessness (and any concomitant sensory specialisations). In this study, we explore the sensory anatomy of representatives from these four lineages of flightless avians—kiwi, the Kakapo, moa, and elephant birds—with a view to better characterising their patterns of sensory use and specialisation.

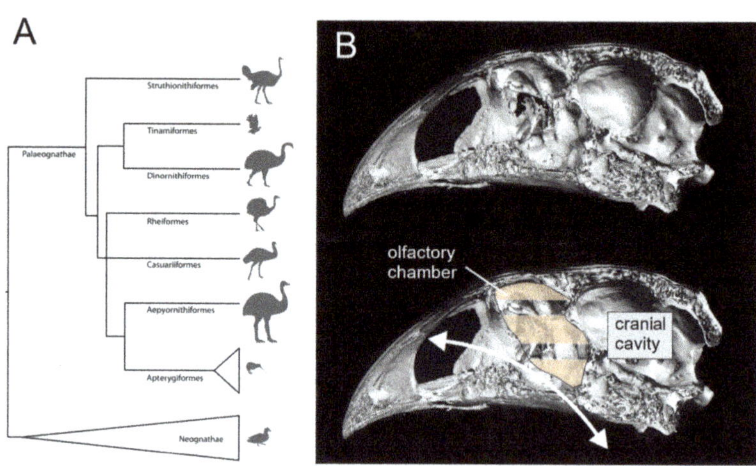

Figure 1. (**A**) phylogenetic tree indicating relationships of palaeognathous birds. (**B**) *Pachyornis elephantopus*, mid-sagittal skull reconstruction, showing nasal air passage (arrow), olfactory chamber, and cranial cavity.

Kakapo have previously been studied in detail in relation to their sensory capabilities and adaptations to nocturnality [9]: enhanced light sensitivity, reduced visual acuity, a wide binocular visual field, and reliance on olfaction. Similarly, the sensory capabilities of kiwi are well known, having a nocturnal lifestyle, very limited vision, enhanced olfaction, and mechanoreceptive function at the tip of the bill and vibrissae at the base of the bill [10–12]. The anatomy of the kiwi olfactory system has also been described in detail at the level of neurological morphology and fine structure of the olfactory mucosa [12], and of the air passages and complex caudal turbinate scroll [13], but the mechanics of airflow in the nasal passages is less clear. The external nares of kiwi are placed at the tip of the bill, unique among avians, and a complex turbinate system, similar to that of mammalian rodents, lies adjacent to the rostral end of the brain in a greatly expanded interorbital septum. The olfactory behaviour of kiwi is described as poking the bill into the substrate, olfactory search with a raised bill arcing around the direction of search [10], and frequent loud 'sniffing'. It is not clear, however, if sniffing, a typical mammalian behaviour, is possible in the low-pressure respiratory system of birds, and it is also unclear whether the very small external nares of kiwi are effective for air entry. In contrast to Kakapo and kiwi, very little is known about the biology of elephant birds [14]: small eyes were a feature, and inferences from endocasts suggest a reduced reliance on vision [4]; herbivorous diet and evolution in a predator-free environment are assumed.

The sensory capabilities of moa are also uncertain and remain debated. Moa eye size is relatively small, and the interorbital septum is expanded by a large chamber in continuity with the nasal air passage [15]—this has been regarded as an olfactory chamber since the earliest accounts of moa anatomy [16,17]—and the bony turbinate system is moderately elaborated [15]. In examinations of moa brain endocast morphology [18,19], however, the olfactory bulb at the rostral end of the forebrain was found to be within the range of extant palaeognaths (kiwi excepted), with olfactory bulb size being regarded as a surrogate for olfactory ability [20,21]. It was thus concluded that moa had an 'apparently poor advanced olfactory capacity', and suggested that the 'olfactory chamber' was a resonating chamber for vocal behaviour [18]. Indeed, it had earlier been suggested on the basis of the flora thought to have been eaten by moa that they had excellent visual acuity and poor olfaction [22]. In contrast, it was instead postulated that the olfactory lobe of the brain had been absorbed into the bulk of the telencephalon as a response to local space constraints, and was—in fact—much larger than the structure identifiable at the rostral end of the brain [23]. Subsequently, the substantial absorption of the olfactory lobe into the telencephalon in kiwi was confirmed [12], leaving open the possibility that the same is true for moa, while questions of relatively poor vision and nocturnality in moa have recently been raised on the basis of the very small optic lobes of the midbrain in two moa species [3,4]. The sensory ecology of moa thus remains unresolved.

Here we present new findings that add to existing evidence on sensory function in flightless bird taxa, focusing on olfaction, vision, and hearing, and consider how these senses may be integrated. We focus on these four taxa—moa, kiwi, elephant birds, and Kakapo as they form or are parts of island radiations on New Zealand and Madagascar, and we can make comparisons among them with the analytical methods and new data that we present. The data on moa concentrate on one species, the Upland Moa (*Megalapteryx didinus*), for which we have the most complete information. We also present new measurements from an elephant bird (*Aepyornis maximus*), Chilean Tinamou (*Nothoprocta perdicardia*), American Rhea (*Rhea americana*), Kakapo (*S. habroptilus*), and Southern Brown Kiwi (*Apteryx australis*).

2. Materials and Methods

2.1. Materials

A list of material examined is given in Appendix A.

2.2. Moa

Moa osteology material was examined with particular focus on incomplete and broken skulls that revealed the olfactory chamber and foramina between the cranial cavity and olfactory chamber. A head of *M. didinis* with attached mummified soft tissue (MNZ S400) [24] includes a complete scleral ring in several fragments, loosely held together for parts of the circumference by soft tissue, currently preserved in alcohol. This is one of three known moa specimens with intact scleral ossicles [25]. An MRI scan of this moa specimen was obtained as documented previously [26], and dimensions of the orbit obtained from this.

2.3. Kiwi

Skulls of kiwi were examined, again with focus on broken material that reveals the internal structure of the olfactory passages. CT and MRI scans of kiwi heads were used to reconstruct brain and olfactory passages. A serial sectioned hatchling kiwi head was used to define details of the nasal glands. Newly deceased kiwi obtained as road-kill were obtained under permit from the Department of Conservation, New Zealand, for dissection.

Computational Fluid Dynamics (CFD)

Simulations were performed using a 3D reconstruction of the nasal and olfactory passages in *Apteryx* (AMNH18456) from segmentation on axial CT slices for CFD simulation in Simscale [27]. The resulting mesh with its intricate internal geometry was too complex for simulations after multiple simplifications, so a mesh was built to the same external dimensions with Meshmixer [28] and openings made at appropriate sizes and positions of external and internal nares. Simulations were run with the following assumptions: incompressible medium, air in standard conditions as the fluid medium, Newtonian viscosity model, and standard meshing algorithm. Pressure differences between external and internal nares were set at 100 Pascals (Pa), within the physiological pressure range for quiet breathing in avians [29]. Kiwi respiratory rate is about 30 breaths per minute [30], so simulations of orthonasal and retronasal flow were captured at 10 s intervals up to 40 s. Other parameters for fluid dynamics including the Reynolds number are calculated by the software on the basis of the size of the model and the input and output values (pressures in this case). Simulation results were visualised with the particle trace algorithm of Simscale.

2.4. Comparative Material

Skulls of all palaeognaths apart from tinamous were examined, together with a range of neognaths. Cranial CT scans of all genera of ratites and three genera of tinamous were surveyed, and about 400 CT scans of neognaths available at online resources [31–34] were reviewed with focus on the interorbital septum and olfactory nerve passages. Kakapo morphology was studied on 3D skull mesh reconstructions and high-resolution CT scan images. Neognaths with known superior olfactory capability [13,20] were specifically sought in these collections, in particular vultures and Procellariiformes.

2.5. Observation of Living Kiwi

Stewart Island Brown Kiwi (Southern Tokoeka) *Apteryx australis lawryi* were observed foraging at Mason Bay, Stewart Island Rakiura, New Zealand. Kiwi in this locality forage by day as well as the typical nocturnal activity.

2.6. Compliance Statement

Use of all specimens conformed to NZ Department of Conservation regulations and permits, and to CITES conventions.

2.7. Analyses

2.7.1. Vision

Visual fields in *M. didinus* were estimated in Blender [35] by loading a mesh of the skull from a 3D CT reconstruction. A virtual 360° light was placed in the expected position of the cornea, and the emitted light projected onto a spherical mesh.

Inference of daily activity pattern (nocturnal versus non-nocturnal) was estimated with measurements of scleral ring and orbit dimensions. An 'optic ratio' of (internal scleral ring diameter)^2/(optic length x external ring diameter) was plotted against the geometric mean of these three measurements according to the method of [36] and added to the data of that study, for a total of 370 taxa, to place birds in nocturnal and diurnal bands (Table S1). A 'flexible phylogenetic discriminant analysis' (fPDA) [37] to derive a posterior probability of nocturnality was done using the method and R codes of [36]. An earlier version of this analysis [37] which adds axial length of the eye to the formula: (log lens diameter)^2 plotted against log (external scleral ring diameter x axial eye length) was also calculated. These analyses were performed for *M. didinus* and *S. habroptilus*.

2.7.2. Hearing

The length of the endosseous cochlear duct (lagena) (ECD) of the inner ear was measured on inner ear labyrinths reconstructed from CT series in *M. didinus*, *Aepyornis maximus*, *Rhea americana* and *Nothoprocta pericardia*. Measurements were made in 3D in Amira, with the axis of measurement lying within the structure. Maximal cranial height over the basisphenoid was measured also. Data were added to the data set of [36] (Table S2) and used for regression of (log10 ECD length) against (log10 braincase height) using the method and R codes of [36]. A phylogenetic regression was performed using the gls function of nlme [38] allowing λ to be fitted using the corPagel function of ape [39] using R version v4.0.1 in RStudio v1.4.1717. Residuals from this regression were used as an index of hearing ability, as verified previously [40] and were plotted against phylogeny.

2.8. Phylogenetic Trees

Estimating λ requires a phylogeny to correct for correlation due to shared ancestry. Most previous studies, including [36] have used supertrees derived from studies by [41] or [42]. However, these trees are unsuitable for our purposes because they all support the reciprocal monophyly of Tinamiformes and the other palaeognaths—which has since been definitively rejected (e.g., [7,8]). They also do not include representatives of the extinct Aepyornithiformes or Dinornithiformes. Since our expanded ECD matrix includes the moa *M. didinus*, elephant bird *Ae. maximus*, and tinamou N. *pericardia*, we needed updated phylogenetic reconstructions that accurately reflected the relationships of these taxa in order to estimated λ.

We constructed phylogenetic trees based on an alignment of published mitochondrial DNA sequences (Table S3). Where available, we downloaded complete mitochondrial genome sequences for the 92 species represented in the ECD matrix. Where mitochondrial genome sequences were not available we either used those of a close relative and/or sequences for only a subset of mitochondrial genes (Table S3). These sequences were aligned using the MUSCLE v3.8.425 [43] algorithm as implemented in Geneious v9.1.6 [44]. We then extracted—where available—the first and second codon positions of the 12 mitochondrial protein coding genes encoded on the leading strand for downstream analysis; third codon positions were excluded to avoid branch compression and other artifacts caused by substitution saturation.

Time-scaled phylogenetic trees were constructed using BEAST v1.8.4 [45]. First and second codon positions were analysed as separate partitions using substitution models—GTR + I + G and TVM + I + G, respectively—determined using ModelFinder as implemented in IQ-TREE v1.6.11 [46,47]. The monophyly of several higher taxa was enforced to match the phylogenomic results published by [48] (see Table S1). Following [6] we calibrated our phylogeny by constraining the age of six nodes: the common ancestor of Neoaves (uniform distribution between 66.5 Ma and 124.1 Ma), the common ancestor of Galloanseres (uniform distribution between 66.5 MA and 83.8 Ma), the divergence of Psittaciformes (uniform distribution between 53.5 Ma and 72.3 Ma), the divergence of Procellariformes (uniform distribution between 60.5 Ma and 72.3 Ma), the common ancestor of all non-ostrich palaeognaths (uniform distribution between 56.0 Ma and 72.3 Ma), and the divergence of *Casuarius* (uniform distribution between 24.5 Ma and 72.3 Ma).

We ran three separate BEAST analyses that were identical except that in each analysis we constrained the relationships between three palaeognath lineages—*Rhea*, *Casuarius*, and a clade comprising *Apteryx* and *Aepyornis*—to match one of the three possible topologies: *Rhea* & *Casuarius* as sister taxa, *Casuarius* as the sister-taxon to *Apteryx* + *Aepyornis*, and *Rhea* as the sister-taxon to *Apteryx* + *Aepyornis*. We did this to test the sensitivity of our downstream results to the order of these divergences, which are not well resolved [49]. For each of these three analyses we performed seven separate Markov Chain Monte Carlo (MCMC) runs. All BEAST analyses used a single lognormal relaxed clock model (with rate multipliers for the two partitions) and a birth-death tree prior. Each MCMC was run for 20,000,000 generations, sampling parameter values every 2000. The first 10% of each chain (1000 samples) was discarded as burn-in, and the remaining 63,000 samples for each analysis were combined. Convergence of parameter values and ESSs > 200 were monitored using Tracer v1.7.1 [50]. From each analysis we randomly selected a sample of 100 representative trees to use for estimating λ. We also estimated λ using a combined sample of all 300 trees, thus averaging across the uncertainty around the branching order among *Rhea*, *Casuarius*, and the clade comprising *Apteryx* and *Aepyornis*.

3. Results

3.1. Olfaction

3.1.1. Moa

The large olfactory chamber and its relation to the cranial cavity and the airway in *Pachyornis elephantopus* is demonstrated in Figure 1B. A complex olfactory nerve exit from the cranium is found in all moa, no adult birds having a single nerve trunk. Multiple nerve branches, as many as 15, penetrate the bony interface between cranium and olfactory chamber, resembling the cribriform plate of mammals; there is variation among individuals and between right and left sides, and typically a small bony mound is seen on the olfactory chamber side of this plate with multiple foramina penetrating and surrounding it (Figure 2C,D). The inner walls of the posterior part of the chamber are grooved by nerve branches radiating from the foramina (Figure 2C,D). This can be seen also in CT scans, and a reconstruction of the proximal nerve branching is shown in *Dinornis robustus* in Figure 2E. In juvenile moa a separate central bone element is found that represents the most caudal part of the olfactory chamber [15] (Figure 2A,B). This an open box-like structure that is entered by large olfactory nerve foramina caudally, has a median septum, and opens rostrally to the rest of the olfactory chamber.

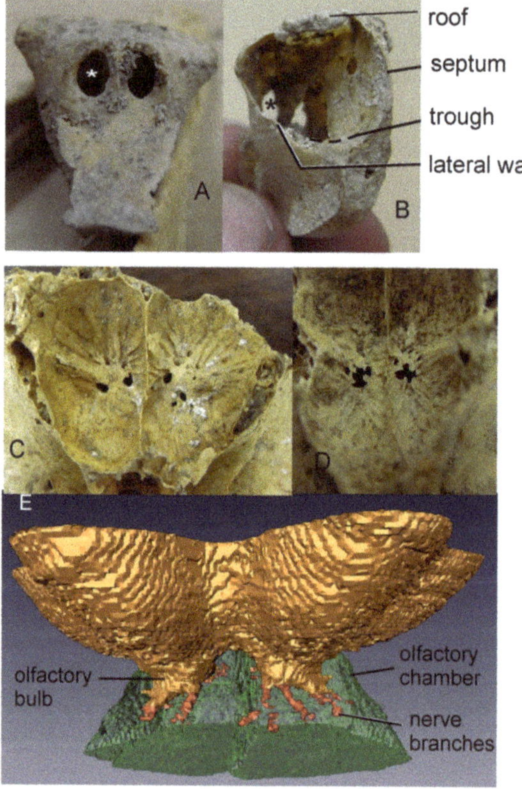

Figure 2. Moa olfactory foramina. (**A**) juvenile *Anomalopteryx didiformis*, ethmoid ossification, caudal view. (**B**) right rostrolateral view. * marks the foramen (**C**) *A. didiformis*, rostral view of foramina with radiating nerve grooves; (**D**) *Dinornis novaezealandiae*, same view. (**E**) *Dinornis robustus*, reconstruction of interface brain-olfactory chamber dorsal view.

3.1.2. Kiwi

The external nares in specimens with an intact rhamphotheca admit a wire of 0.6 mm diameter (Figure 3A), giving a cross-section area of 0.28 mm^2. The internal nares Figure 3B are formed by longitudinal ellipses of area 9.42 mm^2 in adult birds. The complex turbinate system and its relation to the forebrain are shown in Figure 4.

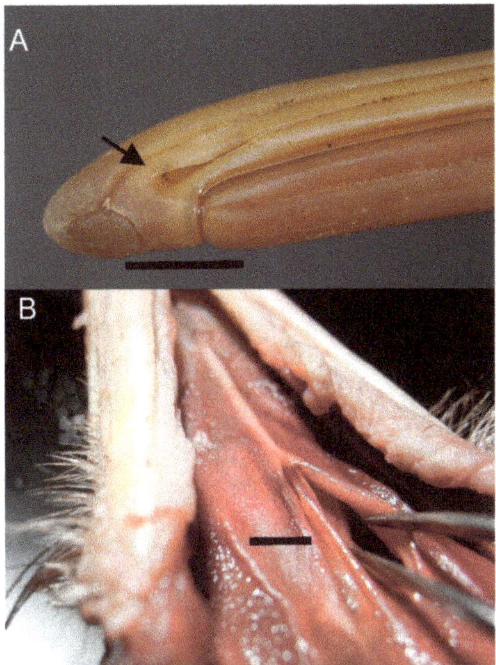

Figure 3. *Apteryx mantelli.* (**A**) external nares (arrow) (**B**) internal nares (in forceps). Scale bars = 1cm.

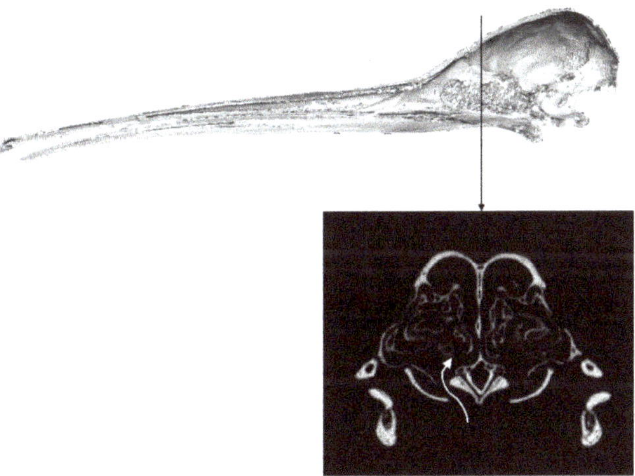

Figure 4. *Apteryx*, CT scan slice at the level of internal nares and turbinate system. The arrow indicates the route through the nares to the olfactory region.

CFD simulations revealed very little flow into the olfactory airways with orthonasal flow under the physiological pressures used. Simulation of retronasal olfaction showed airflow to around the olfactory chamber and nasal airway at all 10 s iterations (Figure 5). With the simulation process used here, when air velocity is very low or zero a particle trace is not generated. With retronasal aeration, vectors of flow are consistent with a retronasal route as the dominant mode of olfaction in kiwi.

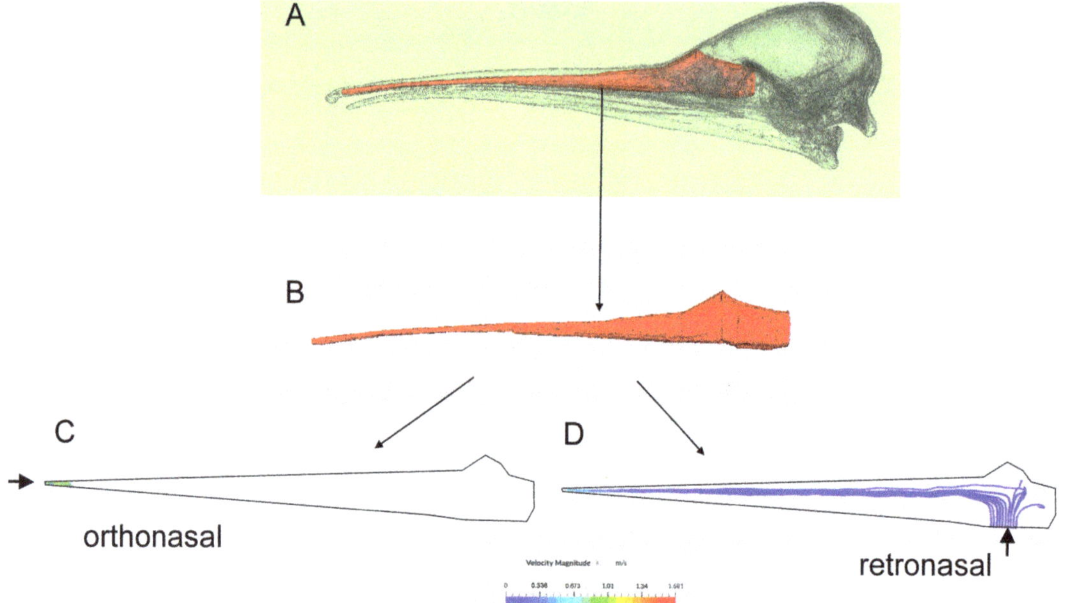

Figure 5. Computational flow dynamics; nasal cavity (**A**) in situ (**B**) isolated (**C**) and (**D**), flow simulations.

Wild kiwi were observed to make snuffling or snorting respiratory noises while foraging, accompanied by head movements as previously described [10]; it was not determined whether these were inspiratory or expiratory noises except on occasions when bubbles appeared at the nares, which was clearly an expiratory phenomenon.

In histological sections the nasal glands of *Apteryx* are situated in a conventional position on the dorso-lateral aspect of the maxilla, immediately rostral to the orbit. Medial and lateral ducts lead forward on either side of the conchal system to reach the vestibule of the nasal passage immediately adjacent to the external naris. Uncommonly for birds [51], the right and left medial ducts converge to form a single duct at the ventral edge of the septum for the rostral two-thirds of the bill, before dividing again close to the naris. The large lacrimal gland duct passes rostrally parallel and ventral to the lateral nasal gland duct, again opening into the vestibule adjacent to that duct. Thus, all three ducts enter adjacent to the naris.

3.1.3. Aepyornithidae

In *Aepyornis maximus* and *Ae. hildebrandti* the forebrain and the olfactory chamber occupy the dorsal third of the space between the eyes; the extent of the chamber is indicated in Figure 6A. The caudal end of the chamber is grooved by radiating olfactory nerve branches in a pattern similar to that of moa (Figure 6B).

3.2. Vision

3.2.1. Moa

The analysis of eye dimensions in *M. didinus* using the method of [36] place it within the band of non-nocturnal birds (Figure 7A). The posterior probability of non-nocturnal state calculated by the fPDA is above 99% with all values of lambda (an optimal lambda of 0.07 was obtained). In the predictive plot of [37], *M. didinus* falls among cathemeral birds (Figure 7B). Visual fields estimated for *M. didinus* are shown in Figure 8. They reveal a small binocular field, and a large blind sector caudally.

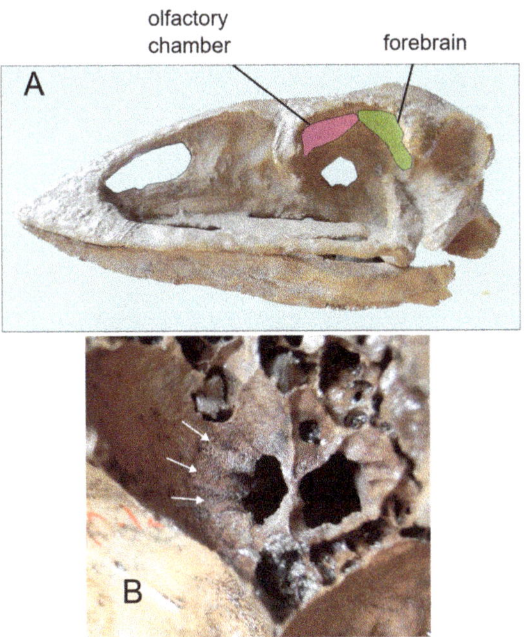

Figure 6. (**A**) *Aepyornis ?hildebrandti*, MNHN MAD6724 olfactory chamber and forebrain outlined (**B**) *Aepyornis maximus* MNHN 1910.12, rostral view of olfactory formaina with radiating nerve grooves (arrows).

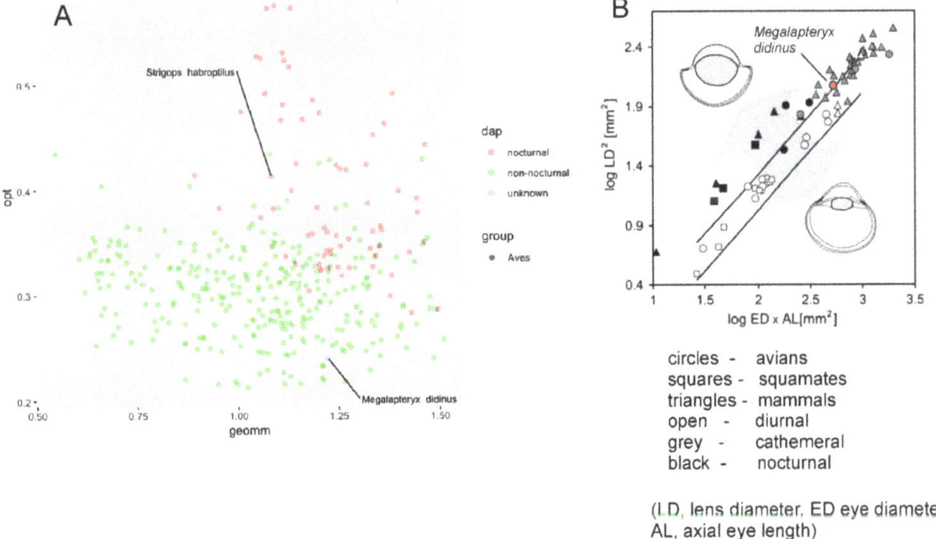

Figure 7. (**A**) plot of optic ratio (opt) against geometric mean of eye measurements (geomm). dap,daily activity pattern. (**B**) plot reproduced with permission from [37], with *Megalapteryx didinus* added.

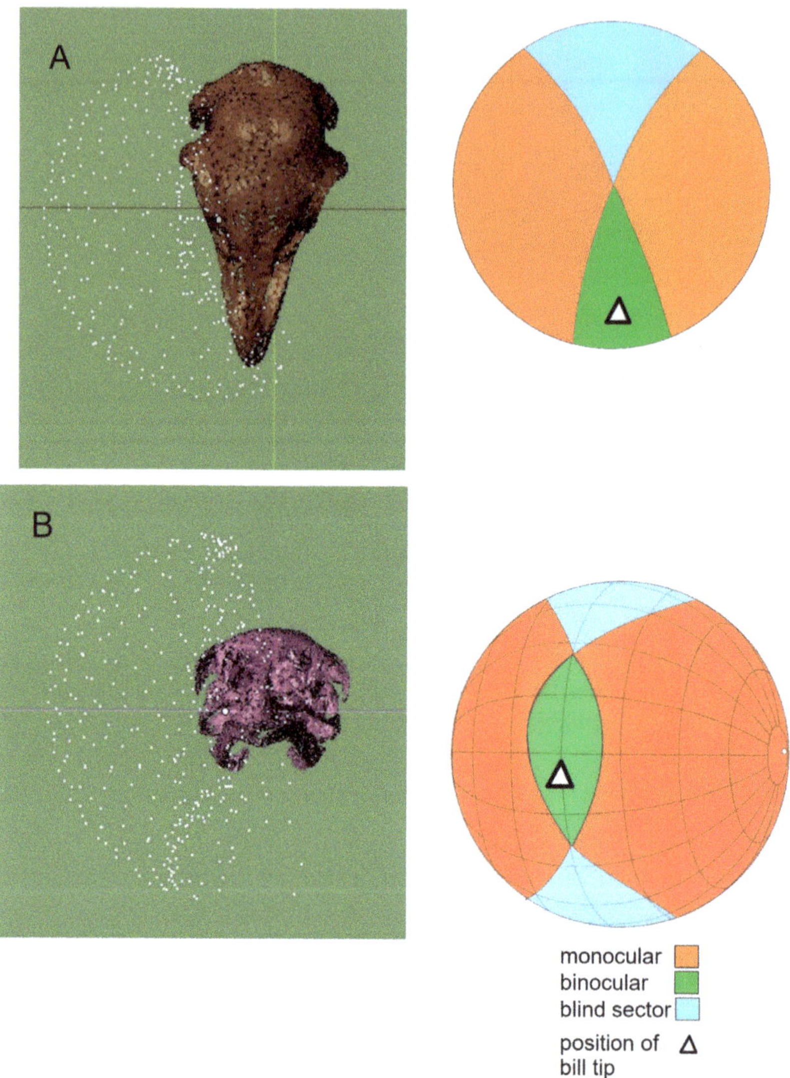

Figure 8. *Megalapteryx didinus*, visual field plot, as reconstructed and graphically after the convention of Martin [1], in (**A**) dorsal and (**B**) frontal views.

3.2.2. Kakapo

In the analysis framework used, *S. habroptilus*, known to be nocturnal but with other parameters of eye anatomy deemed not typical for either diurnal or nocturnal activity [9], falls at the overlap zone in the plot but receives a probability of nocturnality of 0.68.

3.3. Hearing

Residuals of ECD against cranial height in phylogenetic generalised least squares regression are plotted against our phylogeny in Figure 9. A number of interesting results appear: *M.didinus* had the largest spectrum of hearing frequencies of any palaeognath, while *Aepyornis* was at the lower end of the group of birds sampled. *Apteryx* and *Strigops* are both placed as birds with limited frequency hearing. This result for *Strigops* is also shown in [36] within the phylogeny of that study, but is not commented on.

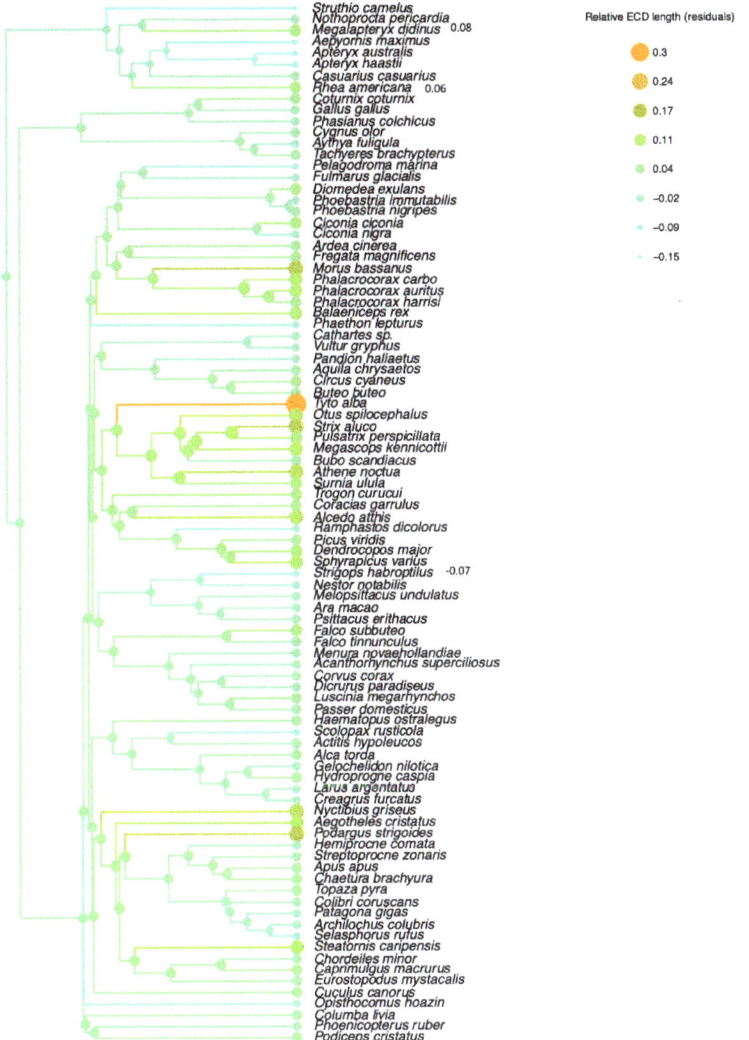

Figure 9. Endosseous cochlear duct residuals plotted to phylogeny (residual values added against taxa of interest).

The inner ear labyrinth of *M. didinus* is shown in Figure 10A. This also demonstrates the cerebellar flocculus, or more accurately, the endocast of the floccular fossa [52]. *M. didinus* has a qualitatively substantial floccular endocast.

3.4. Bill Tip Sensory Organs

Specialised mechanoreceptors known as Herbst corpuscles are found in bony pits in the bill tips of extant palaeogaths and a variety of neognaths that use bill-probing behaviour. Pits are present in all moa species; in *Megalapteryx*, they are most prominent on the oral aspect of the premaxilla (Figure 10B). The trigeminal ganglion endocast in *Megalapteryx* appears prominent, more so than in figures of other palaeognaths [4], and the ophthalmic division of the trigeminal nerve (V_1) is also relatively large, qualitatively (Figure 10C,D).

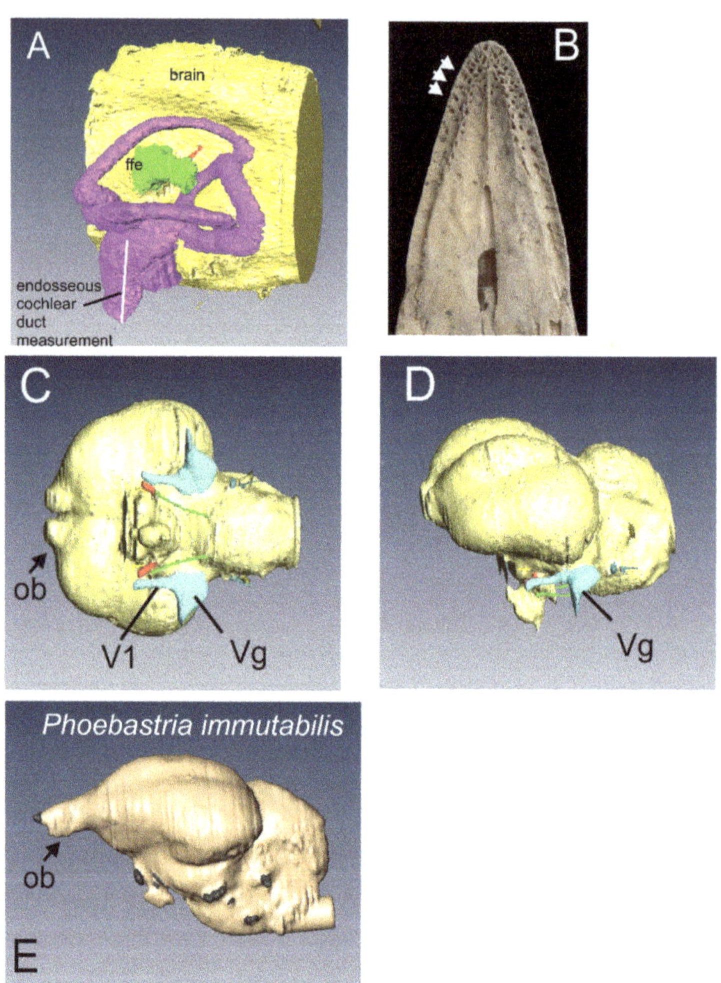

Figure 10. (**A**) *Megalapteryx didinus* left otic labyrinth and adjacent structures. ffe, floccular fossa endocast (**B**) *M. didinus*, bill pits in oral view, photo courtesy of Trevor Worthy (**C**,**D**) *M. didinus*, brain reconstruction in (**C**) ventral and (**D**) left lateral views.ob, olfactory bulb, V1, ophthalmic division trigeminal nerve; Vg, trigeminal ganglion (**E**) An albatross with a well-developed olfactory bulb.

3.5. Comparative Material

3.5.1. Palaeognathae

A small dorsal olfactory chamber is present in non-NZ ratites, directly in contact with the cranial cavity with a very short olfactory nerve trunk in *Struthio*, *Dromaeus*, and *Casuarius*, and with an olfactory nerve that traverses the orbital cavity for 1.5 cm in *Rhea*. A substantial mono-laminar septum is present ventral to this dorsal chamber in these taxa. Tinamous have a single nerve that traverses the orbit for the whole length of the interorbital septum. A substantial mono-laminar septum is present ventral to this dorsal chamber in these taxa.

3.5.2. Neognathae

In neognaths, there are two principal patterns: a small, dorsal olfactory chamber above a mono-laminar septum, and a complete mono-laminar septum. Some species with known olfactory ability [13,20] have marked expansion of the septum by an olfactory chamber, notably in vultures and Procellariiformes. The Turkey Vulture *Cathartes aura* is known for olfactory specialisation, and in this taxon, the whole of the interorbital septum is expanded by a broad olfactory chamber, part of which contains the scrolled caudal concha (Figure 11A). This configuration of olfactory chamber is very similar to that of moa. The Black Vulture *Coragyps atratus* are closely related to the Turkey Vulture and have above average olfaction in terms of the numbers of mitral cells in the olfactory bulb [53], but still reduced in comparison to *Cathartes aura*. The interorbital septum is again expanded by an olfactory chamber (Figure 11B), but to a lesser degree than in *Cathartes*. Among other birds, an extensive olfactory chamber filling much of the interorbital septum is seen in the olfactory specialists *Puffinus grisea*, *Pachyptila desolata*, *Thalassarche chlororhynchus*, *Fulmaris glacialis*, and *Pagodroma nivea*. In the many CT series reviewed, marked expansion of the dorsal septum or any expansion of the ventral part of the septum was only seen in taxa such as the above with known olfactory specialisation.

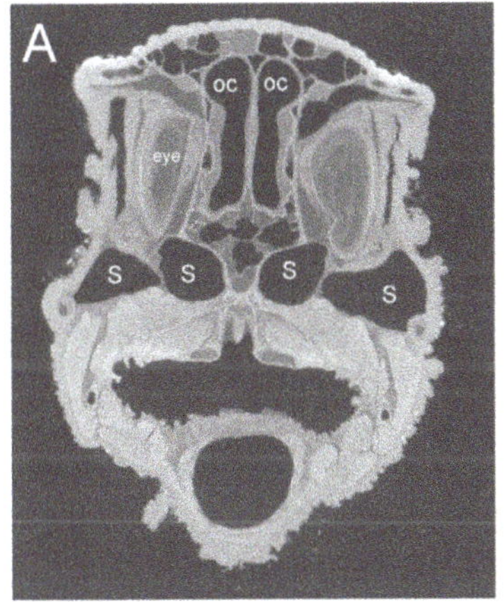

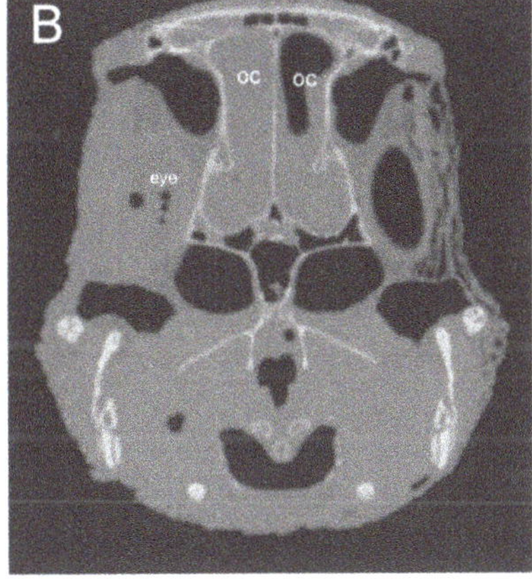

Cathartes aura
(Turkey Vulture)

Coragyps atratus
(Black Vulture)

Figure 11. Axial CT sections of olfactory chambers of vultures. (**A**), Cathartes aura, Turkey Vulture, (**B**), Coragyps atratus, Black Vulture. oc, olfactory chamber; s, sinus.

4. Discussion

We have presented a range of new information on the sensory systems of flightless birds from four different clades, variously linked by geography (NZ), nocturnality (kiwi, Kakapo), large size (moa, aepyornithids), and by the availability of data for our analyses.

We will summarise this new information for each group.

4.1. Olfaction and Sensory Systems in Kiwi

4.1.1. Retronasal Olfaction

The phenomenon of retronasal olfaction—entry of air to the olfactory epithelium via the internal nares—has been studied extensively in mammals, and indeed it has been asserted that retronasal olfaction is a unique attribute of mammals [54], and a separate neural pathway within the brain has been proposed. A little reflection, however, will show that retronasal olfaction must be possible in birds: a variety of aquatic birds, notably gannets and most penguins, do not have patent external nares (Figure 12) but do clearly have olfactory epithelium [13], so the retronasal route is the only possibility.

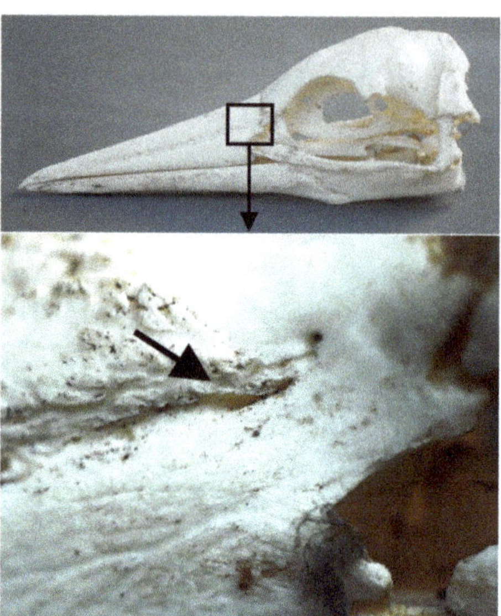

Figure 12. Australasian gannet, *Morus serrator*. skull with site of closed external naris indicated.

The Haagen-Poiseuille equation describes parameters for flow in pipes and other enclosed spaces:

$$\Delta p = 8\mu LQ/(\pi R\hat{\ }4) = 8\pi\mu LQ/A^2$$

where Δp = pressure difference, μ = viscosity, L = length, Q = flow rate, A= area.

When the function for area of pipes is considered, it is clear that resistance to flow is inversely proportional to the square of the area. When the data for the external and internal nares of *Apteryx* are used, the resistance of the external nares is 1200 times that of the internal nares. This, the proximity of the choana to the olfactory epithelium (Figure 4), and the baffle-like complex structure of the caudal concha, which presumably directs the air flow, make retronasal olfaction a likely or probably predominant mode of olfaction in kiwi.

Retronasal airflow has been assessed in a CFD study of pachycephalosaur dinosaurs [55], and the result does not appear to differ significantly from orthonasal flow.

The CFD analysis included here is a simplified simulation of a morphologically very complex airspace, but does support the hypothesis that retronasal air flow is the major mode of ventilation of the kiwi olfactory epithelium. In a more detailed analysis of flow in the intricate rat nasal airways [56], low velocity flow is seen in the olfactory area— this may be an advantage, in allowing longer odorant contact with the epithelium. A significant unknown factor is the speed of diffusion of odorants from the airstream to the olfactory receptors [56]. CFD could be applied widely in the investigation of avian olfaction; studies to date have relied on hypothetical flow distributions. A more advanced simulation with more detailed geometry and phasic inspiration-expiration flow could have offered more data for *Apteryx*.

4.1.2. Sniffing

It is not clear that sniffing, at least in the mammalian sense of short, noisy inspiratory air movement, or rapid cyclical inspiration and expiration, is actually possible in birds. The avian respiratory system operates at low pressures [29], and the extensive air sac network acts as capacitance mechanism that mitigates against any large or sudden pressure alteration on inspiration and expiration. Kiwi have typical large abdominal air sacs, and a non-muscular diaphragm [57]. The mammalian lung is tightly mechanically constrained within the thorax and capable of much greater pressure differences— negative pressures operating for sniffing in humans and rats is typically up to 10 times what avians are capable of [56]. The exit of secretions from nasal glands in birds has been described by [58]: nasal gland secretions are typically cleared via the external nares, and this activity, with expiratory effort, may account for the respiratory noises of kiwi. Mucus secretion from the respiratory epithelium is moved across the mucosal surfaces and leaves the nasal cavity via the internal nares. The bill elevation observed during kiwi olfactory search could be a mechanism for gravitationally directing secretions away from the small external nares. The small nares may be as much an issue for egress of nasal gland and lacrimal duct secretions as for air entry.

4.2. Olfaction in Moa

The data presented here support the hypothesis that moa were olfactory specialists, in spite of the minimally developed olfactory bulb (Figure 10C,D). The branching pattern of olfactory nerves at the cribriform plate and radiation of branches around the wall of the chamber are indicative of their supply to a broad area of olfactory epithelium; the nerve branches in the chamber in *Rhea* as demonstrated by [59], and in Procellariiformes as detailed by [58], are evidence for this. Furthermore, an olfactory chamber expanding the whole or much of the interorbital septum as in moa species is only otherwise found in neognath olfactory specialists such as vultures and Procellariiformes.

The evolutionary history of the avian interorbital septum can be deduced from accounts of developmental anatomy [60,61]. The region between the eyes is originally a tri-laminar structure: the midline septal cartilage is accompanied by an anterior orbital cartilage from each side. In most avians the latter structure disappears with the enlargement of the eye; in those cases where the tri-laminar configuration persists, an olfactory chamber and sinus cavities occupy the space between, and the midline septum may regress. In the early embryo, the olfactory capsules are directly applied to the forebrain; as the facial skeleton grows, the brain stem and olfactory capsules separate. The olfactory bulbs remain in contact with the capsules in fishes, and are connected with the rest of the brain by a peduncle. In amniotes, pedunculated bulbs remain in lepidosaur and archosaur lineages. In the theropod line towards avians, a progressive caudal retreat of an olfactory chamber and bulb toward the rest of the forebrain occur, to reach the situation found in birds where the bulb is sessile rather than pedunculated. A small peduncle may secondarily arise in birds with a large olfactory bulb (Figure 12E). With the enlarging avian eye, the olfactory chamber and bulb may become separated and a large olfactory nerve trunk traversing the orbit or interorbital septum results. There are thus two main patterns found in modern birds: a caudally extended olfactory chamber in apposition to the bulb, as in palaeognaths with the exception of tinamous and *Rhea*, which has a 1 cm single olfactory nerve; or alternatively a separation of bulb and chamber, which is found in most neognaths, with the exception of the olfactory specialists described above. Curiously, in amphibians and turtles, the bulb is contiguous with the telencephalon and a long single olfactory nerve trunk is present [62].

The ethmoid region in birds is made more complex by the variety of terminology that has been used, but an account of this region developed in an attempt to explain the same area in theropod dinosaurs [63] offers a clear explanation of the box-like separate ethmoid ossification seen in juvenile moa. The midline nasal septum is retained, the trough-like floor and sides of the cavity are formed from the anterior orbital cartilages (planum supraseptale), and its roof is the parieto-tectal cartilage [64], ossified as the 'dorsal plate'. This ethmoid structure is found in both neognaths and palaeognaths, where in the latter it may appear on the dorsum of the skull in juveniles, before being covered by the nasal bones.

The evidence that moa are in fact olfactory specialists does contradict the inference from the evident olfactory bulb dimensions. The suggestion that components of the olfactory bulb may be buried in the bulk of the telencephalon [23] remains an open question. The forebrain of moa is shaped differently to other palaeognaths, as noted by its earliest observers [16,65]: wider and blunter at its rostral end. Olfactory bulb size relative to brain size has been shown to follow allometric and phylogenetic trends [20] and empirical evidence of olfactory behaviour is available for only a limited range of birds; available evidence is sometimes not congruent with inference from brain morphology. We quote Graham Martin from his recent book on the sensory ecology of birds: 'how much of the brain is devoted to analysis of olfactory information does not seem to be a good guide to the importance of olfaction in the behaviour of birds' [1]. Much information on the exclusively herbivorous diet of moa has been gathered from preserved gizzard contents and coprolites, using direct examination for seeds and plant remains, and ancient DNA techniques [66], and a profile of the diet ranges for species of six moa genera has been assembled. Where data are sufficient, different diet preferences for sympatric moa species can be defined, indicating dietary selection. This research has also defined a range of plants present in their respective environments that were avoided by moa; both these positive and negative selection processes were presumably driven by olfaction. Moa had the genetic information for ultraviolet (UV) vision [67]; foliage that moa were known to eat has not been tested for its appearance in UV light but this could yield interesting results.

4.3. Hearing in Moa

We show that, with the parameters used here, *Megalapteryx* had more sensitive hearing than any other palaeognath. Without the need for hearing to hunt for prey, and in a relative absence of predators, hearing capacity may have been mainly needed for intraspecific communication. *Rhea* and *Nothoprocta* were included here to achieve a more complete palaeognath phylogeny; *Rhea* also has a more extensive hearing range than other palaeognaths with the exception of *Megalapteryx*; this is interesting, as *Rhea* is the only palaeognath with a complex syrinx [68], and they have a specific vocal profile [69]. The question of the syrinx in moa has been a little mysterious. Oliver [70] produced a figure evidently redrawn from Richard Owen [71], which was stated to be the tympanum of the syrinx of *Emeus crassus*. This attribution has not stood up to scrutiny [15,68] and we agree with that position. However, we offer an alternative explanation. In Owen's original paper (p389) he described working out of matrix an expanded distal tracheal ring attached to an incomplete bronchial ring. His figure of this conforms with the description of a bifurcating lumen with a pessulus that he described as similar to that of a raven. Review of the original description of his raven [72] and its figure does indeed demonstrate a tympanum in the raven with a ventral fused band connecting four rings with a pessulus, similar to his *Emeus* account. This is quite convincing, but if so it is unusual that no other ossification compatible with a moa syrinx has been discovered [68]. We believe Oliver [70] was looking at the wrong drawing in Owen's figures and reworked a thyroid cartilage into the form of a very flattened and quite atypical syrinx. It is tempting to link a possible developed syrinx in *Emeus* with the sensitive hearing of *Megalapteryx* and the parallel with vocalisations as in *Rhea* as adaptations to intraspecific communication. However, interpretations from syringeal morphology must be guarded— 'vocal learners' among birds have the standard developed form of avian syrinx, as in the raven, whereas less vocally specialised birds can have a more elaborate syrinx [73], the adaptation in vocal learners being in more advanced neuromuscular control. Confirmation of a developed syrinx in moa together with that of *Rhea* would probably relegate the 'undeveloped' syrinx of other palaeognaths to a derived state, given the specific similarities between the syringes of *Rhea* and the typical tracheobronchial variety of neognath syrinx.

4.4. Vision in Moa

Nocturnality and limited visual capacity in moa have been suggested or implied by observation of small optic lobes of the brain in recent studies [3,4]. A species for which scleral rings are known, *M. didinus*, is investigated here and found to fit within the published range of cathemeral birds. *M. didinus* had small optic lobes of the brain, similar to all other moa [3,4,18]. Other moa species may of course have been nocturnal, perhaps as a mode of niche partitioning in habitats where several moa species were sympatric. Cathemeral activity does, however, fit best with the pattern of other megaherbivores, which need to eat for a large part of the day to meet energy requirements [74].

The visual field map presented here for moa is differs from that of ostriches [75], in which the bill tip is not included in the visual field. The wide blind sector posteriorly could have presented an avenue of approach for the giant eagle *Hieraaetus* (*Harpagornis*) *moorei*: claw marks described on moa skeletal remains are on the dorsal trunk area [15]. There has been lengthy discussion on the flying abilities of this extinct eagle, given its weight (15 kg) and relatively short wings [15]. The ability of this extinct eagle to prey on moa many times its size and location of attack could suggest that gliding, silent flight on approach to the prey may have been the scenario, as flapping flight for a bird of this size would presumably have been quite noisy, and we have shown that at least one species of moa had sensitive hearing. The significance of a binocular field in birds has been explained as different from that in mammals: the binocular field is not for stereoscopic vision, and only represents a continuity of visual field [76]. Inclusion of the bill tip in the field is associated with foraging and the feeding of chicks. Moa chicks were presumably precocial, as with other ratites with large eggs, and may not have required direct parental feeding. In another study, visibility of the bill tip was related to a pecking foraging strategy; however, in birds relying on tactile foraging, the bill may or may not be included in the visual field [77].

4.5. Bill-Tip Sensory Organ

The presence of bill pits has not been studied formally in moa, although Richard Owen did notice these pits without being aware of their significance [16]. Their presence in moa is consistent with other palaeognaths. Their presence adds to the sensory repertoire used by moa to negotiate their environment; for example, a combination of olfaction and bill sensation would allow foraging among foliage in a context of low light or reduced visual acuity. The bill tip organ is also present in Kakapo [78], as in some other parrots. There are appear to be no comparative data on trigeminal ganglion endocast size, but the prominent structure in *M. didinus* is consistent with bill sensation as a major component of the sensory toolkit [79].

4.6. Floccular Fossa Endocast

The floccular lobe of the cerebellum is part of the central nervous system rather than a sensory organ, but is known to integrate visual and vestibular information and maintenance of a stable gaze [52,80] and thus relevant to those senses. Inferences from its morphology and size have not so far reached definite conclusions, and the significance of floccular fossa morphology remains enigmatic in any predictions regarding extinct taxa. We show here that *M. didinus* has a substantial floccular fossa endocast. This is in contrast to an observation that *D. robustus* had an absent flocculus [81]. The floccular endocast shows a range of morphologies within this single avian radiation [82], and future research in a phylogenetic context may bring another angle to bear on this question, particularly when combined with information about sensory systems and lifestyle.

4.7. Kakapo

The Kakapo *S. habroptilus* also has a herbivorous diet and nocturnal lifestyle, and has specific visual specializations for nocturnality, in that the retina is adapted for light sensitivity and not for visual acuity [9]. *Strigops* is known to use olfactory cues in feeding [83] and has a larger repertoire of olfactory receptor genes [84], but with a relative olfactory bulb size in phylogenetic context which in the middle of the avian range when corrected for allometry and phylogeny [20]. We see here and in [36] that the hearing range of this bird is at the lower end of the range, similar to the findings in kiwi and elephant birds. Our new data on the Kakapo scleral ring modifies the analysis of eye dimensions alone [9], which had described the optics of the eye as resembling those of diurnal birds. This helps to validate our predictive framework for moa and other birds in the data set.

4.8. Aepyornithids

Previous research has shown an olfactory bulb of comparable proportions to other palaeognaths [4], and small eyes are a feature of the skull. We add here the information that a moderately large olfactory chamber was present, and that the hearing range was limited, among the lowest in the birds studied here and by [36]. The optic lobes of the brain were very small [4], of an equivalent size to those of kiwi, suggesting a nocturnal activity pattern. Very little is known of elephant bird ecology [14]; it could be expected that olfaction comprised a major part of the sensory repertoire.

5. Conclusions

We return to the concept of complementary sensory information in birds and trade-offs among different senses. This expectation has been demonstrated more readily in mammals than in birds, where the best documented example is the reduction of other senses in instances of dominant bill sensation- trigeminal hypertrophy taxa [85]. Following [36], we have plotted nocturnality against ECD length from our data where both are available (Figure 13)—(the 'posterior probability of nocturnality' is only relevant to *M. didinus* here) — and a tendency for increased hearing sensitivity is observed in nocturnal birds, *Apteryx* being a notable exception. This plot also indicates *Megalapteryx* ranked for ECD among our dataset. If we follow some of the published suggestions, we could conclude that moa had both diminished visual and olfactory ability. How, then would moa thrive in their herbivorous mode of life in a variety of habitats? The balance is obvious in extreme sensory specialisations, such as olfaction in kiwi and hearing in some owls; in most cases, multiple sources of information must be viewed together. Palaeoneurology explores the limits of inference; in this context, some morphometric studies of brain shape in birds and mammals conclude that brain shape is responsive to dimensions of the eyes and facial skeleton [86,87], and a recent theme in cranial morphometrics is that the various facial, muscular, sensory and central neural structures compete for space within the head, and that this may explain the shapes of structures as much as individual functional needs [88,89]. Thus, as well as sensory complementarity and trade-offs, and the metabolic cost of supporting sensory structures and brain regions to support them [85], we also have to consider competition for the cranial domain. From the data we have assembled here and reviewed in the literature, we can offer tentative hypotheses regarding these trade-offs in our taxa (Figure 14). These are necessarily global assessments from the multiple sources of evidence we have discussed; different patterns of sensory use would presumably apply for feeding, reproductive, intraspecific and interspecific requirements. We have not examined vestibular anatomy as a recent comprehensive study could not reveal any significant functional inferences from this domain [89].

Here, we need to ask what part flightlessness plays in the sensory patterns of the taxa considered here. There is an association between reduction of flight muscles and increase in brain size [90]; this applies to kiwi but not to the other taxa we have addressed: moa have small brains [18] and Kakapo brains are not larger than those of other parrots [9]. Aepyornithids appear to have small brains but the rarity of associated cranial and post-cranial remains makes measurement of brain/body mass uncertain. Kiwi have obvious hypertrophy of the olfactory lobe of the brain, but otherwise, 'whole brain size is a blunt instrument when it comes to assessing avian brain evolution' [2].

In an analysis of potential factors enabling flightlessness in many lineages of island-dwelling birds, absence of predation stood out in the evolution of reduced flight muscles and longer hind limbs [91]. How might this have affected sensory evolution in the island taxa we have considered? A reduction in the need for auditory and visual monitoring for predators may have enabled reductions in those systems and made these birds highly vulnerable to introduced predators: the extinction of moa and aeypornithids was related to human arrival on their respective islands [15,92], Kakapo are currently critically endangered, and kiwi are protected as they are vulnerable to introduced mustelids, rodents, and domestic animals.

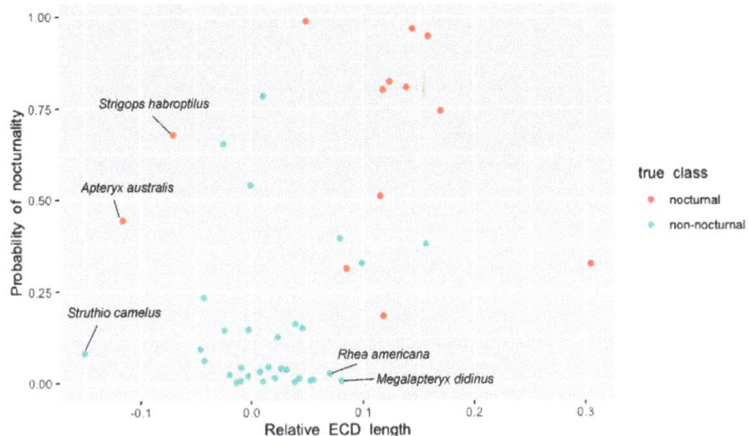

Figure 13. Plot of endosseous cochlear duct residuals against probability of nocturnality from eye measurements.

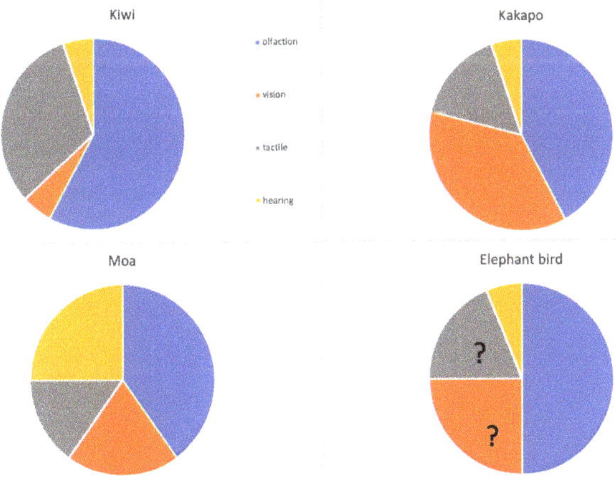

Figure 14. Possible patterns of sensory modalities, based on new date and published information on sensory organs, neural structures and observed behavior. ? = unknown.

Trade-offs among sensory systems are apparent in the extant taxa we have studied, but remain conflicting and incomplete for the extinct birds. To take these questions further, our ongoing research includes geometric morphometrics of endocasts and other domains of cranial morphology, and interrogation of ancient DNA data in all moa genera for signals of positive selection of in genes known to be associated with sensory modalities and nocturnality. The data we have presented here form a baseline for these and other investigations.

Supplementary Materials: The following are available online at https://www.mdpi.com/article/10.3390/d13110538/s1, Table S1 eye data (.csv); Table S2: ECD data (.csv); Table S3: sequence data.

Author Contributions: Concept: P.J., K.J.M.; methodology: P.J., K.J.M.; morphological analyses, P.J.; constructed phylogenetic trees, K.J.M.; original draft preparation, P.J.; reviewing and editing, P.J., K.J.M. All authors have read and agreed to the published version of the manuscript.

Funding: K.J.M. is supported by a Marsden Fund Fast-Start Grant (20-UOO-130).

Institutional Review Board Statement: No animal interventions were used in this research.

Data Availability Statement: microCT series of *M. didinus* and Amira Mesh segmentation file are available on request.

Acknowledgments: For access to collections (for abbreviations see Appendix A) we thank; AIM, Brian Gill, Jason Froggatt; MNZ, Alan Tennyson; MNHN, Ronan Allain and Romain David; Tübingen Zoologische Sammlung, Wolfgang Maier and Erich Weber. We are also grateful to Lars Schmitz for permission to reproduce Figure 7B, Graham Martin for his visual field grid, and for information and discussion, Trevor Worthy, Brian Gill and Martin Wild. Hamish Johnston kindly assisted with data analysis. Contributors to online resources are acknowledged with individual specimens in Appendix A.

Conflicts of Interest: The authors declare no conflict of interest.

Abbreviations: AIM: Auckland Museum; MNZ: Museum of New Zealand, Te Papa Tongarewa; JVC: author's collection (PJ); ZSUT: Zoologische Sammlung Universität Tübingen; AMNH, American Museum of Natural History; SAM, South Australia Museum; TMM, Texas Memorial Museum; UMHN, Utah Museum of Natural History.

Appendix A

Material Examined

Named contributors to online resources and individual requests are gratefully acknowledged.

- Moa:
- *Dinornis novaezealandiae*: MNZ S37876; AIM: LB6400; LB6952; LB7870; LB6952; LB6400; LB6833; LB6401; LB6310; LB 6308; LB6432;
- *Dinornis robustus*: MNZ S28225
- *Anomalopteryx didiformis* MNZ S35274; MNZ S5795; AIM: LB5548; LB5545; LB5979; LB5596; LB5819; LB5796; LB5843; LB5465; LB5485; LB5504; LB5515; LB5519; LB5511; LB5552; LB5553; LB5555; LB5550; LB5593; LB5596; lb5627; LB5793; LB5620; LB5653; LB5684; LB5914; LB5798; LB5819;
- *Emeus crassus* MNZ S470; MNZ S792; AIM: LB6285;
- *Euryapteryx curtus* MNZ S30212; AIM LB6710; LB6637; LB6616; LB6666; LB6246; LB6251; LB6285;
- *Pachyornis elephantopus* AIM: LB5946
- *Pachyornis geranoides* AIM: LB6030; LB6020; LB6021; LB6024; LM6069;
- *Pachyornis australis* MNZ S27896
- *Megalapteryx didinus* MNZ S28206; MNZ S33763; MNZ S400; AIM: LB5904;
- Moa: CT scans: *D. robustus* MNZ S28225, *A. didiformis* MNZ S35274, *E. crassus*, MNZ S470, *E. curtus*, MNZ S30212, *P. australis* MNZ S27896, *M.didinus*, MNZ S28206. Pacific Radiology, Wellington (New Zealand), on a General Electric Discovery CT750 HD scanner, at 80 kV and 40 μA, and reconstructed as axial 0.3mm slices).
- *P. elephantopus* AIM LB5946; Mercy Radiology, Auckland. GE Discovery CT750, 120 kV, 150 mA, 0.625 mm slices.
- *P. elephantopus* MNHN 1875-602, from Dryad Digital Repository https://doi.org/10.5061/dryad.7519042 (accessed on 14 September 2021), C. Torres and J. Clarke

- *M. didinus* AIM LB5904, microCT scan (30.5.2011,Bioengineering Institute, University of Auckland; Skyscan 1172: 100 kV, 100 µA, reconstructed as 1626 slices, voxel size 34.6 µm, image size 1984 × 1984 pixels).
- Moa: MRI scan: *M. didinus*, MNZ S400 Siemens Magnetom Avanto 1.5 Tesla scanner with a Siemens 12 channel head matrix coil and B17 software. Performance per axis details were: maximum amplitude 33 mT/m, minimum rise time 264 microseconds from 0–33 mT/m, maximum slew rate 125 T/m/s.
- Kiwi:
- *Apteryx mantelli*: JVC 386, JVC 387; AIM LB7709; LB7289; LB2182; LB5540; LB7202; LB5539; LB9246; LB14145;
- *Apteryx australis*: AIM LB13427; LB2182
- *Apteryx owenii*: AIM LB9427; LB11246
- Kiwi: CT scan: *Apteryx* species—AMHN18456, http://digimorph.org/specimens/Apteryx_sp/ (accessed on 14 September 2021).
- Kiwi: MRI scan: *Apteryx mantelli*: Centre for Advanced MRI, University of Auckland. Siemens Magnetom Avanto 1–5T, gradient strength 40 (across) and 45 (along) mTm-1, maximum slew rate 200 Tm-1s-1 with a 4-channel wrist coil; 2D turbo spin echo with 0.4 mm in-plane resolution and 1 mm slice thickness; echo time/repetition time/flip angle/averages = 156 ms/5510 ms/1501/6—Jeremy Corfield
- Kiwi, histological series: ZSUT-SAJ78110. *Apteryx australis*. Serial sectioned head of hatchling
- Aepyornithidae:
- *Aepyornis maximus* MNHN 1910.12; *Ae. ?hildebrandti* MNHN MAD6724; *Ae. 'medius'* MNHN1911-27
- microCT scan: *Ae. maximus* MNHN 1910.12; 629 slices at voxel size 138 µm; Romain Allain and Ronan David
- *Struthio camelus*: JVC 343; AIM LB11730
- CT scan: L. Witmer lab: https://people.ohio.edu/witmerl/3D_ostrich.htm; https://youtu.be/gDQ8a0_oH6k (accessed on 14 September 2021).
- *Dromaeus novaehollandiae*: JVC355; AIM541
- CT scan, SAM39373,—Trevor Worthy
- *Rhea americana* TMM M-6721, from Dryad Digital Repository https://doi.org/10.5061/dryad.7519042 (accessed on 14 September 2021), C. Torres and J. Clarke
- *Casuarius casuarius* TMM M-12033, from Dryad Digital Repository https://doi.org/10.5061/dryad.7519042 (accessed on 14 September 2021), C. Torres and J. Clarke
- *Rhea pennata*: AIM LB1216
- *Nothoprocta pericardia*, UMNH 23838, from Dryad Digital Repository https://doi.org/10.5061/dryad.7519042, C. Torres and J. Clarke
- *Morus serrator* (Australasian gannet): JVC201
- *Cathartes aura* (Turkey Vulture); Morphosource 000125045, Jessie Maisano
- *Coragyps atratus* (Black Vulture); http://digimorph.org/specimens/Coragyps_atratus/ (accessed on 14 September 2021)—Tim Rowe
- *Stripogs habroptilus* (Kakapo): Morphosource 000158358, Roger Benson
- *Pachyptila desolata*: Morphosource 000167145, Jeff Zeyl
- *Thalassarche chlororhynchos*: Morphosource 000166936, Jeff Zeyl
- *Fulmaris glacialis*: Morphosource 000032762 Roger Benson
- *Puffinus grisea*: Morphosource 000166694, Jeff Zeyl
- *Phoebastria immutabilis*: http://digimorph.org/specimens/Diomedea_immutabilis/ (accessed on 14 September 2021)—Tim Rowe.

References

1. Martin, G.R. *The Sensory Ecology of Birds*, 1st ed.; Oxford University Press: Oxford, UK, 2017; p. 296.
2. Knoll, F.; Kawabe, S. Avian palaeoneurology: Reflections on the eve of its 200th anniversary. *J. Anat.* **2020**, *236*, 965–979. [CrossRef]

3. Early, C.M.; Ridgely, R.C.; Witmer, L.M. Beyond Endocasts: Using Predicted Brain-Structure Volumes of Extinct Birds to As-sess Neuroanatomical and Behavioral Inferences. *Diversity* **2020**, *12*, 34. [CrossRef]
4. Torres, C.R.; Clarke, J.A. Nocturnal giants: Evolution of the sensory ecology in elephant birds and other palaeognaths inferred from digital brain reconstructions. *Proc. R. Soc. B* **2018**, *285*, 20181540. [CrossRef] [PubMed]
5. Yonezawa, T.; Segawa, T.; Mori, H.; Campos, P.; Hongoh, Y.; Endo, H.; Akiyoshi, A.; Kohno, N.; Nishida, S.; Wu, J.; et al. Phylogenomics and Morphology of Extinct Paleognaths Reveal the Origin and Evolution of the Ratites. *Curr. Biol.* **2017**, *27*, 68–77. [CrossRef] [PubMed]
6. Grealy, A.; Phillips, M.; Miller, G.; Gilbert, M.; Rouillard, J.-M.; Lambert, D.; Bunce, M.; Haile, J. Eggshell palaeogenomics: Palaeognath evolutionary history revealed through ancient nuclear and mitochondrial DNA from Madagascan elephant bird (*Aepyornis* sp.) eggshell. *Mol. Phylogenetics Evol.* **2017**, *109*, 151–163. [CrossRef] [PubMed]
7. Phillips, M.J.; Gibb, G.C.; Crimp, E.A.; Penny, D. Tinamous and Moa Flock Together: Mitochondrial Genome Sequence Analysis Reveals Independent Losses of Flight among Ratites. *Syst. Biol.* **2009**, *59*, 90–107. [CrossRef] [PubMed]
8. Mitchell, K.J.; Llamas, B.; Soubrier, J.; Rawlence, N.J.; Worthy, T.H.; Wood, J.R.; Lee, M.S.Y.; Cooper, A. Ancient DNA reveals elephant birds and kiwi are sister taxa and clarifies ratite bird evolution. *Science* **2014**, *344*, 898–900. [CrossRef]
9. Corfield, J.R.; Gsell, A.C.; Brunton, D.; Heesy, C.P.; Hall, M.I.; Acosta, M.; Iwaniuk, A. Anatomical Specializations for Nocturnality in a Critically Endangered Parrot, the Kakapo (*Strigops habroptilus*). *PLoS ONE* **2011**, *6*, e22945. [CrossRef] [PubMed]
10. Castro, I.; Cunningham, S.J.; Gsell, A.C.; Jaffe, K.; Cabrera, A.; Liendo, C. Olfaction in birds: A closer look at the Kiwi (Apter-ygidae). *J. Avian Biol.* **2010**, *41*, 213–218. [CrossRef]
11. Owen, R. *Memoirs on the Extinct Wingless Birds of New Zealand, with an Appendix on Those of England, Australia, New Foundland, Mauritius and Rodriquez*; John van Voorst: London, UK, 1879.
12. Corfield, J.; Eisthen, H.L.; Iwaniuk, A.N.; Parsons, S. Anatomical specialisations for enhanced olfactory sensitivity Kiwi, *Ap-teryx mantelli. Brain Behav. Evol.* **2014**, *84*, 214–226. [CrossRef]
13. Bang, B. Functional Anatomy of the Olfactory System in 23 Orders of Birds. *Acta Anat.* **1971**, *79* (Suppl. 1), 1–76. [CrossRef] [PubMed]
14. Hansford, J.P.; Turvey, S.T. Unexpected diversity within the extinct elephant birds (Aves: Aepyornithidae) and a new identity for the world's largest bird. *R. Soc. Open Sci.* **2018**, *5*, 181295. [CrossRef] [PubMed]
15. Worthy, T.H.; Holdaway, R.N. *The Lost World of Moa*; Indiana University Press: Bloomington, IN, USA, 2002.
16. Owen, R. On Dinornis (part XXIII): Containing a description of the head and feet, with their dried integuments, of an indi-vidual of the species Dinornis didinus, Owen. *Trans. Zool. Soc. Lond.* **1883**, *11*, 257–261.
17. Parker, T.J. On the cranial osteology, classification and phylogeny of Dinornithidae. *Transactions of the Zoological Society of London.* **1895**, *8*, 373–428. [CrossRef]
18. Ashwell, K.; Scofield, R. Big Birds and Their Brains: Paleoneurology of the New Zealand Moa. *Brain Behav. Evol.* **2007**, *71*, 151–166. [CrossRef] [PubMed]
19. Corfield, J.R.; Wild, J.; Hauber, M.E.; Parsons, S.; Kubke, M.F. Evolution of Brain Size in the Palaeognath Lineage, with an Emphasis on New Zealand Ratites. *Brain Behav. Evol.* **2007**, *71*, 87–99. [CrossRef]
20. Corfield, J.R.; Price, K.; Iwaniuk, A.N.; Gutiérrez-Ibáñez, C.; Birkhead, T.; Wylie, D.R. Diversity in olfactory bulb size in birds reflects allometry, ecology, and phylogeny. *Front. Neuroanat.* **2015**, *9*, 102. [CrossRef] [PubMed]
21. Zelenitsky, D.K.; Therrien, F.; Ridgely, R.C.; McGee, A.R.; Witmer, L. Evolution of olfaction in non-avian theropod dinosaurs and birds. *Proc. R. Soc. B.* **2011**, *278*, 3625–3634. [CrossRef]
22. Atkinson, I.A.E.; Greenwood, R.M. Relationships between moas and plants. *N. Z. J. Ecol.* **1989**, *12*, 67–96.
23. Worthy, T.H. Aspects of the biology of two moa species (Aves: Dinornithiformes). *N. Z. J. Archaeol.* **1989**, *11*, 77–86.
24. Vickers-Rich, P.; Trusler, P.; Rowley, M.J.; Cooper, A.; Chambers, G.K.; Bock, W.J.; Millener, P.; Worthy, T.H.; Yaldwyn, J.C. Morphology, Myology, Collagen and DNA of a Mummified Uplandmoa, Megalapteryx Didinus (Aves: Dinornithiformes) from New Zealand. *Tuhinga Rec. Mus. New Zealand Te Papa Tongarewa* **1995**, *4*, 1–26.
25. Rawlence, N.; Wood, J.; Scofield, R.; Fraser, C.; Tennyson, A.; Wood, J.; Scofield, R. Soft-tissue specimens from pre-European extinct birds of New Zealand. *J. R. Soc. N. Z.* **2013**, *43*, 154–181. [CrossRef]
26. Attard, M.R.G.; Wilson, L.A.B.; Worthy, T.H.; Scofield, R.P.; Johnston, P.; Parr, W.H.C.; Wroe, S.R. Moa diet fits the bill: Virtual reconstruction incorporating mummified remains and prediction of biomechanical performance in avian giants. *Proc. R. Soc. B* **2016**, *283*, 20152043. [CrossRef]
27. Simscale. Available online: https://www.simscale.com (accessed on 14 September 2021).
28. Autodesk Meshmixer. Available online: https://www.meshmixer.com (accessed on 14 September 2021).
29. Brackenbury, J.H. Lung-Air-Sac Anatomy and Respiratory Pressures in the Bird. *J. Exp. Biol.* **1972**, *57*, 543–550. [CrossRef] [PubMed]
30. Beale, G. A radiological study of the kiwi (*Apteryx australis mantelli*). *J. R. Soc. N. Z.* **1985**, *15*, 187–200. [CrossRef]
31. Morphosource. Available online: https://www.morphosource.org (accessed on 14 September 2021).
32. Rowe, T.B. Digimorph. Available online: http://digimorph.org/index.phtml (accessed on 14 September 2021).
33. Goswami, A. Phenome10K: A Free Online Repository for 3-D Scans of Biological and Palaeontological Specimens. Available online: www.phenome10k.org. (accessed on 14 September 2021).
34. Witmer, L.M. WitmerLab. Available online: https://people.ohio.edu/witmerl/projects.htm (accessed on 14 September 2021).

35. Community, B.O. Blender-A 3D Modelling and Rendering Package. Available online: http://www.blender.org. (accessed on 14 September 2021).

36. Choiniere, J.N.; Neenan, J.M.; Schmitz, L.; Ford, D.P.; Chapelle, K.E.J.; Balanoff, A.M.; Sipla, J.S.; Georgi, J.A.; Walsh, S.A.; Norell, M.A.; et al. Evolution of vision and hearing modalities in theropod dinosaurs. *Science* **2021**, *372*, 610–613. [CrossRef]

37. Schmitz, L.; Motani, R. Morphological differences between the eyeballs of nocturnal and diurnal amniotes revisited from op-tical perspectives of visual environments. *Vis. Res.* **2010**, *50*, 936–946. [CrossRef]

38. Pinheiro, J.; Bates, D.; DebRoy, S.; Sarkar, D.; Team, R.C. Nlme: Linear and Nonlinear Mixed Effects Models. 2014. Available online: http://CRAN.R-project.org/package=nlme (accessed on 30 October 2020).

39. Paradis, E.; Schliep, K. ape 5.0: An environment for modern phylogenetics and evolutionary analyses in R. *Bioinform.* **2018**, *35*, 526–528. [CrossRef] [PubMed]

40. Walsh, S.A.; Barrett, P.M.; Milner, A.C.; Manley, G.; Witmer, L.M. Inner ear anatomy is a proxy for deducing auditory capa-bility and behaviour in reptiles and birds. *Proc. R. Soc. B* **2009**, *276*, 1355–1360. [CrossRef] [PubMed]

41. Hackett, S.J.; Kimball, R.T.; Reddy, S.; Bowie, R.C.K.; Braun, E.L.; Braun, M.J.; Chojnowski, J.L.; Cox, W.A.; Han, K.-L.; Harshman, J.; et al. A Phylogenomic Study of Birds Reveals Their Evolutionary History. *Science* **2008**, *320*, 1763–1768. [CrossRef]

42. Jetz, W.; Thomas, G.; Joy, J.B.; Hartmann, K.; Mooers, A.O. The global diversity of birds in space and time. *Nat. Cell Biol.* **2012**, *491*, 444–448. [CrossRef] [PubMed]

43. Edgar, R.C. MUSCLE: Multiple sequence alignment with high accuracy and high throughput. *Nucleic Acids Res.* **2004**, *32*, 1792–1797. [CrossRef]

44. Geneious Prime. Available online: http://www.geneious.com/ (accessed on 7 April 2020).

45. Drummond, A.J.; Rambaut, A. BEAST: Bayesian evolutionary analysis by sampling trees. *BMC Evol. Biol.* **2007**, *7*, 214. [CrossRef] [PubMed]

46. Nguyen, L.-T.; Schmidt, H.A.; Von Haeseler, A.; Minh, B.Q. IQ-TREE: A Fast and Effective Stochastic Algorithm for Estimating Maximum-Likelihood Phylogenies. *Mol. Biol. Evol.* **2015**, *32*, 268–274. [CrossRef]

47. Kalyaanamoorthy, S.; Minh, B.Q.; Wong, T.K.F.; Von Haeseler, A.; Jermiin, L.S. ModelFinder: Fast model selection for accurate phylogenetic estimates. *Nat. Methods* **2017**, *14*, 587–589. [CrossRef]

48. Jarvis, E.D.; Mirarab, S.; Aberer, A.J.; Li, B.; Houde, P.; Li, C.; Ho, S.Y.W.; Faircloth, B.C.; Nabholz, B.; Howard, J.T.; et al. Whole-genome analyses resolve early branches in the tree of life of modern birds. *Science* **2014**, *346*, 1320–1331. [CrossRef]

49. Cloutier, A.; Sackton, T.B.; Grayson, P.; Clamp, M.; Baker, A.J.; Edwards, S.V. Whole-Genome Analyses Resolve the Phylogeny of Flightless Birds (Palaeognathae) in the Presence of an Empirical Anomaly Zone. *Syst. Biol.* **2019**, *68*, 937–955. [CrossRef]

50. Rambaut, A.; Drummond, A.J.; Xie, D.; Baele, G.; Suchard, M.A. Posterior Summarization in Bayesian Phylogenetics Using Tracer 1.7. *Syst. Biol.* **2018**, *67*, 901–904. [CrossRef] [PubMed]

51. Marples, B.J. The Structure and Development of the Nasal Glands of Birds. *Proc. Zool. Soc. Lond.* **1932**, *102*, 829–844. [CrossRef]

52. Walsh, S.A.; Iwaniuk, A.; Knoll, M.A.; Bourdon, E.; Barrett, P.M.; Milner, A.C.; Nudds, R.L.; Abel, R.L.; Sterpaio, P.D. Avian Cerebellar Floccular Fossa Size Is Not a Proxy for Flying Ability in Birds. *PLoS ONE* **2013**, *8*, e67176. [CrossRef]

53. Grigg, N.P.; Krilow, J.M.; Gutierrez-Ibanez, C.; Wylie, D.R.; Graves, G.R.; Iwaniuk, A.N. Anatomical evidence for scent guided foraging in the turkey vulture. *Sci. Rep.* **2017**, *7*, 17408. [CrossRef] [PubMed]

54. Rowe, T.B.; Shepherd, G.M. Role of ortho-retronasal olfaction in mammalian cortical evolution. *J. Comp. Neurol.* **2016**, *524*, 471–495. [CrossRef] [PubMed]

55. Bourke, J.M.; Porter, W.R.; Ridgely, R.C.; Lyson, T.R.; Schnachner, E.R.; Bell, P.R.; Witmer, L.M. Breathing life into dinosaurs: Tackling challenges of soft-tissue restoration and nasal airflowiin extinct species. *Anat. Rec.* **2014**, *297*, 2148–2186. [CrossRef] [PubMed]

56. Zhao, K.; Dalton, P.; Yang, G.C.; Scherer, P.W. Numerical Modeling of Turbulent and Laminar Airflow and Odorant Transport during Sniffing in the Human and Rat Nose. *Chem. Senses* **2006**, *31*, 107–118. [CrossRef] [PubMed]

57. Huxley, T.H. On the Respiratory Organs of Apteryx. *Proc. Zool. Soc. Lond.* **1882**, *50*, 560–569. [CrossRef]

58. Bang, B.G.; Wenzel, B.M. Nasal cavity and olfactory system. In *Form and Function in Birds*; King, A.S., McLelland, J., Eds.; Academic Press: London, UK, 1985; Volume 3, pp. 195–225.

59. Müller, H. Die Morphologie und Entwicklung des Craniums von Rhea americana Linné. *Z. Für Wissensschaftliche Zoologie* **1961**, *165*, 221–319.

60. Bellairs, A.D. The Early Development of the Interorbital Septum and the Fate of the Anterior Orbital Cartilages in Birds. *J. Embryol. Exp. Morphol.* **1958**, *6*, 68–85. [CrossRef]

61. Hüppi, E.; Werneburg, I.; Sánchez-Villagra, M.R. Evolution and development of the bird chondrocranium. *Front. Zool.* **2021**, *18*, 21. [CrossRef]

62. Parsons, T.S. Nasal Anatomy and the Phylogeny of Reptiles. *Evolution* **1959**, *13*, 175–187. [CrossRef]

63. Ali, F.; Zelenitsky, D.K.; Therrien, F.; Weishampel, D.B. Homology of the "Ethmoid Complex" of Tyrannosaurids and its im-plications for reconstruction of the olfactory apparatus of non-avian theropods. *J. Vertebr. Paleontol.* **2008**, *28*, 123–133. [CrossRef]

64. Parker, T.J., II. Observation on the anatomy and development of apteryx. *Philos. Trans. R. Soc. London. Ser. B Biol. Sci.* **1891**, *182*, 25–134. [CrossRef]

65. Starck, D. Die endokraniale Morphologie der Ratiten, besonders der Apterygidae und Dinornithidae. *Morphol. Jahrb.* **1955**, *96*, 14–72.

66. Wood, J.; Richardson, S.; McGlone, M.; Wilmshurst, J. The diets of moa (Aves: Dinornithiformes). *N. Z. J. Ecol.* **2020**, *44*, 3397. [CrossRef]

67. Aidala, Z.; Huynen, L.; Brennan, P.L.R.; Musser, J.; Fidler, A.; Chong, N.; Capuska, G.E.M.; Andersen, M.G.; Tabala, A.; Lambert, D.; et al. Ultraviolet visual sensitivity in three avian lineages: Palaeognaths, parrots, and passerines. *J. Comp. Physiol. A* **2012**, *198*, 495–510. [CrossRef]

68. McInerney, P.L.; Lee, M.S.Y.; Clement, A.M.; Worthy, T.H. The phylogenetic significance of the morphology of the syrinx, hyoid and larynx, of the southern cassowary, *Casuarius casuarius* (Aves, Palaeognathae). *BMC Evol. Biol.* **2019**, *19*, 1–18. [CrossRef]

69. Pérez-Granados, C.; Schuchmann, K.-L. Vocalisations of the Greater Rhea (*Rhea americana*): An allegedly silent ratite. *Bioacoustics* **2021**, *30*, 564–574. [CrossRef]

70. Oliver, W.R.B. *The Moas of New Zealand and Australia*; Dominion Museum, Bulletin 15: Wellington, New Zealand, 1949.

71. Owen, R. On Dinornis (part XVI): Containing notices of the internal organs of some species, with a description of the brain and some nerves and muscles of the head of Apteryx australis. *Trans. Zool. Soc. Lond.* **1871**, *7*, 381–396. [CrossRef]

72. Owen, R. *On the Anatomy of Vertebrates*; Longmans, Green: London, UK, 1866; Volume 2.

73. Garcia, S.M.; Kopuchian, C.; Mindlin, G.B.; Fuxjager, M.J.; Tubaro, P.L.; Goller, F. Evolution of vocal diversity through morphological adaptation without vocal learning or complex neural control. *Curr. Biol.* **2017**, *27*, 2677–2683. [CrossRef]

74. Leader-Williams, N.; Owen-Smith, R.N. Megaherbivores: The Influence of Very Large Body Size on Ecology. *J. Anim. Ecol.* **1990**, *59*, 381. [CrossRef]

75. Martin, G.; Osorio, D.; Masland, D.; Albright, T.D. Vision in birds. In *The Senses: A Comprehensive Reference*; Masland, R.H., Al-bright, T.D., Dallos, P., Oertel, D., Firestein, S., Beauchamp, G.K., Bushnell, M.C., Basbaum, A., Kaas, J.H., Gardner, E.P., Eds.; Academic Press: Cambridge, MA, USA, 2008; Volume 1.

76. Martin, G.R. What is binocular vision for? A birds' eye view. *J. Vis.* **2009**, *9*, 14. [CrossRef]

77. Tyrrell, L.P.; Fernández-Juricic, E. Avian binocular vision: It's not just about what birds can see, it's also about what they can't. *PLoS ONE* **2017**, *12*, e0173235. [CrossRef]

78. Froggatt, J.; Gill, B. Bill morphology reflects adaptation to a fibrous diet in the kākāpō (*Strigops*: Psittaciformes). *N. Z. J. Zool.* **2016**, *43*, 138–148. [CrossRef]

79. Crole, M.R.; Soley, J. Comparative morphology, morphometry and distribution pattern of the trigeminal nerve branches supplying the bill tip in the ostrich (Struthio camelus) and emu (*Dromaius novaehollandiae*). *Acta Zool.* **2016**, *97*, 49–59. [CrossRef]

80. Ferreira-Cardoso, S.; Araújo, R.; Martins, N.; Martins, G.; Walsh, S.; Martins, R.M.S.; Kardjilov, N.; Manke, I.; Hilger, A.; Castanhinha, R. Floccular fossa size is not a reliable proxy of ecology and behaviour in vertebrates. *Sci. Rep.* **2017**, *7*, 2005. [CrossRef] [PubMed]

81. Early, C.M. *Quantitative Assessments of Avian Endocasts as Tools for Inferring Neuroanatomical Traits and Potential Functional Capabilities*; Ohio University: Athens, OH, USA, 2019.

82. Johnston, P. Unpublished obeservations: Endocast Diversity in Moa. 2021.

83. Hagelin, J.C. Observations on the olfactory ability of the Kakapo Strigops habroptilus, the critically endangered parrot of New Zealand. *Ibis* **2003**, *146*, 161–164. [CrossRef]

84. Steiger, S.S.; Fidler, A.E.; Kempenaers, B. Evidence for increased olfactory receptor gene repertoire size in two nocturnal bird species with well-developed olfactory ability. *BMC Evol. Biol.* **2009**, *9*, 117. [CrossRef] [PubMed]

85. Wylie, D.R.; Gutiérrez-Ibáñez, C.; Iwaniuk, A.N. Integrating brain, behavior, and phylogeny to understand the evolution of sensory systems in birds. *Front. Neurosci.* **2015**, *9*, 281. [CrossRef]

86. Selba, M.C.; Bryson, E.R.; Rosenberg, C.L.; Heng, H.G.; DeLeon, V.B. Selective breeding in domestic dogs: How selecting for a short face impacted canine neuroanatomy. *Anat. Rec. Adv. Integr. Anat. Evol. Biol.* **2021**, *304*, 101–115. [CrossRef]

87. Kawabe, S.; Shimokawa, T.; Miki, H.; Matsuda, S.; Endo, H. Variation in avian brain shape: Relationship with size and orbital shape. *J. Anat.* **2013**, *223*, 495–508. [CrossRef]

88. Weisbecker, V.; Rowe, T.; Wroe, S.; Macrini, T.E.; Garland, K.L.S.; Travouillon, K.J.; Black, K.; Archer, M.; Hand, S.J.; Berlin, J.C.; et al. Global elongation and high shape flexibility as an evolutionary hypothesis of accommodating mammalian brains into skulls. *Evolution* **2021**, *75*, 625–640. [CrossRef]

89. Benson, R.B.J.; Starmer-Jones, E.; Close, R.A.; Walsh, S.A. Comparative analysis of vestibular ecomorphology in birds. *J. Anat.* **2017**, *231*, 990–1018. [CrossRef] [PubMed]

90. Isler, K.; van Shaik, C. Costs of encephalization: The energy trade-off hypothesis tested on birds. *J. Hum. Evol.* **2006**, *51*, 228–243. [CrossRef] [PubMed]

91. Wright, N.A.; Steadman, D.W.; Witt, C.C. Predictable evolution toward flightlessness in volant island birds. *Proc. Natl. Acad. Sci. USA* **2016**, *113*, 4765–4770. [CrossRef] [PubMed]

92. Li, H.; Sinha, A.; André, A.A.; Spötl, C.; Vonhof, H.B.; Meunier, A.; Kathayat, G.; Duan, P.; Voarintsoa, N.R.G.; Ning, Y.; et al. A multimillenial climatic context for the megafaunal extinctions in Madagascar and Mascarene Islands. *Sci. Adv.* **2020**, *6*, 42. [CrossRef] [PubMed]

Article

Endocranial Anatomy of the Giant Extinct Australian Mihirung Birds (Aves, Dromornithidae)

Warren D. Handley * and Trevor H. Worthy

Palaeontology Group, Flinders University, GPO 2100, Adelaide 5001, Australia; trevor.worthy@flinders.edu.au
* Correspondence: warren.handley@flinders.edu.au

Abstract: Dromornithids are an extinct group of large flightless birds from the Cenozoic of Australia. Their record extends from the Eocene to the late Pleistocene. Four genera and eight species are currently recognised, with diversity highest in the Miocene. Dromornithids were once considered ratites, but since the discovery of cranial elements, phylogenetic analyses have placed them near the base of the anseriforms or, most recently, resolved them as stem galliforms. In this study, we use morphometric methods to comprehensively describe dromornithid endocranial morphology for the first time, comparing *Ilbandornis woodburnei* and three species of *Dromornis* to one another and to four species of extant basal galloanseres. We reveal that major endocranial reconfiguration was associated with cranial foreshortening in a temporal series along the *Dromornis* lineage. Five key differences are evident between the brain morphology of *Ilbandornis* and *Dromornis*, relating to the medial wulst, the ventral eminence of the caudoventral telencephalon, and morphology of the metencephalon (cerebellum + pons). Additionally, dromornithid brains display distinctive dorsal (rostral position of the wulst), and ventral morphology (form of the maxillomandibular [V_2+V_3], glossopharyngeal [IX], and vagus [X] cranial nerves), supporting hypotheses that dromornithids are more closely related to basal galliforms than anseriforms. Functional interpretations suggest that dromornithids were specialised herbivores that likely possessed well-developed stereoscopic depth perception, were diurnal and targeted a soft browse trophic niche.

Keywords: Cenozoic fossil birds; Galloanserae; dromornithids; brain morphology

Citation: Handley, W.D.; Worthy, T.H. Endocranial Anatomy of the Giant Extinct Australian Mihirung Birds (Aves, Dromornithidae). *Diversity* **2021**, *13*, 124. https://doi.org/10.3390/d13030124

Academic Editor: Eric Buffetaut

Received: 29 January 2021
Accepted: 8 March 2021
Published: 15 March 2021

Publisher's Note: MDPI stays neutral with regard to jurisdictional claims in published maps and institutional affiliations.

1. Introduction

The dromornithids were large flightless birds, collectively known as 'mihirungs', whose fossils are a distinctive component of the Cenozoic avifauna of Australia, and are sometimes comparatively abundant in the Australian Neogene fossil record [1,2]. The greatest diversity of the group occurs during the Miocene [1,3–6], but the family is known from fossils dating from the Palaeogene, with a record consisting of a mould of fossil footprints from the Eocene of Queensland [7], postcranial remains from the late Oligocene Pwerte Marnte Marnte Local Fauna (LF) in the Northern Territory [4], and trackways, probably made by dromornithids, reported from the late Oligocene of Tasmania [1,8].

The fossil record shows that the characteristic morphology of dromornithids had already evolved by the late Oligocene, and that it changed little over the next ~20 million years (Ma) until the group became extinct in the late Pleistocene [6,9]. The first dromornithid named was *Dromornis australis* Owen, 1872, from undated deposits at Peak Downs, Queensland [1,6,10–13]. Eight species in four genera of dromornithids are now recognised: Stirling and Zietz [14] named *Genyornis newtoni* Stirling and Zietz, 1896 from Lake Callabonna, South Australia, from what was originally thought to be late Pliocene to early Pleistocene [14] (p. 177), but more recently proposed to be middle to late Pleistocene [15] (p. 16) deposits. Rich [11] described *D. stirtoni* Rich, 1979, *Ilbandornis lawsoni* Rich, 1979 and *I. woodburnei* Rich, 1979 from the late Miocene Waite Formation, Alcoota, Northern Territory, and *Barawertornis tedfordi* Rich, 1979 from the late Oligocene to early Miocene Carl

Creek Limestone at Riversleigh, Northern Queensland. She also named *Bullockornis planei* Rich, 1979 from the middle Miocene Camfield beds, Bullock Creek, Northern Territory [11] (p. 27), but the genus *Bullockornis* was later synonymised with *Dromornis* [12]. This was supported by Worthy et al. [6] upon revision of cranial material of the Bullock Creek specimens, in conjunction with the description of *Dromornis murrayi* Worthy et al., 2016 from Riversleigh. At the same time, Worthy et al. [6] proposed the eight dromornithid species formed two lineages, where the *Dromornis* lineage is monotypic throughout its range, and includes in a temporal succession *D. murrayi*, *D. planei*, *D. stirtoni*, and *D. australis*. The *Ilbandornis/Barawertornis* lineage comprises the more gracile taxa *B. tedfordi*, *I. lawsoni*, *I. woodburnei*, and *G. newtoni*.

Dromornithids were long considered to be ratites [11,14,16–18]. All ratites exhibit reduced wing morphology and are generally large, flightless birds [15,19,20]. These features are shared with dromornithids, but Olson [21] (p. 104) succinctly opined that "large size and flightlessness do not a ratite make", and pointed out that characteristics of the dromornithid mandible, quadrate and pelvis suggested that they were likely derived from an entirely different group of birds. In more recent times, with the discovery of additional cranial elements, phylogenetic analyses by Murray and Megirian [3] concluded that dromornithids were the sister-group of the Anhimidae, and so were Anseriformes. This conclusion was reinforced by Murray and Vickers-Rich [2], who also found similarities to Anseranatidae, a closer sister-group to anatids within Anseriformes. In a phylogenetic study of the affinities of Pelagornithidae (bony-toothed birds), Mayr [22] suggested that dromornithids were likely stem Galloanseres, i.e., a sister group to Galliformes and Anseriformes. A relationship more distant from Anseriformes was also found by Worthy et al. [23], whereby with inclusion of a representative sample of extant galloanseres, representative Neoaves and palaeognaths, and key fossil taxa, *Dromornis* was found to have a stem-galliform relationship. Most recently, Worthy et al. [24,25] used an expanded taxon set, and resolved dromornithids as the sister group to the flightless gastornithids of Eurasia and North America, forming a galloansere clade termed Gastornithiformes Stejneger, 1885. However, the relationship of this clade to either galliforms or anseriforms within Galloanserae, was poorly resolved.

As with gastornithids [26,27], there exists convincing evidence for a herbivorous diet in dromornithids, as some specimens have been preserved with gastroliths [28] (p. 79). Individual stones, presumed to be gastroliths, are common in the fossiliferous silty unit of the Waite Formation producing the Alcoota LF wherein dromornithids are abundant ([2] (p. 262), [29] (p. 164), [30] (Figure 8A)). Undoubted gastroliths in the form of complete or partial gizzard stone sets, are known from several specimens of the Pleistocene dromornithid *Genyornis newtoni*, collected on recent expeditions by THW et al.

Handley et al. [31] demonstrated significant male dominated sexual dimorphism in the largest of all dromornithids, the Miocene species *Dromornis stirtoni*, and revealed those birds identified as male had a mean mass of 528 kg based on tibiotarsus circumference metrics. The tibiotarsus was preferred for estimating body mass in large birds, especially in dromornithids, after statistical evaluations of several mass estimation algorithms applied across a large sample, showed that femoral metrics likely overestimated body mass for them [31,32]. These dromornithids, along with the giant aepyornithid morphotype *Vorombe titan* (Andrews, 1894), from the Holocene of Madagascar [33], likely comprise the largest birds to have ever evolved.

Dromornithid cranial anatomy has previously been comprehensively described [1,3,6], and while Murray and Megirian ([3] (Figure 5)), repeated in Murray and Vickers-Rich ([2] (Figures 78–80)), briefly described a physical endocast for *Dromornis planei*, there exists no information regarding the specific shape and size of the dromornithid brain across the two lineages proposed by Worthy et al. [6].

Worthy et al. [6] identified that from the Oligocene through the late Miocene, the shape of crania of dromornithids changed, with a foreshortening of the length relative to the height of the cranium (e.g., Figure 1). How the shape of the dromornithid brain changed to accommodate these temporal changes in cranial anatomy and whether there exists quantifiable differences in endocranial anatomy between the *Dromornis* and *Ilbandornis* lineages have yet to be appropriately assessed.

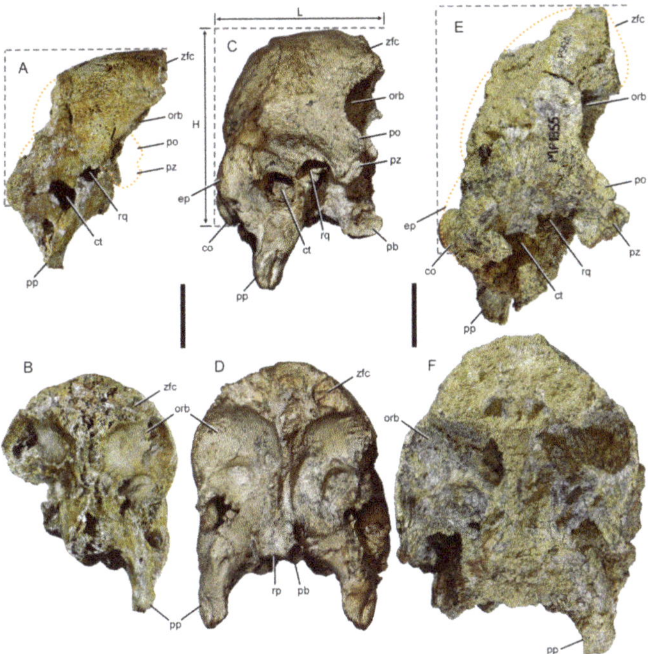

Figure 1. Time series of *Dromornis* specimens showing a progressive increase in cranial height (H) relative to length (L) over ~20–8 Ma: (**A,B**), *Dromornis murrayi* (QM F57984) Oligo–Miocene (~25–23 Ma); (**C,D**), *D. planei* (NTM P9464-106) middle Miocene (~15–12 Ma); (**E,F**), *D. stirtoni* (NTM P5420) late Miocene (~9–7 Ma). Views. Right lateral view (**A,C,E**); Rostral, (**B,D,F**). *D. murrayi* lateral view (**A**) shows the more complete left hand side (**B**) which has been flipped for comparison. Missing cranial areas are shown by orange stippled lines on (**A**) and (**E**). Abbreviations: ct, cavum tympanicum; ep, exoccipital prominence; mm, millimetres; orb, orbit; pb, processus basipterygoidei; po, processus postorbitalis; pp, processus paroccipitalis; pz, processus zygomaticus; rp, rostrum parasphenoidalis; rq, recessus quadratica; zfc, zona flexoria craniofacialis. Scale bars equal 40 mm.

The objectives of this study are to assess dromornithid endocast material spanning the late Oligocene to the late Miocene. This will: (1) allow a comprehensive description of morphological characteristics of the dromornithid brain and its principal innervation for the first time; (2) inform our understanding of how dromornithid brains differ morphologically from those of other galloanseres; (3) assess how *Dromornis* and *Ilbandornis* differ in endocranial anatomy; (4) assess how brain shape responded to significant changes in cranial anatomy through time; (5) identify potential functional constraints shaping the evolution of dromornithid endocranial anatomy across this period; and (6) aid in resolving the phylogenetic position of dromornithids.

2. Materials and Methods

2.1. Abbreviations

Institutions

ANSTO, Australian Nuclear Science and Technology Organisation, Lucas Heights, Sydney, New South Wales, Australia, QM, Queensland Museum, Brisbane, Queensland, Australia, QVM, Queen Victoria Museum and Art Gallery, Launceston, Tasmania, Australia;

NTM, Museum of Central Australia, Alice Springs, Northern Territory, Australia, SAHMRI, South Australian Medical and Health Research Institute, Adelaide, South Australia, SAM, South Australian Museum, Adelaide, South Australia, MV, Museums Victoria, Melbourne, Victoria, Australia.

2.2. Geological and Temporal Data for Fossil Specimens

The fossil materials used in this study were sourced from three localities. Two crania of *Dromornis murrayi* (QM F57984; QM F57974), and a fossil endocast (QM F50412) came from Riversleigh World Heritage Area in north-western Queensland, Australia (Figure 1; SI Figures S1–S3). The fossil endocast (QM F50412) was not scanned and does not contribute to numerical analysis, but is figured for comparative purposes (see SI Figure S3). *Dromornis murrayi* (QM F57984) is derived from the Hiatus site (Queensland Museum Locality 941), Hal's Hill, D Site Plateau, forming part of the Riversleigh Faunal Zone A deposits (e.g., "System A" of [34,35] and "Faunal Zone A" of [36,37]). Hiatus site comprises "pure" limestone formed in an aquatic setting, and has proved difficult to successfully date radiometrically, due to the lack of speleothem or flowstone material often included in palaeo-cave deposits elsewhere at Riversleigh [38]. The Hiatus fauna is considered late Oligocene to early Miocene (~25–23 Ma) in age, based on biocorrelation (i.e., vertebrate stage of evolution; see [34–39]). The second specimen of *D. murrayi* (QM F57974), and the fossil endocast (QM F50412), come from Cadbury's Kingdom site, considered Faunal Zone B and early Miocene (~23–16 Ma) in age [37–39].

The second site complex is located at Bullock Creek in the Northern Territory of central Australia: one cranium, respectively, of *Dromornis planei* (NTM P9464-106) and *Ilbandornis woodburnei* (QVM:2000:GFV:20) were studied from this site (Figure 1; SI Figures S4 and S5). The Camfield Beds exposed at Bullock Creek, are fossiliferous freshwater conglomeratic limestone deposits that contain the Bullock Creek LF, which includes several aquatic and "stream-bank" species [30], and forms the type locality for the Camfieldian Land Mammal Age [40]. Fossils from the site are generally well preserved [6], and are considered to be middle Miocene (~14–12 Ma) in age based on biocorrelation, specifically the stage of evolution of diprotodontid *Neohelos* spp. [14,30,38,40,41].

The third site complex is located at Alcoota Station, approximately 110 km northeast of Alice Springs in the Northern Territory of central Australia [42]: two crania of *Dromornis stirtoni* (NTM P5420; NTM P3250) from Alcoota LF were studied (Figure 1; SI Figures S6 and S7). The Alcoota LF derives from unconsolidated fluviatile clays and silts of the Waite formation, previously interpreted as lacustrine deposits [29]. The sediments are now considered to be overbank silts accumulated via debris flow, wherein fossils are concentrated in extensive bonebeds with little or no association [42–45]. Specimens are generally poorly preserved, likely due to repeated fluctuations in moisture content of the siltstone matrix, causing fracturing and compaction of fossils over time [30]. Alcoota LF is believed to be late Miocene (~9–7 Ma) in age based on biocorrelation [1,30,40,44,46], and is the type locality for the Waitean Land Mammal Age [40]. Alcoota LF is unique in that it preserves the only late Miocene vertebrate community known from Australia outside of Riversleigh [2,30].

Four crania of extant basal galloansere birds were also included: the phasianid *Gallus gallus* (SAM B34041), a megapodiid *Leipoa ocellata* (SAM B11482), an anhimid *Anhima cornuta* (MV B12574), and the anseranatid *Anseranas semipalmata* (SAM B48035).

2.3. Nomenclature

We follow the anatomical nomenclature in Baumel et al. [47] for osteology, innervation, and external or brain surface anatomy (see Figure 2, Figure 3A–D, Figure 4A–D; SI Figures S4K–N and S5K–L). Therefore, cranium is the term preferred for that part of the skull enclosing the brain, rather than neurocranium. Descriptions of the internal architecture of the avian brain follow Jarvis et al. [48,49] (Figure 6). At first mention, osteological, innervation, and brain surface anatomy is described using Latin nomenclature, with the anglicised equivalent in parenthesis, or mentioned immediately subsequent. Thereafter, we use anglicised equivalents where appropriate.

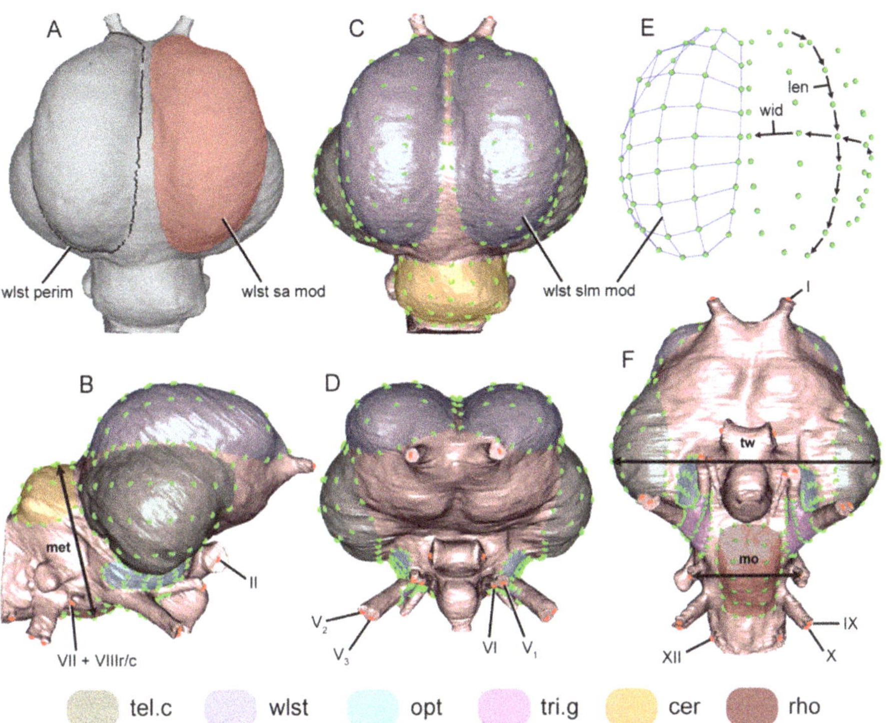

Figure 2. Modular Surface Area (**A**) and Measurement data (**E**), along with landmark modules used to capture endocast shape (see Section 2.6.2, Section 2.6.3 and SI Section S3.2). Brain surface Slm modules (green dots) and innervation Lm module (red dots), are mapped onto the endocast of *Dromornis planei* (NTM P9464-106), and Slm modules are colour-shaded to facilitate anatomical identification (see also SI Figures S10 and S11). Acquisition of Modular Surface Area data are illustrated by a dorsal endocast (**A**), showing the right side wulst surface area module selected (wlst sa mod–pink), prior to computation of the surface area value for the defined region. The endocast (**A**) also shows the previously defined left side wulst module, circumscribed by a perimeter polyline (wlst perim) for which the surface area value was computed (see Section 2.6.3 above). Measurement values are illustrated by a 3D shape plot (**E**) showing wulst Slm modules (wlst slm mod), left side Slms (green dots) are linked (blue) to provide perspective. Distance (vector) values were calculated between individual Slms forming the modular width (**wid**) and length (**len**) measurements, vector values were then combined to form the total measurement value; Measurements were also calculated between individual Slms for; (**met**), metencephalon (cerebellum + pons) total height; (**tw**), endocast total width; (**mo**), medulla oblongata total width (see Section 2.6.2 above). Views: (**A,C,E**), Dorsal; (**B**), right lateral; (**D**), rostral; (**F**), ventral. Abbreviations: cer, cerebellum; Lm, landmark; mod, module; opt, optic lobe; perim, perimeter; rho, rhombencephalon; sa, surface area; Slm, semilandmark; tel.c, caudal telencephalon; tri.g, trigeminal ganglion; wlst, wulst; I, left olfactory nerve; II, right optic nerve; V_1, ophthalmic nerve; V_2, maxillary nerve; V_3, mandibular nerve; VI, abducent nerve; VII + VIIIr/c, rami of the facial nerve (VII), and the rostral (VIIIr) and caudal (VIIIc) vestibulocochlear nerves; IX, glossopharyngeal nerve; X, vagus nerve; XII, ramus of the hypoglossal (XII) nerve (XIIc in the extant galloanseres (Section 3.1.6, Figure 4D; SI Figures S10C and S11C). Endocasts are not to scale. [Note: for the modular Slm suite mapped onto the endocast of *Leipoa ocellata* (MV B12574), see SI Figures S10 and S11).

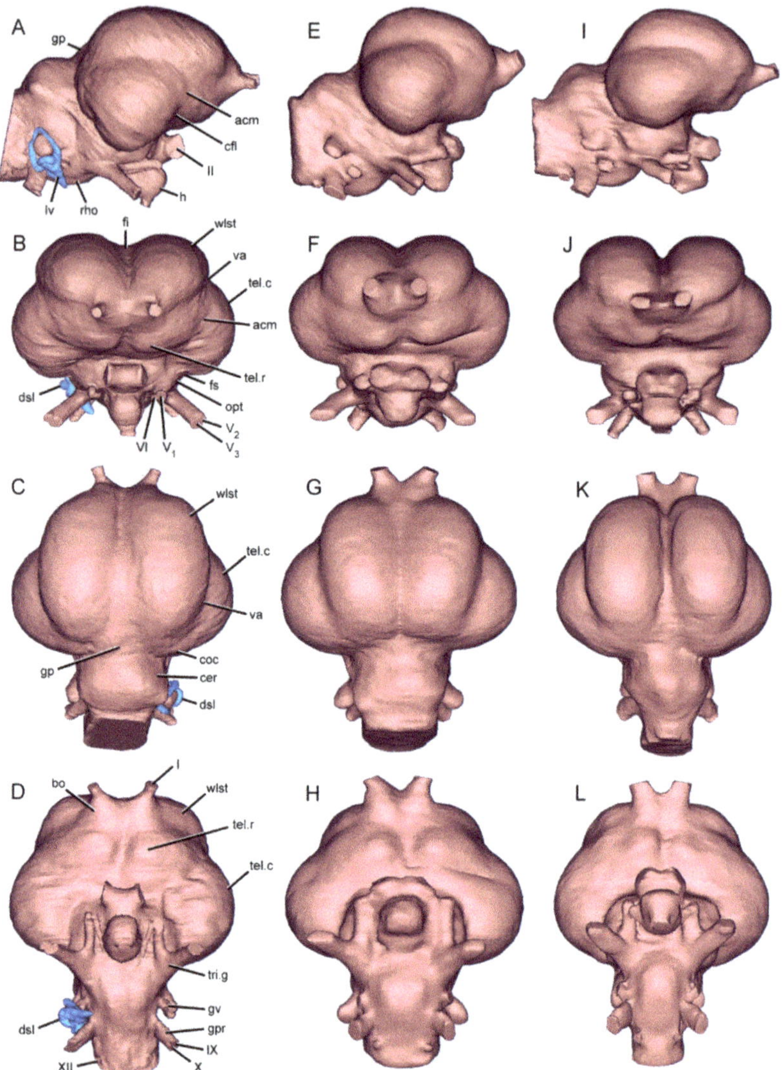

Figure 3. Dromornithid endocasts: (**A–D**), *Dromornis planei* (NTM P9464-106); (**E–H**), *D. murrayi* reconstruction (QM F57984 + QM F57974; see SI Figure S9); (**I–L**), *Ilbandornis woodburnei* (QVM:2000:GFV:20). Views. Right lateral (**A,E,I**); rostral (**B,F,J**); dorsal (**C,G,K**); ventral (**D,H,L**). Trigeminal nerves (V_1, V_2, V_3) are truncated where exiting the cranium. Abbreviations: acm, arteria cerebralis medialis; bo, olfactory bulb; cer, cerebellum; cfl, cerebrum fovea limbica; coc, cerebrum pars occipitalis; dsl, lateral semicircular duct; fi, fissura interhemispherica; fs, fissura subhemispherica; gp, glandula pinealis; gpr, proximal ganglion; gv vestibular ganglion; h, hypophysis; lv, vestibular organ (semicircular ducts + cochlea [blue]; see also Section 4.1.4); mm, millimetres; opt, optic lobe; rho, rhombencephalon; tel.c, caudal telencephalon; tel.r, rostral telencephalon; tri.g, trigeminal ganglion; va, vallecula telencephali; wlst, wulst; I, olfactory nerve; II, optic nerve; V_1, ophthalmic nerve; V_2, maxillary nerve; V_3, mandibular nerve; VI, abducent nerve; IX, glossopharyngeal nerve; X; vagus nerve; XII, ramus of the hypoglossal nerve. Endocasts are not to scale.

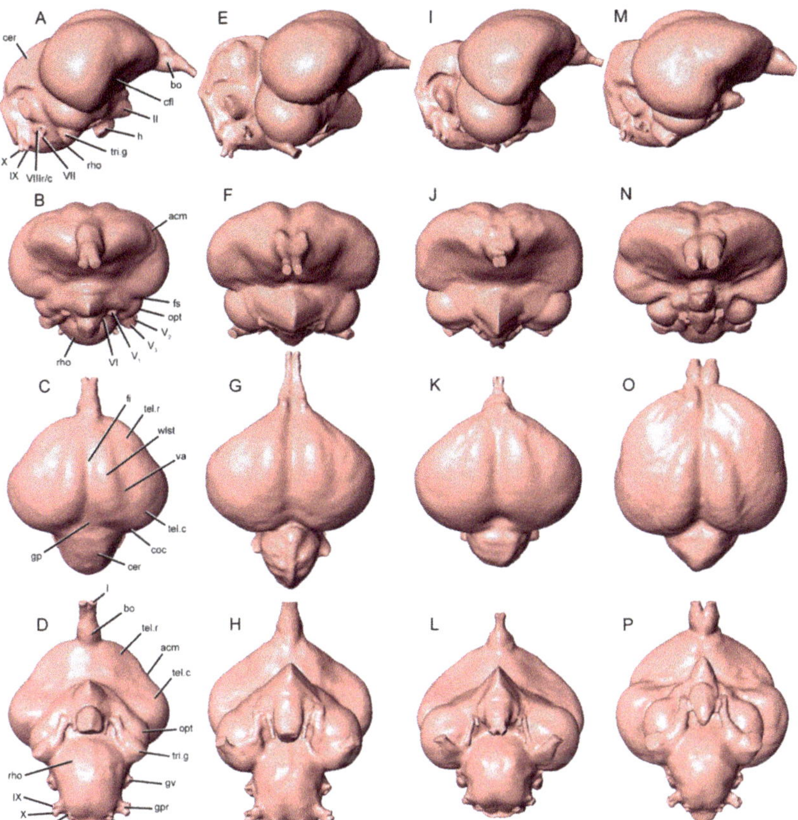

Figure 4. Galloansere endocasts (**A–D**), *Anhima cornuta* (MV B12574); (**E–H**), *Gallus gallus* (SAM B34041); (**I–L**), *Leipoa ocellata* (SAM B11482); and (**M–P**), *Anseranas semipalmata* (SAM B48035). Views: Right lateral (**A,E,I,M**); Rostral (**B,F,J,N**); Dorsal (**C,G,K,O**); Ventral (**D,H,L,P**). The full extension of the olfactory (**I**) nerves in *G. gallus* are shown in dorsal (**G**) view, but were cropped in lateral (**E**) and ventral (**H**) views to fit the plate. Abbreviations: acm, arteria cerebralis medialis; bo, olfactory bulb; cer, cerebellum; cfl, cerebrum fovea limbica; coc, cerebrum pars occipitalis; fi, fissura interhemispherica; fs, fissura subhemispherica; gp, glandula pinealis; gpr, proximal ganglion; gv vestibular ganglion; h, hypophysis (note the caudoventral hypophysis has been trimmed to facilitate access to rostroventral rhombencephalon surfaces); opt, optic lobe; rho, rhombencephalon; tel.c, caudal telencephalon; tel.r, rostral telencephalon; tri.g, trigeminal ganglion; va, vallecula telencephali; wlst, wulst; I, olfactory nerve; II, optic nerve; V_1, ophthalmic nerve; V_2, maxillary nerve; V_3, mandibular nerve; VI, abducent nerve; VII, facial nerve; VIIIr/c, rostral and caudal rami of the vestibulocochlear nerve; XIIr, rostral and; XIIc, caudal rami of the hypoglossal nerve. Endocasts are not to scale.

Much disparate terminology has been used for the description of surface morphology of the avian cranium and brain, and in some instances with no consensus for precedence of any particular term over another (e.g., see [47,48], and references therein). We agree with Early et al. [50] who considered the term 'wulst' (wlst, Figure 2, Figure 3B–D and Figure 4C) appropriate to describe the external dorsal eminences of the internal hyperpallium [49] (Figure 6), and lobus opticus (optic lobe; opt, Figures 2, 3B and 4B,D) to describe the external ventral eminences of the internal tectum opticus (optic tectum), or tectum mesencephali (mesencephalon). Additionally, we use cerebrum pars frontalis (rostral telencephalon; tel.r, Figure 3B,D and Figure 4C,D; see also Section 2.5), to describe the surface topology of dorsorostrolateral mesopallium and nidopallium [49] (Figure 6), rostrad of the arteria cerebralis medialis (medial cerebral artery; acm, Figure 3A–B and Figure 4B,D) dorsoventral traversal of the hemispherium telencephali (telencephalic hemisphere); cerebrum pars parietalis (caudal telencephalon; tel.c, Figure 2, Figure 3B–D and Figure 4C–D), to describe the surface topology of the mesopallium, nidopallium and arcopallium [49] (Figure 6), forming part of the caudolateral telencephalon pallial complex caudad of the traversal of the medial cerebral artery (see Section 2.5 below); ganglion trigeminale (trigeminal ganglia, tri.g, Figures 2, 3D and 4A,D) to describe ganglia of the trigeminal nerve (V) complex (see Section 3.1.3 below), inserting on the ventral surface of the optic lobe.

2.4. Modelling

One cranium each of the following species were micro-computed tomography (µCT) scanned using the Skyscan 1076 µCT instrument (Bruker microCT) at Adelaide Microscopy, University of Adelaide: *Gallus gallus*, 17.0 micrometre (µm) resolution, at 59 kilovolts (kV) and 167 microamps (µA); *Leipoa ocellata*, 17.4 µm resolution, at 48 kV and 139 µA; *Anhima cornuta*, 34 µm resolution, at 100 kV and 100 µA; and *Anseranas semipalmata*, 34.8 µm resolution, at 100 kV and 90 µA. Skyscan µCT acquisition data were reconstructed using NRecon v1.6.10.4 (Bruker microCT) and compressed using ImageJ v1.51w [51] software.

The crania of *D. murrayi*, *D. stirtoni* and *I. woodburnei* were medical X-ray CT scanned using the Siemens Somatom Force CT instrument located at the SAHMRI facility in Adelaide. Acquisition CT data were captured at a slice thickness of 0.4 mm, but with the application of an oversampling technique allowing the acquisition of twice the number of slices per detector row, an effective resolution of 240 µm was achieved for all specimens, excluding *D. stirtoni* (NTM P3250) which was CT scanned at a resolution of 320 µm. Acquisition data were reconstructed by M. Korlaet of Dr Jones and Partners using Siemens proprietary software.

The cranium of *D. planei* was scanned at the ANSTO nuclear facilities in Sydney using the DINGO neutron CT instrument, located in the OPAL reactor beam hall on thermal beam HB2. Neutron CT acquisition data were captured at low-intensity mode at 95 µm resolution and were reconstructed by Dr. J. Bevitt of ANSTO. All CT data used were isotropic.

2.4.1. Three-Dimensional (3D) Surface Model Construction

Three-dimensional (3D) surface model construction was conducted via segmentation using Materialise Mimics v18 software, and 3D surface stereolithograph (STL) endocast models were produced from reconstructed CT data to represent the shape of the brain (Figures 3 and 4). These included the base and immediate stem of the major nerves passing from the cranium into the brain (see also Section 3.1, Section 4.4.1 and SI Section S3.1.1). Surface STL models were exported to Materialise Mimics 3-Matic v10 software for reconstruction and remeshing.

2.4.2. Model Reconstructions

In many fossils, structures are often lost or damaged by taphonomic processes over time, or during recovery. Where specimens are somewhat bilaterally symmetrical, as is the case of endocasts, damaged or missing structures may be digitally reconstructed based on preservation of one side, or parts of a particular endocast. The incomplete endocasts for the two fossil specimens of *D. stirtoni* NTM P5420 and NTM P3250 constrained their interpretation. Thus, a two-dimensional (2D) reconstruction representative of the species was derived using both endocast models (see SI Figure S8 and Section S2.1). Similarly, endocasts for specimens of *D. murrayi* were, respectively, damaged and incomplete, where QM F57984 preserves only the left hand side (LHS) dorsolateral endocast, and QM F57974 preserves only the ventral endocast. Consequently, a single 3D endocast model was compiled from CT data of the two specimens of *D. murrayi* (see SI Figure S9 and Section S2.2).

2.4.3. Remeshing

Remeshing of 3D STL surface models is required to optimise the quality of the triangles comprising the surface mesh, and to reduce the file size of models for landmarking (see below). Remeshing operations were carried out in Materialise 3-Matic v10 and passing of remeshed STL objects to Polygon File Format (PLY) was conducted in MeshLab v2016.12 [52].

2.5. Landmarking

To capture the shape of the endocasts, we defined seven morphological zones using semilandmark (Slm) patches, or modules (see Figure 2; SI Figures S10 and S11, Section S3.2). This approach allowed us to assess whether each morphological zone differed in the same way between taxa and along a lineage, or whether the zones differed in contrasting ways across specimens.

Additionally, the dorsal endocast of dromornithids is dominated by wulst structures (e.g., Figures 2 and 3), which are distinct with respect to those of the extant galloanseres (e.g., wlst, Figure 3 versus [vs] Figure 4, and Section 3.2.2 below). We prefer the explanation that the expansion, or hypertrophy, of the wulst in dromornithids has effectively masked the surface morphology of the rostrodorsal telencephalon (see Section 3.2.1 below). This required the segregation of telencephalic hemispheres into rostral and caudal regions, defined by the traversal of the medial cerebral artery (see acm, Figure 4B, and SI Sections S3.2.1.3–S3.2.1.5), such that telencephalic regions caudal of the medial cerebral artery might be compared across all specimens.

Digital landmarking of 3D endocast surface models was conducted in IDAV Landmark v3.6 [53] using 20 fixed (type 1) and 460 semi- (type 3) landmarks [54] for a total of 480. The set of landmarks (Lms) and semilandmarks (Slms) comprising the modules (Figure 2; SI Figures S10 and S11) were the basis for all subsequent shape assessment. Endocast landmarking protocols and descriptions of the full Modular landmark (Lm) suite are given in Supplementary Information (SI, Section S3).

2.6. Data

We used Slm modules to capture the shape of discrete regions of the brain (see Section 2.5 above). Derived from those shape data (see Section 2.6.1), we computed Measurement data (see Section 2.6.2), and acquired Surface Area data based on the 'footprint' of Slm modules as defined (see Section 2.6.3), with the exception of trigeminal ganglion data, where Measurement data were computed from trigeminal ganglia Slm modules (e.g., Figure 10C and Figure 11C), and Surface Area data were computed from the truncated faces of the maxillomandibular (V_2+V_3; see Figure 2E, Figure 3B and SI Figure S5L) branch of the trigeminal (V) nerve complex (see Figures 3D and 4A,D). Collectively, those three forms of data informed the systematic assessment of morphological differences between the endocasts of individual dromornithid specimens (see Section 2.7.2, Section 3.3 and SI Section S4), and the comparison of dromornithid endocast morphology with those of the extant galloanseres (see Section 3.4 and SI Section S5).

2.6.1. Modular Lm Data

Three-dimensional digital shape data derived from the Modular Lm suite (Figure 2; SI Figures S3, S10 and S11) were used for all assessments of shape (see Section 2.7 below).

2.6.2. Measurement Data

Measurement data were calculated between Lm and Slm locations along specific transects for each specimen, using the 'interlmkdist' function in Geomorph v3.1.3 [55]; see also Section 2.7 below. Measurements for the length and width of each modular structure, capturing the directional 'curve' over a 3D surface, were calculated by adding together the distances between each Slm forming the measurement vector (see Figure 2E). Data for each paired structure (i.e., wulst, rostral and caudal telencephalon, optic lobe, and trigeminal ganglion modules) were combined, and mean Measurement values calculated (see Table 1A). Additionally, measurements describing gross endocast morphological 'vector' distances were calculated between two Lm or Slm locations (see Figure 2D,F;

Table 1A). Size-standardised mean Measurement log shape ratios were calculated by the log shape ratios method [56], where species Measurement values were divided by species endocast total volume values and $\log_{10}$ transformed [57] (p. 99), [58] (p. 117). The log shape ratio approach produces size-standardised shape variables from univariate data and is analogous with the Procrustes superimposition method for multivariate Lm data (i.e., GPA; see Section 2.7.1 below), where both methods correct for size while retaining shape variation [59] (p. 1389). Measurement log shape ratios are presented in text, in Table 1B, and plotted in SI Figures S12 and S13.

Table 1. **A**, Mean Measurement values calculated between Slm locations for each species. **B** and **D**, respectively, Size-standardised Mean Measurement and modular Surface Area log shape ratios. Log shape ratios were calculated by the log shape method (see Section 2.6.2). **C**, Mean modular Surface Areas computed directly from endocast surfaces (see Section 2.6.3). All bilateral structure data (i.e., wulst, rostral and caudal telencephalon, optic lobe, and trigeminal ganglion modules) were combined and mean values calculated. Abbreviations: *A. cornuta*, *Anhima cornuta* (MV B12574); *A. semipalmata*, *Anseranas semipalmata* (SAM B48035); Cer, cerebellum; *D. murrayi*, *Dromornis murrayi* reconstruction (QM F57984 + QM F57974); *D. planei*, *Dromornis planei* (NTM P9464-106); Endo Surf, endocast total surface area; Endo Vol, endocast total volume; *G. gallus*, *Gallus gallus* (SAM B34041); *I. woodburnei*, *Ilbandornis woodburnei* (QVM:2000:GFV:20); L, length; *L. ocellata*, *Leipoa ocellata* (SAM B11482); Med.Ob, medulla oblongata; Meten, metencephalon; mm, millimetres; mm^2, square millimetres; mm^3, cubic millimetres; Opt, optic lobe; Rho, rhombencephalon; Slm, semilandmark; Tel.c, caudal telencephalon; Tel.r, rostral telencephalon; TH, total height; Tri.g F, trigeminal ganglion maxillomandibular ($V_2 + V_3$) face; TW, total width; W, width; Wlst, wulst.

Measurement	A. Mean Measurement Values (mm)						
	G. gallus	*L. ocellata*	*A. cornuta*	*A. semipalmata*	*D. murrayi*	*D. planei*	*I. woodburnei*
Wlst L	12.47	14.32	16.79	20.50	55.90	67.65	51.32
Wlst W	4.81	5.52	6.34	7.92	27.63	34.43	28.87
Tel.r L	5.59	9.59	10.37	17.73	N/A	N/A	N/A
Tel.r W	4.02	5.45	8.13	11.93	N/A	N/A	N/A
Tel.c L	15.46	12.75	17.30	17.52	47.26	49.71	41.88
Tel.c W	12.91	13.40	18.43	22.18	41.14	40.61	30.52
Opt L	15.08	18.25	13.37	15.87	16.96	19.11	16.72
Opt W	5.98	7.81	4.67	4.85	6.75	5.96	8.02
Tri.g L	5.96	4.32	7.95	9.60	13.47	13.76	12.30
Tri.g W	4.83	3.76	3.03	3.32	9.56	9.20	9.18
Cer L	12.73	10.95	17.53	15.27	17.79	20.85	21.24
Cer W	10.71	9.91	13.63	17.55	37.87	45.99	33.21
Rho L	9.69	10.71	13.64	14.54	25.65	26.82	23.47
Rho W	7.66	9.51	12.39	10.03	18.19	18.02	13.83
Tel.c TW	21.07	22.25	28.21	31.16	67.34	72.83	57.52
Meten TH	15.43	16.26	20.34	22.04	38.85	40.68	34.56
Med.Ob TW	12.94	11.66	14.33	15.50	37.87	40.59	28.91
Endo Vol (mm^3)	3733.84	4519.27	8031.85	10881.35	95577.71	122859.93	60289.34
B. Mean Measurement Log Shape Ratios							
Wlst L	0.127	0.163	0.157	0.186	0.362	0.415	0.368
Wlst W	−0.287	−0.251	−0.266	−0.227	0.056	0.122	0.119
Tel.r L	−0.221	−0.011	−0.052	0.123	NA	NA	NA
Tel.r W	−0.365	−0.257	−0.158	−0.049	NA	NA	NA
Tel.c L	0.220	0.113	0.170	0.118	0.289	0.281	0.280
Tel.c W	0.142	0.134	0.198	0.220	0.229	0.193	0.143
Opt L	0.210	0.268	0.058	0.075	−0.156	−0.134	−0.119
Opt W	−0.193	−0.101	−0.398	−0.440	−0.556	−0.640	−0.438
Tri.g L	−0.194	−0.358	−0.167	−0.143	−0.256	−0.276	−0.252
Tri.g W	−0.285	−0.418	−0.587	−0.604	−0.405	−0.452	−0.379
Cer L	0.136	0.046	0.176	0.058	−0.135	−0.096	−0.015
Cer W	0.061	0.003	0.067	0.119	0.193	0.248	0.180
Rho L	0.017	0.037	0.067	0.037	0.024	0.013	0.029
Rho W	−0.085	−0.015	0.025	−0.124	−0.125	−0.159	−0.201
Tel.c TW	0.355	0.354	0.382	0.368	0.443	0.447	0.418
Meten TH	0.219	0.218	0.240	0.218	0.204	0.194	0.197
Med.Ob TW	0.143	0.074	0.088	0.065	0.193	0.193	0.119

Table 1. *Cont.*

Measurement	A. Mean Measurement Values (mm)						
	G. gallus	*L. ocellata*	*A. cornuta*	*A. semipalmata*	*D. murrayi*	*D. planei*	*I. woodburnei*
Module	C. Mean modular Surface Area Values (mm^2)						
Wlst	58.50	65.71	90.65	117.40	1353.96	1851.49	1170.99
Tel.r	19.62	47.39	82.55	216.04	NA	NA	NA
Tel.c	144.52	133.10	249.41	297.48	1213.28	1297.48	852.94
Opt	78.90	118.28	52.33	71.57	139.37	164.03	145.83
Tri.g F	1.342	1.335	5.114	6.237	14.537	13.477	10.431
Cer	124.10	103.59	205.25	192.07	678.37	816.82	528.19
Rho	63.21	86.91	136.69	135.38	394.24	585.79	290.76
Endo Surf (mm^2)	1544.33	1682.29	2404.61	2985.52	13199.82	15874.02	10200.52
	D. Mean Modular Surface Area Log Shape Ratios						
Wlst	0.176	0.136	0.079	0.078	0.655	0.720	0.708
Tel.r	−0.299	−0.006	0.039	0.343	NA	NA	NA
Tel.c	0.569	0.443	0.519	0.482	0.608	0.566	0.571
Opt	0.306	0.391	−0.159	−0.137	−0.332	−0.332	−0.196
Tri.g F	−1.464	−1.556	−1.169	−1.197	−1.314	−1.418	−1.342
Cer	0.502	0.334	0.434	0.292	0.355	0.365	0.363
Rho	0.210	0.258	0.258	0.140	0.119	0.221	0.103

2.6.3. Surface Area Data

Surface Area data for each endocast module as defined here (see SI Section S3.2) were computed directly from the surface of each endocast using MeshLab v2016.12 (see Figure 2A). Two forms of Surface Area data were acquired: (1) total endocast Surface Area; and (2) modular Surface Area in square millimetres (mm^2), from which mean Surface Area values for all bilateral modules (i.e., wulst, rostral and caudal telencephalon, optic lobe, and trigeminal ganglion face) were computed (see Table 1C). Size-standardised mean Surface Area log shape ratios were calculated by the log shape ratios method, where species Mean Surface Area values were divided by species endocast Surface Area values and log$_{10}$ transformed (see Section 2.6.2 above). Surface Area log shape ratios are presented text, in Table 1D, and plotted in SI Figure S14.

Additionally, due to taphonomic processes over some 14 Ma, the caudoventral endocast of *Dromornis planei* had suffered somewhat of a rostrocaudal ventral rotation, along with a subtle rostrally orientated 'twisting' of the caudoventral endocast with respect to dorsal endocast surfaces. This can be appreciated in the slight caudal displacement of LHS optic lobe margins in the *D. planei* endocast, when observed from the ventral aspect (see Figures 2F and 3D below), and in the *D. planei* cranium itself (see Figure 1D above). During landmarking, optic lobe and trigeminal ganglion module margins on the ventral *D. planei* endocast were situated with respect to existing morphological boundaries (see SI Section S3.2.1.9), without attempting to adjust margins for the subtle taphonomic distortion present in the endocast. Subsequent univariate Measurement and Surface Area data for the optic lobe and trigeminal ganglia formed the primary focus for downstream morphological assessments, as the computation of mean log shape ratio values from those paired modular data (see Section 2.6.2 above), accommodated for the subtle bilateral misalignment of ventral midbrain regions in the *D. planei* endocast.

2.7. Analyses

All data analyses and visualisations (Figure 2, Figure 5 and SI Figure S15), excluding SI Figures S12–S14 (OriginPro v2018b.95.1.195, OriginLab Corporation), were conducted in R v3.6.1 [60] using RStudio v1.2.5019 [61], and package Geomorph v3.1.3.

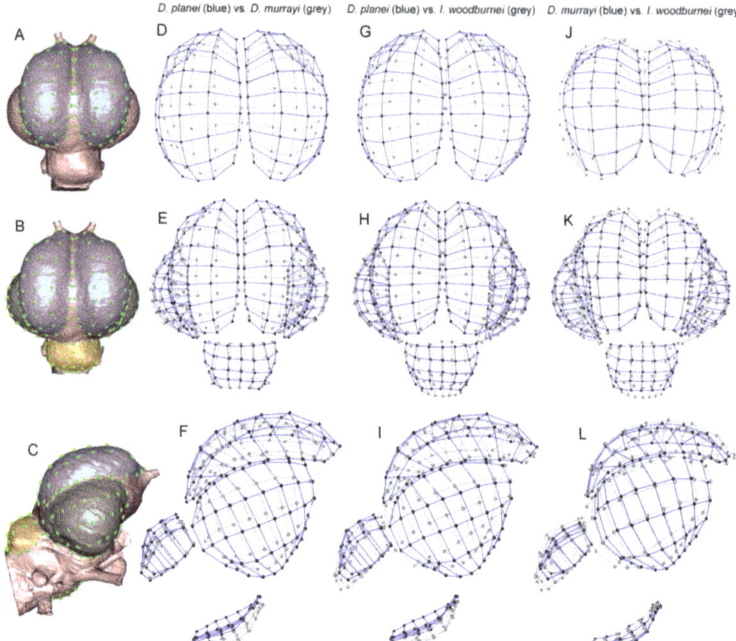

Figure 5. Three-dimensional modular shape variation plots for dromornithid specimens (see Section 2.7.2). (**A**), *Dromornis planei* (NTM P9464-106) endocast showing the caudodorsal view of the wulst (grey) Slm modules represented in plots (**D**), (**G**), and (**J**); (**B**), *D. planei* (NTM P9464-106) endocast showing the dorsal view of the wulst (grey), caudal telencephalon (green) and cerebellum (yellow) Slm modules represented in plots (**E,H,K**); (**C**), *D. planei* (NTM P9464-106) endocast showing the right lateral view of the wulst (grey), caudal telencephalon (green), cerebellum (yellow), and rhombencephalon (brown) Slm modules represented in plots (**F,I,L**). Modular Slms (green dots) are mapped onto the inset endocasts, and modules are colour-shaded to assist identification (see Figure 2; SI Figures S10 and S11). Shape variation plots are arranged by column (see column header): (**D–F**), *D. planei* (NTM P9464-106–blue) vs. *D. murrayi* reconstruction (QM F57984 + QM F57974–grey); (**G–I**), *D. planei* (NTM P9464-106–blue) vs. *Ilbandornis woodburnei* (QVM:2000:GFV:20–grey); (**J–L**), *D. murrayi* reconstruction (QM F57984 + QM F57974–blue) vs. *I. woodburnei* (QVM:2000:GFV:20–grey). Abbreviations: Slm, semilandmark; vs., versus.

2.7.1. Generalized Procrustes Analysis (GPA)

Generalized Procrustes Analysis [62,63] is how shape variables, or Procrustes coordinates, are derived from specimen landmark data which are translated, scaled, and optimally rotated using a least-squares criterion [64,65]. During superimposition, Slms on curves and surfaces were slid along tangent directions and tangent planes, respectively [66–69], and locations of Slms were optimised by minimising bending energy [66,67]. Aligned Procrustes coordinates were used for subsequent assessments of shape differences between specimens. The advantage of this approach allows for specimen data to be translated, scaled, and rotated about a common centroid, and shape differences between specimens may be examined within a shared shape space (see below).

2.7.2. Three-Dimensional Modular Shape Variation Plots

To better understand the extent of particular morphological variation between species of dromornithid, we used 3D shape variation plots to visualise the modular shape variation between individual specimens. One dromornithid species represented by black dots and blue links, is superimposed over another represented by grey dots and grey links, visualising the extent and direction of modular shape variation between the two specimens. In this manner, the morphological differences between the endocasts of each species of dromornithid were described (see Figure 5, Section 3.3 and SI Section S4).

3. Results

3.1. Dromornithid Innervation That Differs from the Extant Galloanseres (see Figures 3 and 4; SI Figures S4 and S5)

3.1.1. Nervus Olfactorius

The olfactory (I) nerve transmits rostrocaudally into the bulbus olfactorius (olfactory bulb; bo, Figures 3D and 4D), through the bony foramen n. olfactorii (folf, SI Figures S4K–S4L, S5K) of the rostrodorsal cranium. The dromornithid olfactory bulb is best described by the more complete RHS lateral view of *D. murrayi* (Figure 3E). The margins of the olfactory bulb are pronounced both dorsally and ventrally, but caudodorsally masked by the rostral eminence of the wulst. The caudomediolateral transmission of the olfactory bulb margins are shown by the ventral view of *D. murrayi* and *D. planei* (Figure 3D,H, respectively) as transitioning into the rostroventral endocast without reduction in mediolateral width, as seen most notably in *A. cornuta* and *A. semipalmata* (e.g., Figure 4C,D,O,P, respectively).

3.1.2. Nervus Opticus

The optic (II) nerves form the chiasma opticum caudally, and pass through the os laterosphenoidale, forming the caudomedial wall of the orbit via the foramen opticum (fopt, SI Figures S4K,L and S5K), meeting rostrally at the septum interorbitale. In dromornithids, the interorbital septum rostral of the foramen opticum is robust, and the optic (II) nerves divides rostrolaterally into two well defined branches (i.e., II; Figures 3A and 4A).

3.1.3. Nervus Trigeminus

The trigeminal (V) nerve complex comprises three divisions. The medial or ophthalmic branch carries nervus ophthalmicus (ophthalmic [V_1] nerve; Figures 3B and 4B), transmitting to the trigeminal ganglia (see Section 3.2.5 below) on the ventral surfaces of the optic lobe (see Section 3.2.4 below), through the foramen n. ophthalmici (foph; SI Figures S4K,L and S5K). The foramen n. ophthalmici opens into the "lacerate (presphenoid) fossa" *sensu* [6] (Figure 1D), located ventrolaterad from the foramen opticum, between the os laterosphenoidale, os basisphenoidale, os parasphenoidale and septum interorbitale (os laterosphenoidale complex) of the caudomedial wall of the orbit. In dromornithids, the foramen n. ophthalmici is paired with the foramen n. abducentis, which transmits n. abducens (abducent [VI] nerve, see below). The lateral or maxillomandibular (V_2+V_3) branch of the trigeminal ganglion complex, carries n. maxillaris (maxillary [V_2] nerve) and n. mandibularis (mandibular [V_3] nerve), both of which enter the skull rostroventrolaterally at the foramen n. maxillomandibularis (fmx, SI Figures S4K,L and S5K), a single opening between the prootic and laterosphenoid bones of the skull. In dromornithids, the maxillomandibular branch is distinctive in that it is markedly elongate, compared to the extant galloanseres assessed, and transmits the maxillary (V_2) and mandibular (V_3) cranial nerves minimally 20 mm caudoventrolaterally (in *Dromornis*), after entering the skull at the foramen n. maxillomandibularis (see also V_2 + V_3, Figure 3B, SI Figure S5L; and Section 3.2.5 below).

3.1.4. Nervus Abducens

The abducent (VI) nerve inserts on the rostroventral rhombencephalon and is transmitted caudoventrally through the osseus canalis n. abducentis, after entering the skull at the foramen n. abducentis. In dromornithids, the foramen n. abducentis is paired with the foramen n. ophthalmici, forming a single bi-lobal foramen in the rostromedial os laterosphenoidale structures of the orbit (fa, SI Figures S4K,L and S5K, VI, Figure 3D and SI Figure S5L; and above).

3.1.5. Nervus Glossopharyngeus

The glossopharyngeal (IX) nerve inserts caudoventrolaterally on the rhombencephalon (rho; Figure 3A), and forms the rostral component of the combined root ganglion proximale (proximal ganglion; gpr, Figure 3D) with n. vagus (vagus [X] nerve, see below).

The proximal ganglion is enclosed in the fovea ganglii vagoglossopharyngealis in the lamina parasphenoidalis of the fossa cranii caudalis, between the exoccipital and opisthotic bones. The glossopharyngeal (IX) nerve enters the cranium caudoventrolaterad from the vagus (X) nerve at the foramen n. glossopharyngeus (fg; SI Figure S4M,N), situated in the fossa parabasalis. In dromornithids, the glossopharyngeal (IX) and vagus (X) nerves separate caudoventrolaterad of the eminence of the proximal ganglion from the rhomben-cephalon surface (Figure 3D). The separation of the glossopharyngeal (IX) and vagus (X) nerves in dromornithids is similar to, but occurs to some extent further distally, than the condition seen in *Anhima cornuta* (Figure 4D), *Gallus gallus* (Figure 4H), and *Anseranas semipalmata* (Figure 4P), but distinct to that seen in *Leipoa ocellata* (Figure 4L).

3.1.6. Nervus Hypoglossus

The hypoglossal (XII) nerves typically comprise rostral (XIIr) and caudal (XIIc) rami (e.g., Figure 4D,H,L,P), which in dromornithids are represented by one ramus at either side of the caudoventrolateral medulla oblongata (e.g., XII, Figure 3D,H,L). This condition is distinct to that seen in *A. cornuta* (Figure 4D), *G. gallus* (Figure 4H), *L. ocellata* (Figure 4L), and *A. semipalmata* (Figure 4P), which display both rostral (XIIr) and caudal (XIIc) hy-poglossal rami. In dromornithids, the nerves appear to transmit through a single canalis n. hypoglossi and bifurcate in close proximity with the external surface of the os exoccipitale, at the paired foramen n. hypoglossi (fh, SI Figure S4M,N).

3.2. Characteristics of Dromornithid Endocast Morphology

3.2.1. Rostral Telencephalon

The external morphology of the rostral telencephalon as defined here (see SI Figures S10 and S11 and Section S3.2.1.3), evident rostrodorsally of the medial cerebral artery (acm, see Figure 3A,B and Figure 4B,D) in both anseriforms, and somewhat less so in the galliforms, is notably absent in dromornithids. We prefer the explanation that the rostral telencephalon in dromornithids have been masked by hypertrophy of the wulst (see below), and that the external remnants of rostral telencephalon morphology are only visible, in rostral aspect, as twin eminences ventromediolaterad of the olfactory bulb, on either side of the rostromedial surface of the dromornithid endocast (e.g., tel.r; Figure 3B,D).

3.2.2. Wulst

The wulst in dromornithids are greatly hypertrophied and dominate the entire dor-sal endocast morphology (see wlst, Figure 3B,D). They extend rostromediolaterally to effectively mask the olfactory bulbs (see Section 3.1.1 above), and extend rostroventrally over the most rostral eminence of the rostral telencephalon, substantially overhanging the rostroventral surface of the brain when viewed from the ventral aspect (Figure 3D,H,L). The wulst extend rostrolaterally and mask the rostrodorsal telencephalon (see above, Figure 3), obscuring the rostrodorsal endocast morphology visible in the extant galloanseres (e.g., Figure 4). The wulst extend mediolaterally across the entire dorsal forebrain, to the lateral vallecula transition zones delimiting the boundaries between the mediolateral wulst, and the dorsolateral caudal telencephalon (see Section 3.2.3 below). The vallecula transition zones are well defined, as the wulst are strongly dorsolaterally expanded in those areas (va, Figure 3B,C). Caudodorsally, the wulst grade into the cerebrum pars occipitalis (coc, Figure 3C) in the region of the medial glandula pinealis (gp, Figure 3A,C), rostrolaterad of the dorsomedial cerebellum (Figure 3C). Notably, dromornithid wulst structures are located somewhat rostrally on the dorsal endocast (e.g., Figure 3C,G,K), such that they do not overlap the rostromedial eminence of the cerebellum when viewed from the lateral aspect (e.g., Figure 3A,E,I); this is similar to the condition in *A. cornuta* (Figure 4A,C), *G. gallus* (Figure 4E,G), *L. ocellata* (Figure 4I,K), and distinct to the condition seen in the anseriform *A. semipalmata* (Figure 4M,O).

3.2.3. Caudal Telencephalon

The caudal telencephalon in dromornithids are well defined and are delimited from the wulst by the vallecula transition zone dorsolaterally (va, Figure 3B,C). The mediolateral hypertrophy of the caudal telencephalon begins approximately where the medial cerebral artery traverses the telencephalic hemisphere dorsoventrolaterally (acm, Figure 3A,B). They extend ventrolaterally, approximately level (dorsoventrally) with the fissura subhemispherica (fs, Figure 3B), and return medially, somewhat acutely, to grade into the ventrolateral optic lobe (see below). Caudally, the cerebrum pars occipitalis (coc, Figure 3C) forming part of the dorsal caudolateral caudal telencephalon, grades into the dorsorostrolateral pons and medulla oblongata structures forming the overall metencephalon complex rostromediad of the cerebellum, in the vicinity of the glandula pinealis dorsolaterally, and medially at the rostromediolateral metencephalon.

3.2.4. Optic Lobe

The optic lobes in dromornithids are somewhat visually inconspicuous structures in comparison to those in the extant galloanseres (opt, Figure 3B vs Figure 4B,F,J,N). They are defined by a slight lateral hypertrophy of the ventromedial endocast ventrolaterad of the fissura subhemispherica, and rostrally by transition into the caudolateral chiasma opticum and tractus opticus structures. Caudally, the optic lobes grade into the ventromediolateral metencephalon complex (see below, Figures 3 and 4).

3.2.5. Trigeminal Ganglia

Trigeminal ganglia receive the three divisions of the trigeminal nerve (V, see Section 3.1.3 above), and insert on the ventral surface of the optic lobe (tri.g, Figures 3D and 4A,D). The trigeminal ganglia form part of the Modular Slm suite, therefore, are described here (see SI Figures S10 and S11, Section 2.6.1 and Section S3.2.1.9). In dromornithids, the medial portion of the trigeminal ganglion carrying the ophthalmic nerve (V_1) separates from the lateral branch carrying the maxillary (V_2) and mandibular (V_3) nerves and exhibits a small ganglionic bridge between the two primary eminences (Figure 3D,H,L). The characteristics of dromornithid trigeminal ganglia are distinctive in that the maxillomandibular branch transmits minimally 20 mm caudoventrolaterally (in *Dromornis*), after entering the skull at the foramen n. maxillomandibularis.

3.2.6. Cerebellum

The cerebellum, as reflected in the endocast in dromornithids, is compressed rostrocaudally and expanded mediolaterally (Figure 3C,G,K), and more so in species of *Dromornis*. From the lateral aspect (see Figure 3A,E,I), the dorsal rostroventral surface forms a shelf somewhat level with the dorsal lateral semicircular duct (dsl, Figure 3B,D) of the vestibular organ (lv, Figure 3A), before turning sharply ventrally in the vicinity of the dorsolateral auricula cerebelli, to grade into the caudodorsal medulla spinalis at the osseous foramen magnum. Overall, the exposed dromornithid cerebellum and the associated ventral rhombencephalon (medulla oblongata + pons; see below), compared with the extant galloanseres, form a comparatively distinctive hind brain in these birds.

3.2.7. Rhombencephalon

Rhombencephalon is the collective term describing the structures of the medulla oblongata and pons, forming the caudoventrolateral areas of the hindbrain. In dromornithids, the ventral rhombencephalon surface is somewhat flat rostrocaudally and mediolaterally (i.e., not as ventrally curved as in other galloansere specimens, e.g., Figure 4), and extends further rostrocaudally than it does mediolaterally.

3.3. Key Morphological Differences between Species of Dromornis and Ilbandornis

In the following, we present a summary of results for Modular Lm data, Measurement data, and Surface Area data forms. Full descriptions of results are given in SI (see SI Sections S4 and S5).

3.3.1. Wulst Modules

The overall size and surface area of the wulst increased between the late Oligocene and the middle Miocene in species of *Dromornis*, particularly rostrodorsally, and in the ventrolateral displacement of the vallecula transition zones between the dorsolateral wulst and the dorsal caudal telencephalon (see SI Section S4.1.1). The wulst in *D. stirtoni* likely had a comparable dorsal profile to other species of *Dromornis*, in that the rostrocaudal and mediolateral profile is distinctively hypertrophied (see SI Section S4.2.1). The late Oligocene *D. murrayi* and middle Miocene *I. woodburnei* dromornithids have similar length ratios, but the mediolateral width ratio is markedly smaller in *D. murrayi*. Similarly, results for Surface Area data show that *D. murrayi* has a smaller wulst surface area ratio than *I. woodburnei* (see SI Section S4.4.1).

With respect to the middle Miocene dromornithids, *D. planei* shows greater rostrodorsal hypertrophy of the wulst, and the dorsolateral margin of the vallecula transition zones are more ventrolaterally located than in *I. woodburnei*. *D. planei* has greater overall rostrocaudal length and caudal mediolateral width ratios than *I. woodburnei*. These results are consistent with those for Surface Area data, which show *D. planei* has a larger surface area ratio than *I. woodburnei* (see SI Section S4.3.1). *Ilbandornis woodburnei* displays a deeper fissura interhemispherica transition zone, placing the medial wulst margins closer together than in both species of *Dromornis* assessed.

3.3.2. Caudal Telencephalon Modules

The overall size and surface area of the caudal telencephalon reduced between the late Oligocene and the middle Miocene in species of *Dromornis* (see SI Section S4.1.2), and between species of *Dromornis* and *Ilbandornis* (see SI Sections S4.3.2 and S4.4.2). However, the characteristic shape of the structures are maintained within the late Oligocene through middle Miocene species of *Dromornis*. The caudal telencephalon of the late Miocene *D. stirtoni* is mediolaterally hypertrophied and is similar to the other species of *Dromornis* (see SI Section S4.2.2). The main differences in the caudal telencephalon between species of *Dromornis* and *I. woodburnei*, are that caudoventral margins project further ventrally in species of *Dromornis*, this is reflected in caudal telencephalon dorsoventral width ratios being greater in species of *Dromornis* than in *Ilbandornis*.

3.3.3. Cerebellum Module

There was limited rostrodorsal hypertrophy in the cerebellum between the late Oligocene and the middle Miocene in species of *Dromornis*. The dorsal surface of the cerebellum in the late Miocene *D. stirtoni* has a similar rostrocaudal shape to the other species of *Dromornis*, but the dorsal cerebellum appears more ventrally orientated in *D. stirtoni* (see SI Section S4.2.3). Species of *Dromornis* and *Ilbandornis* differ notably in the shape of the cerebellum, where species of *Dromornis* have mediolaterally wider and rostrocaudally shorter cerebellum profiles.

The middle Miocene species *D. planei* and *I. woodburnei* have a larger cerebellum than the late-Oligocene *D. murrayi* (see SI Sections S4.1.3, S4.4.3). Additionally, *I. woodburnei* displays a rostrocaudally longer and mediolaterally narrower cerebellum profile than both species of *Dromornis* (see SI Sections S4.3.3 and S4.4.3). It is notable, however, that there exist rostrocaudolateral profile similarities between the cerebellum of the middle Miocene *D. planei* and *I. woodburnei* specimens, that differ from the late Oligocene *D. murrayi*. Where both middle Miocene species display relative hypertrophy of the rostrodorsal through dorsomedial cerebellum, affecting a rostrocaudally steeper caudal transition to the dorsal medulla spinalis, compared to the older *Dromornis murrayi*.

3.3.4. Rhombencephalon Module

There was a dorsal displacement of the rhombencephalon surface between the late Oligocene and the middle Miocene in species of *Dromornis*, and the development of an overall 'flatter' mediolateral and rostrocaudal rhombencephalon profile in *D. planei*. As noted for the cerebellum above, the rhombencephalon structure in *D. stirtoni* appears to be more ventrally situated than that of *D. planei*. These observations suggest that the apparent ventral displacement of the dorsal cerebellum and ventral rhombencephalon surfaces in *D. stirtoni*, may reflect a compensatory ventral rotation of the hindbrain complex, with respect to the rostrodorsal rotation evident in the forebrain (see SI Section S4.2.4). In effect, the brain of *D. stirtoni* appears rotated about the median plane, whilst the position of the dorsomedial surfaces of the brain have been maintained, effectively foreshortening the overall rostrocaudal length of the *D. stirtoni* endocast compared with other species of *Dromornis*.

The overall length of the rhombencephalon between species of *Dromornis* and *Ilbandornis* are remarkably similar. However, in species of *Dromornis*, the rhombencephalon describes a more 'flat' rostrocaudal profile, along with greater mediolateral width ratios compared to *I. woodburnei*. The surface area ratio of the rhombencephalon in *D. planei* is markedly larger than that of *I. woodburnei* (see SI Section S4.3.4), and the late-Oligocene *D. murrayi* has a larger ratio compared with the middle Miocene *I. woodburnei* (see SI Section S4.4.4). These trends are similar to those noted in the comparisons of *D. planei* and *I. woodburnei* (see SI Section S4.3.4, and above), and suggest that the size of the cerebellum may have increased, particularly in the rostromedial zone, across both *Dromornis* and *Ilbandornis* lineages between the late Oligocene and the middle Miocene. However, this observation may only be tested by assessment of yet to be discovered cranial material of the *Ilbandornis* lineage from the late Oligocene.

3.4. Key Morphological Differences between Dromornithids and the Extant Galloanseres

3.4.1. Innervation

Unlike in dromornithids, the olfactory zones of extant galloanseres are wholly external to the rostral telencephalon, as revealed by their slight constriction immediately rostrad of the rostral telencephalon, particularly evident in *A. semipalmata* (Figure 4M,O,P) and *A. cornuta* (Figure 1A,C,D). The hypoglossal (XII) nerves have a single origin at either side of the caudoventral medulla oblongata in dromornithids, this condition is unlike the extant galloanseres, which display rostral (XIIr) and caudal (XIIc) rami (Figure 4D,H,L,P). Results for Measurement data show the trigeminal ganglia of *A. semipalmata* and *A. cornuta* are wider than those of all other galloanseres. Surface Area data show the galliforms *L. ocellata* and *G. gallus* have the smallest surface area ratios for the face of the maxillomandibular ($V_2 + V_3$) branch of the trigeminal (V) nerve, closely followed by the dromornithids. The anseriforms *A. semipalmata* and *A. cornuta*, respectively, have the largest maxillomandibular surface area ratios for all galloanseres assessed (see SI Section S5.1).

3.4.2. Wulst Modules

Wulst modules in extant galloanseres are much hypotrophied in comparison with all dromornithids (e.g., Figure 3 vs Figure 4). Results for Measurement data show *G. gallus* has the shortest and narrowest wulst ratios, followed by *L. ocellata*, and the anseriforms *A. cornuta* and *A. semipalmata*, respectively. Dromornithids have substantially larger wulst surface area ratios than all galloanseres, but when endocast absolute size is accounted for, *G. gallus* and *L. ocellata* have the largest wulst surface area ratios among the extant galloanseres, and the anseriforms *A. cornuta* and *A. semipalmata* the smallest, respectively (see SI Section S5.2).

3.4.3. Rostral Telencephalon Modules

Evidence of the rostral telencephalon is conspicuously absent on the external morphology of dromornithid endocasts, and only present rostrally as twin eminences ventromediolaterad of the olfactory bulb, on either side of the rostromedial surface of the endocast (tel.r, Figure 3B,D). Consequently, detailed comparisons were not possible between the rostral telencephalon of dromornithids and those of the extant galloanseres, and we present results for the extant galloanseres only. Measurement data show *G. gallus* has the shortest and narrowest rostral telencephalon ratios, followed by *L. ocellata*, and the anseriforms *A. cornuta* and *A. semipalmata*, respectively. Similarly, results for Surface Area data show that *G. gallus* has the smallest surface area ratio, and *A. semipalmata* the largest surface area ratio of all extant galloanseres (see SI Section S5.3).

3.4.4. Caudal Telencephalon Modules

Results for Measurement data show that all dromornithids have relatively greater rostrocaudal length ratios compared with the extant galloanseres. However, all extant galloanseres overlap in dorsoventral width ratios. Dromornithids have the greater mediolateral endocast total width ratios of all galloanseres, and the galliforms have somewhat similar endocast total width ratios as the anseriform *A. semipalmata*.

Results for Surface Area data show that the dromornithids have the largest surface area ratios of all specimens, *L. ocellata* has the smallest surface area ratio of all specimens, and the phasianid *G. gallus* has the largest surface area ratio of the extant galloanseres (see SI Section S5.4).

3.4.5. Optic Lobe Modules

Results for Measurement data show the extant galliforms have the largest optic lobe length and width ratios, followed by the anseriforms and dromornithids. Results for Surface Area data (see SI Section S5.5) show that the largest taxa have the smallest overall optic lobe ratios, and reveal the optic lobe in *I. woodburnei* is larger than those of both species of *Dromornis*. Surface Area ratios for the anseriforms *A. cornuta* and *A. semipalmata* are somewhat similar to those of the dromornithids, and the galliforms *G. gallus* and *L. ocellata* have the largest surface area ratios of all specimens assessed.

3.4.6. Cerebellum Module

Results for Measurement data show the cerebellum, as exposed in the endocast in dromornithids, are notably rostrocaudally shorter and mediolaterally wider than those of all extant galloanseres, with the width ratio for *A. semipalmata* most approaching those of the dromornithids. Rostrad of the foramen magnum, all extant galloanseres display a gradual rostrolateral divergence of cerebellum mediolateral margins, prior to grading into the cerebrum pars occipitalis regions of the caudal telencephalon. This condition is not evident in dromornithids, which display a more abrupt mediolateral divergence of caudodorsal cerebellum margins, which are more pronounced in species of *Dromornis* than in *I. woodburnei*. Results for Surface Area data show that the dromornithids have the largest ratios of all galloanseres. Among the extant galloanseres, *A. semipalmata* has the smallest surface area ratio for all specimens. *Leipoa ocellata* overlaps with those of the dromornithids, and is most similar to that of *I. woodburnei*. *Anhima cornuta* has a larger surface area ratio than all dromornithids, and the *G. gallus* has the largest surface area ratio of all galloanseres (see SI Section S5.6).

3.4.7. Rhombencephalon Module

Results for Measurement data show that all galloanseres have somewhat similar rhombencephalon length ratios, with the exception of *A. cornuta*, which has the largest rostrocaudal length ratio of all galloanseres. Rhombencephalon width ratios for extant galloanseres are all larger than in dromornithids, and likely reflect the greater ventral projection, or eminence, of the rhombencephalon in the extant specimens. The overall mediolateral width ratio of the hindbrain in species of *Dromornis* are greater than in all galloanseres assessed, and that of *I. woodburnei* overlaps with those of the extant galloanseres. Results for Surface Area data show that the dromornithids *I. woodburnei* and *D. murrayi* have the smallest rhombencephalon surface area ratios, followed by *A. semipalmata* and *G. gallus*. The dromornithid *D. planei* has a somewhat similar surface area ratio as *G. gallus*, and *A. cornuta* and *L. ocellata* have the largest surface area ratios among all galloanseres (see SI Section S5.7).

4. Discussion

Dromornithid endocranial anatomy is described here in detail for the first time, using specimens from fossil sites in Australia spanning ~20–8 Ma. In the following, (1) we summarise the morphological characteristics of the dromornithid brain, and discuss these features with respect to those of exemplars of other major galloansere clades; (2) we review potential lineage-specific morphological characteristics identified in dromornithid endocranial anatomy; (3) we comment on morphological changes observed to have occurred in the endocranial anatomy of the *Dromornis* lineage over time; (4) we assess potential functional implications of the dromornithid endocranial condition.

4.1. Comparisons of Endocranial Characteristics of Dromornithids and Extant Galloanseres
4.1.1. Olfactory Bulb

The olfactory bulb in dromornithids is pronounced both dorsally and ventrally in the oldest (*Dromornis murrayi*) species. However, in middle Miocene specimens (*D. planei* and *Ilbandornis woodburnei*), its dorsal morphology is masked by the rostral eminence of the wulst (see Section 4.1.3 below). There appears no reduction in the size of the olfactory bulb in the younger dromornithids, as from the ventral aspect, lateral margins of the organ transition into the rostroventral endocast without reduction in mediolateral width. In extant galloanseres, the olfactory bulb of *Anseranas semipalmata* displays hypertrophy in excess of that seen in *Anhima cornuta*, which is somewhat more than those of *Leipoa ocellata* and *Gallus gallus*. The olfactory zones of these galloanseres appear wholly external rostrad of the rostral telencephalon, as the olfactory bulb margins constrict somewhat prior to grading caudally into the rostral telencephalon, a condition distinct from those of dromornithids.

Taken together, the evidence shows that the wulst extends rostromediolaterally to effectively mask the olfactory bulbs in dromornithids, and so contrary to a first assessment that might consider that dromornithids had hypotrophied olfactory bulbs, we consider them to be relatively no smaller than in the other galloanseres.

4.1.2. Trigeminal Ganglia

Trigeminal ganglia transmit the three divisions of the trigeminal (V) nerve complex and insert on the ventral surface of the optic lobe. They are distinctive in dromornithids, in that the maxillomandibular ($V_2 + V_3$) branch passes minimally 20 mm caudoventrolaterally through the unusually thick cranium (in *Dromornis*), after entering it at the foramen n. maxillomandibularis. The extended transmission of the maxillomandibular branch through the cranium in dromornithids accommodates for the unusually thick honeycomb-like trabecular bone separating the cortical bone defining the endocranial capsule, from the outer cortical cranial surface (e.g., SI Figure S8I). In the extant galloanseres, the transmission of these nerves from the foramen n. maxillomandibularis is markedly shorter. Notably, however, in the phasianid galliform *G. gallus* (see Figure 4E,F,H), the transmission of the maxillomandibular ($V_2 + V_3$) branch through the cranium, although somewhat shorter than that observed in dromornithids, is longer than observed in the other extant galloanseres. Moreover, CT data for *G. gallus* reveals that this is also accompanied by trabecular matrix surrounding the endocranial capsule through which the nerves transmit [Pers. Obs. Authors].

In general, the shape of the avian cranium, particularly dorsally, exhibits a close relationship to the shape of the brain within [68–74]. However, patterns of the brain size 'lagging behind' the body, accompanying an increase in the relative thickness of trabecular bone surrounding the endocranial capsule have been recognised in Haast's eagle *Hieraaetus moorei* (Haast, 1872) by Scofield and Ashwell [75], who showed that the eagle's "ten-fold" increase in body size was only accompanied by a "doubling or tripling" of endocast volume. This demonstrated lag of neuroanatomy behind rapid hypertrophic skeletal changes in a taxon, may relate to strong selection for body size, i.e., adaptation to a novel trophic niche, or artificial selection in the form of "stringent" human mediated selection for desirable phenotypes (e.g., as in the case of *G. gallus* see [76], and references therein). Consequently, as dromornithids became larger through the course of their evolution (e.g., [6,31]), it is likely the increase in physical size of the cranium, was accommodated for by an increase in trabecular bone enclosing the 'lagging' endocranial capsule [75] (Figure 5a).

Similarly, the *G. gallus* skull used for CT scanning for this project was almost certainly from a domestic chicken, and may demonstrate increasing trabecular thickness in the cranium, mediated by human selection for body size. To assess these observations more comprehensively, additional data in the form of wider sampling across galloanseres in particular, but across Neornithes in general is required, targeting taxa with demonstrated temporal increases in body size.

Results for Measurement data show the trigeminal ganglia of *A. semipalmata* and *A. cornuta* are relatively wider than those of all other galloanseres, and Surface Area data show the galliforms *L. ocellata* and *G. gallus* have the smallest surface area ratios for the face of the maxillomandibular ($V_2 + V_3$) branch of the trigeminal (V) nerve, closely followed by the dromornithids. The anseriforms *A. semipalmata* and *A. cornuta*, respectively, have the largest maxillomandibular surface area ratios for all galloanseres assessed (see Section 3.4.1). The eminence of the hypoglossal (XII) nerves which typically comprise a rostral (XIIr) and caudal (XIIc) branch in extant galloanseres, is represented by one nerve root at either side of the caudoventrolateral medulla oblongata in dromornithids, with bifurcation into rostral and caudal branches occurring within the os exoccipitale, close to the caudomediolateral surface of the cranium.

4.1.3. Wulst

The morphology of the wulst is the most distinguishing feature uniting dromornithids, forming massively hypertrophied structures. The wulst extends rostromediolaterally to effectively cover the olfactory bulbs and extends rostroventrally over the most rostral eminence of the rostral telencephalon, substantially overhanging the rostroventral surfaces of the brain. The dromornithid wulst extend strongly rostrolaterally, masking the rostrodorsal telencephalon (see Section 4.1.4 below), and extends mediolaterally across the entire dorsal forebrain. The structure of the wulst in dromornithids is unlike any seen in the extant galloanseres. Results show the dromornithids all display much larger width and length, and surface area ratios, than all the extant galloanseres. When endocast absolute size is accounted for, *G. gallus* and *L. ocellata* have the largest wulst surface area ratios among the extant galloanseres, and the anseriforms *A. cornuta* and *A. semipalmata* the smallest (see Section 3.4.2).

Among taxa that were not included in these analyses, some palaeognaths display hypertrophy of the wulst. For example, Corfield et al. [49] (Figure 1B–E) figured endocasts of the extinct NZ moa *Dinornis novaezealandiae* Owen, 1843 and *Anomalopteryx didiformis* (Owen, 1843), and those of extant ratites like emu (*Dromaius novaehollandiae*) and ostrich (*Struthio camelus*). Additionally, Ashwell and Scofield [77] (Figure 6G–I) figured the dorsal endocasts of several NZ moa: *D. robustus* (Owen, 1846), *A. didiformis*, *Euryapteryx curtus gravis* (Owen, 1870), and *Emeus crassus* (Owen, 1846) as well, all of which show wulst characteristics similar to those seen in dromornithids, wherein the vallecula, especially in the larger moa taxa, visibly extend rostrocaudally across the entire dorsolateral telencephalic hemispheres. These characteristics suggest that such characteristic hypertrophy of the wulst in large flightless birds (see also [78] (Figures 1 and 2), [79] (Figure 2a), [80] (Figures 1 and 3), [81] (Figure 1), [74] (Figures 5.3 1–6), may represent a parallel convergent modification towards enhanced visual proficiency and stereoscopic capability (see also Section 4.4.2.1 below). However, ratite taxa clearly display a lesser degree of wulst hypertrophy than that evident in dromornithids. Even the oldest dromornithid cranial fossils (e.g., *D. murrayi* from the ~20 Ma sites of Riversleigh), display greater hypertrophy of the wulst than seen in any ratite, indicating a long dromornithid "ghost lineage" must have existed prior to any substantive fossil evidence of dromornithids in Australia (see also [6] (p. 19)). Evidence for such, comprise trackways reported from the late Oligocene of Tasmania [1], postcranial remains from the late Oligocene Pwerte Marnte Marnte LF in the Northern Territory [4], and a mould of fossil footprints from the Eocene Redbank Plains Formation of Queensland [7].

4.1.4. Rostral Telencephalon

Evidence of the rostral telencephalon in dromornithids is only present rostroventrally as twin eminences ventromediolaterad of the olfactory bulb, on either side of the rostromedial endocast (tel.r; Figure 3B,D). It is possible that these eminences are expanded cerebrum tuber ventrolaterale structures, as evident in *A. cornuta* (Figure 4A,B,D). However, their interpretation as remnant rostral telencephalon eminences is favoured, as in all dromornithid endocasts modelled, there exist pronounced paired eminences in this rostromedial zone that are not present to the same degree in any avian endocast modelled or observed in the literature [Pers. Obs. Authors]. Further support for this, is that the positioning of these rostral eminences in dromornithids, agrees with the angle and position of rostral telencephalon eminences in specimens, when endocasts are aligned to putative "alert posture" [82–85], with reference to the horizontal positioning of the lateral semicircular canal of the vestibular organ. Additionally, the rostrocaudal transition angle of the vallecula, describing the dorsal margins of the caudal telencephalon, agree with the extension of the visible rostral eminences of the dromornithid rostral telencephalon, should the dorsolateral curve of the rostral telencephalon not be masked by hypertrophy of the rostromediolateral wulst. In support of this interpretation, the apparent rostral extension of the vallecula across the dorsolateral surface of moa endocasts figured by Ashwell and Scofield ([77] Figures 5E–5I and 6G–6I) and Corfield et al. ([49] Figure 1b D–E), and similarly in the brains of extant flightless ratites (see [78] (Figures 1 and 2), [79] (Figure 2a), [77] (Figure 6), [49] (Figure 1B,C), [80] (Figures 1 and 3), [81] (Figure 1), [74] (Figures 5.3 1–6)), suggest that the evolution of a rostromediolaterally hypertrophied wulst in large flightless birds may effectively mask rostral telencephalon morphology. The accommodation of this apparently major change in rostrodorsal endocranial morphology in dromornithids, would have necessitated a dorsomedial displacement of the olfactory bulb, which we think is a possibility, as this condition is somewhat similar to that seen in moa (e.g., [77] (Figures 5e–l and 6g–l), [49] (Figure 1b D–E)).

Detailed comparisons of the dromornithid rostral telencephalon were not possible. However, results of comparisons among extant galloanseres show the phasianid *G. gallus* has the smallest rostral telencephalon, followed by the megapodiid *L. ocellata*, and the basal anseriform *A. cornuta*. The anseriform *A. semipalmata* has the most hypertrophied rostral telencephalon of the extant galloanseres (see Section 3.4.3).

4.1.5. Caudal Telencephalon

The caudal telencephalon is strongly defined in all galloanseres assessed. Results show dromornithids have relatively greater rostrocaudal length and mediolateral endocast total width ratios than the extant galloanseres. Similarly, the dromornithids have the largest caudal telencephalon surface area ratios, and the megapodiid *L. ocellata* has the smallest surface area ratio of all specimens. The phasianid *G. gallus* has the largest surface area ratio for the extant galloanseres (see Section 3.4.4).

4.1.6. Optic Lobe

The optic lobe in dromornithids appear somewhat indistinct, and not as well delimited as in the extant galloanseres. This is confirmed by results revealing the extant galliforms have the largest optic lobe length and width ratios, followed by the anseriforms and dromornithids. These patterns are also reflected by results for Surface Area data which show the dromornithids have the smallest optic lobe surface area ratios, followed by those of the anseriforms. The largest surface area ratios for all galloanseres assessed are shown by the galliforms *G. gallus* and *L. ocellata*. Overall, results show that when endocast size is accounted for, the largest overall taxa display the smallest overall optic lobe ratios, and reveal relatively greater hypertrophy of the optic lobe in the *Ilbandornis* lineage compared with species of *Dromornis* (see Section 3.4.5).

Additionally, during the process of segmenting the endocast models used for this project, we retained trigeminal ganglion structures (see Figure 3D,H,L) inserting on the ventral surfaces of the optic lobe, primarily to gain insight of dromornithid somatosensory capabilities (e.g., Section 4.4.1 below). This, however, resulted in the ventromedial surfaces of optic lobe Slm modules being constrained by the morphology of the trigeminal ganglia. Retrospectively, it may have been more useful to segment the trigeminal (V) complex out of the final endocast models, enabling access to the full extent of the ventral optic lobes and allowing consideration of optic lobe morphology with respect to those of the wulst and optic foramen (*sensu* [50]; see also [86], and (Section 4.4.2.3 below). This aspect of dromornithid cranial morphology forms part of continuing work on the brains of these giant birds.

4.1.7. Cerebellum

Dorsal cerebellum margins in dromornithids are characteristically rostrocaudally compressed and mediolaterally expanded. Results show that the exposed cerebellum in dromornithid endocasts are rostrocaudally shorter and mediolaterally wider than those of all extant galloanseres. Results for Surface Area data show that the galliform *G. gallus* has the largest cerebellum ratio, and the anseriform *A. semipalmata* the smallest ratio of all galloanseres assessed. *Dromornis planei* has the largest cerebellum surface area ratio of the dromornithids, and the ratios for *A. cornuta* and *L. ocellata* overlap with those of dromornithids. Rostrad of the foramen magnum, all extant galloanseres display a gradual rostrolateral divergence of cerebellum mediolateral margins, prior to grading into the cerebrum pars occipitalis regions of the caudal telencephalon. This condition is not evident in dromornithids, which display a much more abrupt mediolateral divergence of the caudodorsal cerebellum margins, and this is more pronounced in species of *Dromornis* than in *Ilbandornis*. Progressing rostrally, dromornithid cerebellum margins describe a somewhat parallel rostrocaudal transition (see Section 3.4.6).

4.1.8. Rhombencephalon

Dromornithids display relatively flat ventral rhombencephalon surfaces rostrocaudally and mediolaterally, whereas, in comparison, all extant galloanseres show much ventrally hypertrophied rhombencephalon surfaces (rho; Figure 4A,B,D). These trends are reflected by results for Measurement data showing all galloansere species, with the exception of *A. cornuta*, have similar rhombencephalon length ratios. Width ratios for the extant galloanseres are all larger than those of the dromornithids, and likely reflect the greater ventral projection, or eminence, of the rhombencephalon in the extant taxa. The overall width ratio of the hindbrain in species of *Dromornis*, are greater than in all other species, and that of *I. woodburnei* overlaps with those of the extant galloanseres. Results for Surface Area data show that the dromornithids *I. woodburnei* and *D. murrayi* have the smallest surface area ratios, followed by the anseranatid *A. semipalmata* and phasianid *G. gallus*. *Dromornis planei* has a somewhat similar surface area ratio as *G. gallus*, and *A. cornuta* and *L. ocellata* have the largest surface area ratios among all galloanseres assessed (see Section 3.4.7).

4.2. Endocranial Morphology Distinguishing Lineages within Dromornithids

The examination of dromornithid endocasts has revealed morphological features of the dromornithid brain that may provide support for the two-lineage hypothesis (*sensu* [6]). We define these distinctive endocranial attributes below, with a focus on the morphological differences between *Ilbandornis woodburnei* and the two species of *Dromornis* with adequate preservation for comparison (i.e., *D. planei* and *D. murrayi*): (1) the medial boundaries of the wulst in the rostrocaudal fissura interhemispherica zone of the dorsal endocast in *I. woodburnei*, are notably closer together than in species of *Dromornis*; (2) the most ventral eminence of the *I. woodburnei* caudal telencephalon, in the zone of the fissura subhemispherica, are conspicuously less ventrally pronounced than in species of *Dromornis*; (3) the rostrodorsal cerebellum in *I. woodburnei* defines a more rostrally projecting mediolateral curve than in species of *Dromornis*, which display a flatter rostral dorsomediolateral margin; (4) the caudodorsal part of the cerebellum in *I. woodburnei* projects further caudally in the region of the dorsal medulla spinalis, whereas this region is caudally shorter but mediolaterally flatter in species of *Dromornis*; (5) the entire hindbrain (rhombencephalon, medulla oblongata and metencephalon complex [cerebellum + pons]) in *I. woodburnei*, is rostrocaudally longer and mediolaterally narrower than in species of *Dromornis*.

These morphological traits are potential lineage-specific endocranial attributes, but require confirmation through a shared presence in yet to be discovered crania of *Ilbandornis lawsoni* and *Barawertornis tedfordi*, or in presently undescribed material of *Genyornis newtoni*.

4.3. Temporal Changes in the Endocranial Morphology of the Dromornis Lineage

Across the ~10 Ma period represented by specimens of Oligo–Miocene *D. murrayi*, and the middle Miocene *D. planei*, the orientation of the brain within the skull appears to have remained much the same, despite foreshortening of the cranium. Other than regional changes in endocast shape, for example, the rostrodorsal hypertrophy of the wulst, accompanying other trends described above, the brains of these species are generally similar, and distinctively dromornithid. The relatively major increase in the overall size of species of *Dromornis*, reflected by crania of *D. murrayi* and *D. planei* figured here (e.g., Figure 1A–D; SI Figures S1A–P and S4A–H), and postcranial fossils described elsewhere [1,6], appear to have not been substantial enough to affect changes in the position and orientation of the brain in *D. planei* relative to that of *D. murrayi*. However, by the late Miocene, some ~6 Ma after the occurrence of *D. planei*, the cranium of the *Dromornis* lineage had evolved to become even more foreshortened and dorsoventrally deeper, as manifested in *Dromornis stirtoni* (Figure 1E; SI Figures S6 and S7; see also [6]). These morphological changes appear to have affected the orientation of the brain.

With regard to *D. stirtoni*, it is unfortunate that the state of preservation of specimens prevented the level of endocast shape assessment achieved for the other dromornithids. This was primarily due to the taphonomic characteristics of the only site preserving these giant birds (see Section 2.2 above). In turn, this limited the biological and functional inferences derived from their exceptional cranial architecture. However, we have shown from the similarity of preserved features of the *D. stirtoni* endocast models with those of *D. planei* (see SI Sections S2.1 and S4.2), that the brain of *D. stirtoni* does not depart greatly from the only other known dromornithid endocast morphology. However, the altered endocranial alignment to 'fit' the brain in the foreshortened cranium, resulted in rostroventral endocast surfaces in *D. stirtoni* being rotated rostrodorsally around the medial caudal telencephalon into a more dorsally orientated position. Additionally, it appears the brain of *D. stirtoni* has experienced a measure of dorsoventral compression and mediolateral expansion. These changes in the forebrain are accompanied by a more ventrally orientated hindbrain, which may reflect a compensatory ventral rotation of the hindbrain complex in the species, although the 'life position' of the midbrain in the skull of *D. stirtoni* has not changed appreciably from that of *D. planei*.

The reasons for this unusual rotation and apparent subtle compression of the *D. stirtoni* endocast may lie in the highly derived state of cranial morphology attained by this, the largest of the dromornithids, by the late Miocene. The cranium of *D. stirtoni* is unique in that the rostrocaudal cranial length is effectively about half the cranial depth, and represents the terminal state of a concerted trend in cranial foreshortening, along with an increase in bill size, of the most extreme avian cranial specialisation known [3,6]. This trend extends from the oldest known species of *Dromornis*, the Oligo–Miocene *D. murrayi*, through the middle Miocene *D. planei*, to the most derived late Miocene taxon *D. stirtoni*.

4.4. Functional Implications of Dromornithid Endocranial Morphology

Jerison [87] (p. 8) proposed the "Principle of Proper Mass" which specifies particular sensory specialisations in the vertebrate brain are correlated with concomitant hypertrophy of the neural tissue processing related information, and that the relative mass of functional neural tissue implies the relative importance of those functions in the species. Subsequent studies showed that overall brain size was in fact increased by independent hypertrophy of particular brain regions (see [88–91], and references therein), supporting Jerison's [87] observations, which became to be known as the "mosaic" model of brain evolution (*sensu* [90]).

That mosaic evolution characterises some, but not all, of avian brain composition has been demonstrated by several works (see [49,92], and references therein). It is acknowledged that the brain is not strictly compartmentalised into regions that process exclusive neuronal input, but rather includes levels of interconnectivity across the whole structure [92]. It is clear that particular brain nuclei share greater levels of neuronal connectivity associated with specific functions; that a hypertrophied brain region reflects a greater level of "information-processing power", and that these patterns are somewhat indicative of functional specialisation ([49,90,93–95] and references therein). Most recently, Early et al. [50] showed there exists statistically significant correlations between the external surface area of the wulst and optic lobe, and the volume of underlying brain regions.

4.4.1. Innervation

Characteristic mosaic correlations in the trigeminal system have previously been shown in several vertebrate taxa (see [96], and references therein). In birds, the trigeminal (V) nerve system comprises the medial portion carrying the ophthalmic (V_1) nerve which innervates the orbit and nasal cavity, the rostral palate and the tip of the upper bill, and forms a major sensory pathway for the skin of the head and maxillary rostrum. The maxillary (V_2) branch innervates the maxillary rostrum and infraorbital regions, and the mandibular (V_3) division innervates the entire lower bill and several mandibular and interramal regions [97–100]. The trigeminal nucleus receives exclusively proprioceptive information from the descending tract, and the principal sensory nucleus of the trigeminal system [96]. This includes not only projections from ophthalmic (V_1) and maxillomandibular ($V_2 + V_3$) nerves, but taste information from the tongue is conveyed, within the lingual branch of the maxillomandibular ($V_2 + V_3$) ramus, by the facial (VII) nerve to the trigeminal principal sensory nucleus, which also receives input from glossopharyngeal (IX) and hypoglossal (XII) nerves [93,98,101–105]. In short, the trigeminal (V) nerve comprises the largest somatosensory cranial innervation complex, and transmits epicritic sensation from the entire facial region and mastication musculature [98,105,106]. Dubbeldam [107] proposed that differences in the trigeminal principal sensory nucleus were indicative of the functional demands of specific feeding behaviours. Gutiérrez-Ibáñez et al. [96] reported hypertrophy of the trigeminal principal sensory nucleus in species dependent on tactile input for feeding, and that bill morphology and the concentration of mechanoreceptors in the bill and tongue, strongly correlate with feeding behaviour.

The glossopharyngeal (IX) and vagus (X) nerves share the large proximal ganglion. The glossopharyngeal (IX) components of this complex comprise somatic, "special" and visceral afferent fibres. The special fibres connect with the palatine branch of the facial (VII) nerve at the cranial cervical ganglion, and are associated with sensory taste and tactile information [98,101,108,109]. The general visceral efferent fibres of the glossopharyngeal (IX) nerve supply the oesophagus and crop, exhibit size variability across taxa that show greater "distensibility" of the oesophagus [98], and are notably hypotrophied in taxa that have no crop (i.e., owls and hawks). The glossopharyngeal (IX) nerve complex bifurcates, after separation from the vagus (X) nerve at the proximal ganglion, and transmits, in two main afferent branches of the lingual and the laryngopharyngeal nerves, as the descending oesophageal nerves, innervating the tongue and the laryngeal muscles, respectively.

The vagus (X) nerve complex is the most extensive of the sensory and motor cranial nerves, wherein there are two groups of motor fibres. The first consists of "general" visceral efferent fibres which innervate the muscles and glands of the thoracoabdominal viscera, including the heart and lungs, etc., and is associated with circulation, respiration and digestion control [98]. The second consists of "special" visceral efferent fibres innervating the muscles of the pharynx and the larynx, reached via branches of the glossopharyngeal (IX) nerve ([98]; for a contrary opinion, see Wild [103] who argued vagus (X) projections are exclusively cardiovascular and pulmonary in function).

The morphology of dromornithid ventral cranial innervation, in the form of the maxillomandibular ($V_2 + V_3$), glossopharyngeal (IX) and vagus (X) nerves, and results presented here for trigeminal ganglia, are more similar to the species of galliforms. Dromornithid trigeminal ganglia appear less hypertrophied than those of the more derived, and trophic specialist anseriforms too, this is supported by results for Measurement data. Additionally, Surface Area data describing the truncated face of the maxillomandibular ($V_2 + V_3$) branch of the trigeminal (V) nerve complex, show that the dromornithids have somewhat similar surface area ratios as the galliforms, and which collectively, are much smaller than those of the anseriforms (see Sections 3.4.1 and 4.1.2). The morphology of dromornithid ventral innervation, better represented by the endocast of *Dromornis planei* (see Figure 3A–D), suggests that dromornithids were likely possessed of somewhat better tactile capability than the extant galliforms (e.g., Figure 4H–L), and were likely employing greater levels of sensory input from the bill, palate, and tongue in their trophic behaviour, than the extant galliforms do.

Dromornithids, especially species of *Dromornis*, have extremely large, deep bills, with dorsally prominent, mediolaterally compressed culmens [1,3,6]. The herbivorous diet of dromornithids is well established ([1], and references therein), but musculature for operation of the bill is "surprisingly limited" [6] (p. 19), and suggests that these birds were not capable of a particularly forceful bite [6], *contra* [3] (p. 88). For example, there is no temporal fossa on the side of the cranium for insertion of mandibular musculature, which is thus limited to the fused postorbital-zygomatic processes, and hyper-developed insertions on the orbital wall of the cranium (i.e., ma, SI Figure S4K,L). Moreover, the culmen, while large, has a lightly constructed osseous core that was only partially covered in rhamphotheca (at least it was in species of *Dromornis*), was highly vascularised and likely highly innervated, a combination of features conferring relatively weak biting ability [6]. The large size of the dromornithid culmen, combined with indications that they are not strengthened for food manipulation, suggest that they were primarily utilised for display, and that the distinctive morphology was likely driven by sexual selection, or by thermoregulatory requisites.

Bill architecture suggests that dromornithids were likely not consuming coarse browse requiring strong bite forces, as were some species of moa (see [19], and references therein). This contention can be tested by observations of the collections of gastroliths used to process such food. We were able to analyse total volume of gastroliths and size of stones in complete or partial gizzard stone sets from specimens of *Genyornis newtoni* to infer characteristics of diet [110–113], and showed (see SI Figure S15 and Table S1) that dromornithids selected gastroliths of much smaller diameter, and accumulated them in surprisingly small overall volumes compared with *Dinornis* moa, although the dromornithids were somewhat larger birds. Moreover, the data reveal that *G. newtoni* selected smaller stones than did the smaller extant emu (*Dromaius novaehollandiae*) which occasionally selected remarkably large gastroliths (see also [110] (Table 2), [111] (p. 26)). In fact, when overall body size was accounted for, gastrolith size ratios for the emu were larger than those of *Dinornis robustus*, although accumulated in smaller quantities. Notably, the stout-legged moa (*Euryapteryx curtus gravis*) has very similar gastrolith size and gizzard mass ratios to those of *G. newtoni*. Stout-legged moa are hypothesised to have exploited a diet of soft leaves and fruit, in dry scrubland and mosaic environments, in contrast to the generalist tree and shrub browsing *D. robustus* moa, the gizzard contents of which have been shown to comprise much coarser, low-quality fibrous leaf and twig material (e.g., [15,114–116] and references therein). Complete dromornithid gizzard stone sets are rarely preserved, but a complete gizzard set is known for the Oligo–Miocene *Dromornis murrayi*, preserved in situ in limestone (see [28] (p. 79)), and which shows only small stones. The gizzard sets measured for the Pleistocene species *Genyornis newtoni*, show they selected similarly small stones. Thus, if gizzard mass and gastrolith size preferences for *G. newtoni* are representative of dromornithids in general, these results suggest that the fibrosity of browse the dromornithids were targeting was likely similar to that of stout-legged moa (i.e., new growth, soft leaves, and fruit), requiring less vigorous mechanical processing in the gizzard [112]. Such a diet would require specialised visual pathways for the identification and selection of specific food resources. We have shown that dromornithids may have had well-developed somatosensory and sensorimotor capabilities, and so were likely adapted for precise and selective visual browsing.

4.4.2. Visual Pathways

There are three principal visual pathways in birds: (1) the thalamofugal pathway transmits visual signals from the retina via the optic lobe, to the principal optic nucleus of the dorsal thalmus, and thence to the wulst; (2) the tectofugal pathway transmits via the optic lobe to the nucleus rotundas of the thalmus, and proceeds to the entopallium of the caudal telencephalon; and (3) the third visual pathway transmits via the optic lobe through retinal recipient nuclei in the accessory optic system and pretectum, and projects to several regions of the brain, including the cerebellum (see [49,117–120], and references therein).

4.4.2.1. Wulst

The wulst are composed of two main regions, the larger "visual" region, located dorsally and extending caudodorsally, which receives retinal projections, and a smaller rostral somatosensory region, receiving "substantial" somatosensory and kinesthetic input [94,106,121–123]. The thalamofugal pathway, incorporating the wulst, has been shown to be primarily involved in binocular vision capability, and global stereopsis ([94,120,124–126] and references therein). Iwaniuk et al. [94] showed that the size of the wulst was significantly correlated with more frontally orientated orbits and broader binocular fields [120,127], and argued changes in the relative size of the wulst were indicative of increases in somatosensory and motor processing capabilities [48,124,128,129]. Additionally, wulst structures are hypertrophied in species that forage using tactile information from the bill [120,124,130,131].

Dromornithids may have had a well-developed thalamofugal pathway as they display particularly hypertrophied wulst structures. Potential indications of the kinds of adaptive selection driving dromornithid wulst morphology, may lie in the extraordinarily similar morphology of avian predators such as Barn owls (*Tyto alba*), and the Australian Tawny frogmouth (*Podargus strigoides*), which have a similarly hypertrophied wulst structures too (e.g., [124] (Plate 26), [120] (Figure 2), [79] (Figure 2c), [49] (Figure 1bK), [132] (Figure 3A,B), [74] (Figures 5.3 34 and 35)). Barn owls are nocturnal, and possess exceptional low-light visual proficiency and binocularity, or stereopsis [133–137]. Similar specialisations typical of low-light and stereoptic visual proficiency have been recognised in the Tawny frogmouth, which have highly developed visual systems [120,124], and are thought to possess stereoscopic vision [125]. Stereoptic proficiency has been shown to facilitate accuracy in nocturnal prey capture in caprimulgid taxa [125], and spatial, or "topographical cues" associated with feeding activities in *Columba* and *Gallus* [126], and corvids [131,138]. Furthermore, it has been advanced that taxa which use tactile information for foraging (see also Section 4.4.2.2 below) show somewhat hypertrophied rostral wulst structures [120,124], likely as the mandibular (V_3) division of the trigeminal (V) cranial nerve, innervating the lower bill (see Section 4.4.1 above), terminates in the rostrodorsal mesopallial regions of the brain [99,130,139,140]. As such, the development of stereopsis in birds has been linked with the presence of a well-developed wulst, evidence of which is proposed as compelling indications for the presence of stereoscopic vision specialties in fossil material [125] (p. 220).

Therefore, future analyses of dromornithid orbit size, optic foramen and skull orientation may provide additional insights into the hypertrophy noted in wulst structures for these birds. Cranial anatomy unambiguously shows that dromornithids had large, forward facing eyes (see Figure 1; SI Figures S1, S4 and S6; see also [6] (Table 1)). However, inference as to whether dromornithid retinal topography was structurally adapted (i.e., corneal diameter, retinal cell density/type, etc.), for sensitivity to low light conditions, may only be made by interpretation of orbit shape and size from skeletal remains.

In a large study assessing the relationship between corneal diameter and axial length of the avian eye, Hall and Ross [141] showed that species adapted to light-limited (scotopic or crepuscular) habitats, have larger corneal diameters and axial lengths, relative to those active during well-lit (photopic or diurnal) conditions. Hall [142] showed that there exists a close relationship between corneal diameter and axial length of the eye, and that metrics describing the osseous structures of the orbit were "well associated" with photic activity in birds. In support of these observations, several additional studies have shown that in nocturnal birds, eye shape increases relative to skull length, they display larger orbit diameters relative to depth, and orbits are more frontally orientated (e.g., [94,143,144] and references therein, but see also [131]). Animals that exploit low-light environments have evolved in one of two ways: (1) by enlargement and orientation of the visual system (i.e., increasing orbit size and binocular overlap); or (2) they develop enhanced sensitivity of somatosensory and tactile systems (e.g., [144], and references therein). Among ratites, kiwi are the only confirmed nocturnal taxon, and have evolved reduced eye size and distinct endocranial morphology associated with the somatosensory and tactile systems strategy (i.e., (2) see [79,145]). All other ratites are diurnal, as were the extinct NZ moa [77], and inspection of the shape of their brains (e.g., [75] (Figure 2), [79] (Figure 2a,b), [77] (Figure 6g–l), [145] (Figure 1b A–E), [80] (Figures 1 and 3), [81] (Figure 1), [74] (Figures 5.3 1–6)), show rostromediolaterally hypertrophied wulst structures in these large flightless birds, which may represent evolution for enhanced visual proficiency (i.e., strategy (1) above). However, no ratites display the massively hypertrophied state of the wulst seen in dromornithid dorsal endocasts.

Martin [131] advanced that binocular vision in birds may be used for the inspection of food items held in the bill, for bill control during the process of foraging and food provision to chicks, and not primarily for stereopsis. He also argued that binocular vision in the control of locomotion is a secondary function, as spatial information may be provided by each eye independently. Considering dromornithid cranial morphology displays large, forward facing orbits, and their dorsal endocranial morphology is dominated by the wulst, these birds probably adopted the strategy of enlargement and orientation of the visual system to develop good stereoscopic vision and accurate depth perception. This would preadapt them to being specialised browsers capable of selecting individual fruit and leaves from within complex browse. Additionally, such combined features of cranial and endocranial morphology also raises the possibility that dromornithids were adaptively selected for low-light visual proficiency along the nocturnal-diurnal gradient (scotopic *sensu* [142,146]). However, we prefer the more 'parsimonious' explanation that features of dromornithid cranial and endocranial anatomy are more likely associated with foraging and locomotion within complex diurnal environments.

4.4.2.2. Cerebrum

The rostral and caudal telencephalon—the nido- and mesopallial structures of the cerebrum—are recognised to form a complex with "integrative" functions (see [93], and references therein). Thus, we describe functional interpretations for the cerebrum as a whole.

The internal structure of the dorsal and rostrocaudolateral cerebrum incorporates four main subdivisions: the hyperpallium, incorporating the wulst, the mesopallium, incorporating the rostro- and caudodorsal telencephalon, the nidopallium, incorporating the rostro- and caudolateral telencephalon, and the arcopallium incorporating the caudoventral telencephalon. The rostromediolateral telencephalon incorporates the medial striatopallidal complex (striatum + pallidum), overlain rostrocaudally and dorsolaterally by the nidopallium, which is similarly overlain rostrocaudally and dorsolaterally by the mesopallium (see [48] (Figure 1C), [49] (Figure 6)). Dubbeldam and Visser [147] showed that the caudolateral nidopallium receives arcopallial afferents sourced through the medial nidopallium, and contains a complex pattern of terminal fields. They identified a strong connection between the striatopallidal complex and the mediolateral nidopallium, arguing that in the mallard (*Anas platyrhynchos*) there exists two major telencephalic circuits: one relaying through the pallidum, and the other through the striatum (e.g., [48] (Figure 3)), and that these afferent circuits play a major role in feeding behaviour. The telencephalon has been associated with a wide range of behaviours including feeding, taste, tactile sense, taste discrimination, vocalisation, and with high levels of cognition and complex tasks ([49], and references therein). Furthermore, stereotyped species-specific behaviour [93,148], pecking accuracy [149], and the processing of visual information such as brightness, colour, and pattern discrimination [118], have been attributed to processes within the caudolateral telencephalon. Pettigrew and Frost [130] showed that the maxillary (V_2) division of the trigeminal (V) nerve, which innervates the upper bill (see Section 4.4.1 above), transmits to extensive terminal fields in the region of the rostrodorsal mesopallium (e.g., [150] (Figure 3b), see also [139]). Similarly, Dubbeldam et al. [99] showed that ascending maxillary (V_2) and mandibular (V_3) trigeminal (V) projections transmitted rostrodorsally, via the nucleus basalis, to mesopallial terminal fields [140]. These sensorimotor and somatosensory projections were related to the "detection" of food particles, particularly in low-visibility feeding in anseriforms [151–154], and food grasping in columbiforms [140,155] and passeriforms [156]. Located in the caudal telencephalon, the medial spiriform nucleus receives projections from both somatosensory and visual parts of the wulst (see [121]; and Section 4.4.2.1 above), along with arcopallial projections associated with "motor planning information" from the caudoventral telencephalon, and projects to the cerebellum (see Section 4.4.2.4 below), forming the "telencephalic–cerebellar loop" (*sensu* [150] (p. 9)).

A notable caveat when interpreting differences in caudal telencephalon morphology, is that changes may arise from differential hypertrophy of any of several internal structures, each of which vary in function (e.g., [157], and above). Results show that the caudal telencephalon in dromornithids are strongly hypertrophied, and along with the apparently hypotrophied rostral telencephalon, they are morphologically similar to galliforms in the condition of the cerebrum (see Sections 4.1.4 and 4.1.5). Extant galliforms are generally visual foragers and exploit less somatosensory information associated with trigeminal (V) projections than tactile specialists like anseriforms [152–154]. Results show surface area ratios for the face of the maxillomandibular ($V_2 + V_3$) branch of the trigeminal (V) nerve complex in dromornithids are more similar to those of the galliforms, than they are to the anseriforms, which are distinctly larger (see Section 3.4.1). These morphological trends suggest that the tactile capability of dromornithids was likely reasonably well developed, at least comparable with the extant galliforms, but they were certainly not tactile sensory specialists.

4.4.2.3. Optic Lobe

The optic lobe forms a crucial component of the visual pathway system. Hellmann et al. [158] (p. 395) characterised the optic lobe as a "relay station" for transmitting ascending visual output to the forebrain (see Section 4.4.2.1 above), projecting descending output to the premotor regions of the hindbrain (see Section 4.4.2.4 below), and comprise multiple "retinotopically organised, and functionally specific" cell types. So called "optic flow" (*sensu* [159]), are retinal stimuli generated by self-motion through stationary environments (see [150,160] and references therein). Optic flow stimuli are analysed by recipient nuclei in the accessory optic system and pretectum, wherein the lentiformis mesencephali, or pretectal nucleus, responds to "moving large-field" visual stimuli and generates optokinetic response for the control of posture, eye movement stabilisation, and compensatory movement during locomotion ([117,150,160–164] and references therein), facilitated by processes within the cerebellum [87,165].

Optic lobe structures in dromornithids are somewhat hypotrophied in comparison to those of the extant galloanseres. Iwaniuk et al. [166] showed that the optic lobe of the nocturnal Night parrot (*Pezoporus occidentalis*) was relatively reduced compared with those of diurnal parrots, but orbit size did not differ between them. As noted above (see Section 4.4.2.1), eye shape, corneal diameter and retinal topography are better predictors low-light sensitivity than eye size alone (e.g., [86,141,142]). What is more, Kay and Kirk [167] (p. 238) showed variation in 'retinal summation', or the ratio of photoreceptor cells in the eye which interact with a single ganglion cell in the optic (II) nerve, is indicative of the level of visual sensitivity, acuity, and photic activity. This implies that large eye size, accompanied by small optic nerve size, is indicative of nocturnal habitus.

These forms of data are often not accessible from fossil material, and as we have shown, there is much variation in overall optic lobe size across known diurnal species of galloanseres assessed here. This implies that without corroborating evidence of eye structure and optic nerve absolute size, the size of the optic lobe alone is likely not a reliable proxy for inferring photic activity for fossil taxa.

Given that optic lobe structures are primarily relay stations, their absolute size is not definitively correlated with either nocturnal or diurnal behaviour (see also [168]) although this can resolved to an extent by comparisons of orbit and foramen opticum measurements. That dromornithids have large, forward facing eyes (see Section 4.4.2.1), robust optic nerves (see Section 3.1.2), and enlarged morphology of trigeminal (V) innervation (see Section 4.4.1), further supports our argument that dromornithids were likely diurnal taxa with a strong reliance on stereopsis and trigeminal (V) somatosensory input. These results also suggest that caution is required when deriving functional or behavioural inference from comparisons of optic lobe absolute size, without accounting for the collective morphology of somatosensory "circuits" (e.g., [150] (p. 9)), such as those of the thalamofugal, tectofugal and third visual pathways (see Section 4.4.2).

4.4.2.4. Hindbrain

The cerebellum has long been associated with motor integration and posture control in birds [87]. Visual signals are projected through the third visual pathway, via the retinal-recipient nuclei of the optic lobe, to the cerebellum [87,117,165,169–171], where they facilitate obstacle avoidance responses. Additionally, Pakan and Wylie [165] suggested that folia VI–VIII of the cerebellum may be involved in "steering" functions, and Iwaniuk et al. [172] showed that VI and VII folia are hypertrophied in birds they classified as "strong fliers", and showed some evidence to support correlation of hypertrophy of the cerebellum rostral lobe with "strong hindlimbs" in birds. Additionally, due to the dura mater, or dural envelope, which encloses the brain in life, a surface endocast derived from the internal osseous endocranial capsule, may not entirely reflect actual brain morphology, although this is much less of a problem in birds than in other archosaurs (e.g., [71,173] (p. 263), [74] (p. 61)). This is apparent in the lack of folia definition in dromornithid cerebellum morphology presented here, but which was almost certainly present in life. Additionally, it is recognised that an endocast can only capture surface structures of the brain, and that where a brain region is partially occluded by another structure, the endocast will underestimate total size of that region. This is the case for the cerebellum where it is occluded rostrally by the cerebrum. Thus, analyses of the cerebellum only relate to exposed parts of this region, but we consider these were analogous between compared taxa. Therefore in our analyses, the overall shape of the exposed dromornithid cerebellum is captured, and the broad metrics we use to describe the cerebellum illustrate well the differences in dorsal hindbrain shape across specimens.

Results for the dromornithid hindbrain, show the shapes described by the dromornithid metencephalon (cerebellum and rhombencephalon), are certainly distinct from those of the extant galloanseres. Measurement data show that the dromornithid cerebellum and rhombencephalon differ from the extant galloanseres in both length and width ratios. However, results for Surface Area data show the dromornithids overlap with the extant galloanseres in metencephalon surface area ratios. So the dromornithid hindbrain is not particularly dissimilar in overall size to those of the extant taxa, and apparent visual differences between caudal endocasts across specimens may lie in the particularly distinct hypertrophy of the dromornithid forebrain, several ratios of which certainly do differ from those of the extant galloanseres (e.g., Section 4.1.3 above). Additionally, the trend for hypertrophy of the rostrodorsal cerebellum noted in both middle Miocene dromornithid taxa, compared to that of the late Oligocene species (e.g., Section 3.3.3), may represent selection for enhanced neural capability with respect to hindlimb sensorimotor processes, facilitated within the cerebellum rostral lobe [172], as overall body size in dromornithids increased from the late Oligocene through the middle Miocene. Such morphological trends suggest that dromornithids likely maintained the capacity for capable locomotion through complex environments associated with third visual pathway processes in the hindbrain, at least comparable with the extant galliforms assessed here.

5. Conclusions

In this study, we comprehensively described dromornithid endocranial morphology for the first time, comparing endocasts for three species of *Dromornis* and *Ilbandornis woodburnei* to those of four species of extant basal galloanseres spanning the galliform–anseriform dichotomy. We compared endocasts of the Oligo–Miocene *Dromornis murrayi*, the middle Miocene *D. planei*, and the late Miocene *D. stirtoni*, to describe endocranial changes associated with neurocranial foreshortening in a temporal series along the *Dromornis* lineage. The endocasts of species of *Dromornis* were compared to that of the middle Miocene *Ilbandornis woodburnei*, revealing minimally five morphological differences between *Ilbandornis* and *Dromornis* (e.g., relating to the medial boundaries of the wulst in the rostrocaudal fissura interhemispherica zone; the caudoventral regions of the caudal telencephalon; the morphology of the rostrodorsal and caudodorsal cerebellum; and the morphology of the ventral cerebellum + pons hindbrain complex), collectively supporting the two-lineage hypothesis. Overall, dromornithid brains display morphology supporting hypotheses that they are more closely related to basal galliforms than anseriforms, in that the rostral positioning of the wulst on the dorsal endocast; the form of the maxillomandibular (V_2 + V_3), glossopharyngeal (IX) and vagus (X) nerves, and the morphology of trigeminal ganglia, are more similar to the species of galliforms assessed. Functional interpretations suggest that dromornithids were specialised herbivores that likely possessed well-developed stereoscopic depth perception and targeted a soft browse trophic niche.

The retention of a relatively large cerebellum and associated hindbrain morphology in dromornithids, such as those shown by the volant galloanseres assessed, raises the question as to why dromornithids maintained the capacity for capable movement through complex environments (see above). That these functional attributes were selected for during early dromornithid evolution is a possibility.

During the transition from the early Cenozoic through the Eocene, Australia was blanketed by predominantly warm to cool-temperate rainforest, which began to open into sclerophyllous vegetation on higher ground only during the Oligocene (see [174]). These forests became progressively drier during the transition from the Oligocene through the late Miocene, eventuating in the establishment of sclerophyllous fire-sensitive woodlands as the dominant continental floras during the Miocene ([1], see also [175]). Dromornithid evolution has a long history, and the key characteristics of the dromornithid brain were likely assembled when they first evolved into large flightless birds sometime during the

Palaeogene. During those times, dromornithids would have occupied highly complex forested environments, where the capacity for acute stereoscopic vision, the taxon's trophic preferences established, and the capability for navigating complex environments were maintained from their flighted ancestors. Dromornithids likely co-opted these traits in adapting to and exploiting the steadily drying Australian environment through their known temporal range, traits that persisted in the last dromornithid taxon of the late Pleistocene.

During the middle Miocene through to the late Pleistocene, dromornithids were likely increasingly restricted to riparian zones, or sclerophyllous woodland remnants proximal to water courses or temporary water bodies. Dromornithid fossils from these times are almost exclusively recovered from deposits derived from ancient water courses or shallow lakes. For example, their bones are relatively abundant in the fluviatile deposits of Bullock Creek LF, and Alcoota LF in the Northern Territory (see Section 1); and are common in several lacustrine sites around Lake Callabonna in the Eyre basin of Southern Australia, where dromornithid bones are found within the clays of ancient lake margins wherein birds were mired, evidently while in search of water during dry periods (e.g., [176] (p. 208), [15] (p. 3)). Consequently, from the late Miocene in particular, dromornithids may have become increasingly 'tethered' to constricting sclerophyllous woodland remnants in a drying environment, circumstances which, along with temporal overlap with newly arrived humans during the late Pleistocene [177], potentially contributed to the eventual extinction of the group.

The cranial and endocranial morphology of dromornithids is unlike any attained in the evolution of birds, and represents distinct morphological adaptations to progressively changing Australian Cenozoic environments. The specialist browsing diet of these giant birds resulted in the unique juxtaposition of large body size, deep, narrow, and elongate bills, with frontally directed eyes, affording stereoscopic vision on crania bearing minimal musculature, and like much of the idiosyncratic Australian fauna of the past, dromornithids represent combinations of unique adaptations now lost.

Supplementary Materials: The following are available online at https://www.mdpi.com/1424-281 8/13/3/124/s1, Supplementary Information (SI) document [178–180], and an 'Endocasts and R code' folder, including endocasts used, R scripts and supporting data files.

Author Contributions: T.H.W. and W.D.H. designed this study, wrote and edited the manuscript. W.D.H. assembled and analysed the data. All authors have read and agreed to the published version of the manuscript.

Funding: This work was supported by the Australian Research Council DE130101133, and DP180101913 (T.H.W), and an Australian Government Research Training Program Scholarship (W.D.H).

Institutional Review Board Statement: Not applicable.

Informed Consent Statement: Not applicable.

Data Availability Statement: All primary data generated in this study are available in SI and Table 1 above. Image-stack CT data are available on request, see ReadMe.txt file (SI).

Acknowledgments: We thank Michael Archer and Suzanne Hand (University of NSW, Sydney), and Queensland Museum (Brisbane, Qld) for access to *Dromornis murrayi* specimens (QM F57984, QM F57974 and QM F50412 endocast); Adam Yates (Museum of Central Australia, Alice Springs, NT) for access to the *Dromornis planei* (NTM P9464-106) specimen; The Queen Victoria Museum and Art Gallery (Launceston, Tas) for access to the *Ilbandornis woodburnei* (QVM:2000:GFV:20) specimen. Museums Victoria (Melbourne, Vic) enabled access to the *Anhima cornuta* (MV B12574) specimen. We thank Philippa Horton and Maya Penck (South Australian Museum, Adelaide, SA) for access to *Gallus gallus* (SAM B34041), *Leipoa ocellata* (SAM B11482), and *Anseranas semipalmata* (SAM B48035) specimens. Adelaide Microscopy (University of Adelaide, SA) enabled access to the Skyscan μCT instrument. We thank Joseph Bevitt (ANSTO, Sydney, NSW) for neutron scanning of *D. planei*, and Mishelle Korlaet (SAHMRI, Adelaide, SA) for CT scanning of *D. stirtoni*, *D. murrayi* and *I. woodburnei* specimens. W.D.H thanks Gavin Prideaux, Stig Walsh, Em Sherratt, and two anonymous reviewers for their comments which substantially improved earlier versions of this manuscript.

Conflicts of Interest: The authors declare no conflict of interest.

References

1. Vickers-Rich, P. The Mesozoic and Tertiary History of Birds on the Australian Plate. In *Vertebrate Palaeontology of Australasia*; Vickers-Rich, P., Monaghan, J.M., Baird, R.F., Rich, T.H., Eds.; Pioneer Design Studios and Monash University Publications Committee: Melbourne, Australia, 1991; pp. 721–808.
2. Murray, P.F.; Vickers-Rich, P. *Magnificent Mihirungs: The Colossal Flightless Birds of the Australian Dreamtime*; Indiana University Press: Bloomington, IN, USA, 2004; p. 410.
3. Murray, P.F.; Megirian, D. The skull of dromornithid birds: Anatomical evidence for their relationship to Anseriformes. *Rec. South Aust. Mus.* **1998**, *31*, 51–97.
4. Murray, P.F.; Megirian, D. The Pwerte Marnte Marnte Local Fauna: A new vertebrate assemblage of presumed Oligocene age from the Northern Territory of Australia. *Alcheringa* **2006**, *30*, 211–228. [CrossRef]
5. Boles, W.E. The Avian Fossil Record of Australia: An Overview. In *Evolution and Biogeography of Australasian Vertebrates*; Merrick, J.R., Archer, M., Hickey, G.M., Lee, M.S.Y., Eds.; Auscipub: Sydney, Australia, 2006; pp. 387–411.
6. Worthy, T.H.; Handley, W.D.; Archer, M.; Hand, S.J. The extinct flightless mihirungs (Aves, Dromornithidae): Cranial anatomy, a new species, and assessment of Oligo-Miocene lineage diversity. *J. Vertebr. Paleontol.* **2016**, *36*, e1031345. [CrossRef]
7. Vickers-Rich, P.; Molnar, R.E. The foot of a bird from the Eocene Redbank Plains Formation of Queensland, Australia. *Alcheringa* **1996**, *20*, 21–29. [CrossRef]
8. Mayr, G. *Paleogene Fossil Birds*; Springer: Berlin, Germany, 2009; p. 262.
9. Saltré, F.; Rodríguez-Rey, M.; Brook, B.W.; Johnson, C.N.; Turney, C.S.; Alroy, J.; Cooper, A.; Beeton, N.; Bird, M.I.; Fordham, D.A. Climate change not to blame for late Quaternary megafauna extinctions in Australia. *Nat. Commun.* **2016**, *7*, 10511. [CrossRef] [PubMed]
10. Owen, R. [Part 19 of Owen's memoir on *Dinornis*, read June 4, 1872]. *P. Zool. Soc. Lond.* **1872**, *1872*, 682–683.
11. Rich, P.V. The Dromornithidae, an extinct family of large ground birds endemic to Australia. *Bur. Nat. Resour. Geol. Geophys. Bull.* **1979**, *184*, 1–194.
12. Nguyen, J.M.T.; Boles, W.E.; Hand, S.J. New material of *Barawertornis tedfordi*, a dromornithid bird from the Oligo-Miocene of Australia, and its phylogenetic implications. *Rec. Aust. Mus.* **2010**, *62*, 45–60. [CrossRef]
13. Worthy, T.H.; Yates, A. Connecting the thigh and foot: Resolving the association of post-cranial elements in the species of *Ilbandornis* (Aves: Dromornithidae). *Alcheringa* **2015**, *39*, 1–21. [CrossRef]
14. Stirling, E.C.; Zietz, A.H.C. Preliminary notes on *Genyornis newtoni*: A new genus and species of fossil struthious bird found at Lake Callabonna, South Australia. *T. Roy. Soc. South Aust.* **1896**, *20*, 171–211.
15. Wells, R.T.; Tedford, R.H. *Sthenurus* (Macropodidae: Marsupialia) from the Pleistocene of Lake Callabonna, South Australia. *B. Am. Mus. Nat. Hist.* **1995**, *225*, 3–111.
16. Stirling, E.C.; Zietz, A.H.C. *Genyornis newtoni*—A fossil struthious bird from Lake Callabonna, South Australia: Description of the bones of the leg and foot. *T. Roy. Soc. South Aust.* **1896**, *20*, 191–211.
17. Wetmore, A. A classification for the birds of the world. *Smithson. Misc. Collect.* **1960**, *139*, 1–37.
18. Rich, P.V. Changing continental arrangements and the origin of Australia's non-passeriform continental avifauna. *Emu Austral Ornithol.* **1975**, *75*, 97–112. [CrossRef]
19. Worthy, T.H.; Holdaway, R.N. *The Lost World of the Moa: Prehistoric Life of New Zealand*; Indiana University Press: Bloomington, IA, USA, 2002; p. 760.
20. Phillips, M.J.; Gibb, G.C.; Crimp, E.A.; Penny, D. Tinamous and moa flock together: Mitochondrial genome sequence analysis reveals independent losses of flight among ratites. *Syst. Biol.* **2010**, *59*, 90–107. [CrossRef]
21. Olson, S.L. The Fossil Record of Birds. In *Avian Biology*; Farner, D.S., King, J.R., Parkes, K.C., Eds.; Academic Press: New York, NY, USA, 1985; Volume 8, pp. 79–238.

22. Mayr, G. Cenozoic mystery birds–on the phylogenetic affinities of bony-toothed birds (Pelagornithidae). *Zool. Scr.* **2011**, *40*, 448–467. [CrossRef]
23. Worthy, T.H.; Mitri, M.; Handley, W.D.; Lee, M.S.Y.; Anderson, A.; Sand, C. Osteology supports a stem-Galliform affinity for the giant extinct flightless bird *Sylviornis neocaledoniae* (Sylviornithidae, Galloanseres). *PLoS ONE* **2016**, *11*, e0150871. [CrossRef]
24. Worthy, T.H.; Degrange, F.J.; Handley, W.D.; Lee, M.S.Y. The evolution of giant flightless birds and novel phylogenetic relationships for extinct fowl (Aves, Galloanseres). *Roy. Soc. Open Sci.* **2017**, *4*, 170975. [CrossRef]
25. Worthy, T.H.; Degrange, F.J.; Handley, W.D.; Lee, M.S.Y. Correction to 'The evolution of giant flightless birds and novel phylogenetic relationships for extinct fowl (Aves, Galloanseres)'. *Roy. Soc. Open Sci.* **2017**, *4*, 171621. [CrossRef] [PubMed]
26. Andors, A.V. Reappraisal of the Eocene groundbird *Diatryma* (Aves: Anserimorphae). *Nat. Hist. Mus. Los Angeles County Sci. Ser.* **1992**, *36*, 109–125.
27. Angst, D.; Lécuyer, C.; Amiot, R.; Buffetaut, E.; Fourel, F.; Martineau, F.; Legendre, S.; Abourachid, A.; Herrel, A. Isotopic and anatomical evidence of an herbivorous diet in the Early Tertiary giant bird *Gastornis*. Implications for the structure of Paleocene terrestrial ecosystems. *Naturwissenschaften* **2014**, *101*, 313–322. [CrossRef]
28. Archer, M.; Godthelp, H.; Hand, S.J.; Attenborough, D. *Australia's Lost World: Riversleigh, World Heritage Site*; New Holland Publishers: Sydney, Austrilia, 1991; p. 264.
29. Woodburne, M.O. The Alcoota Fauna, central Australia: An integrated palaeontological and geological study. *B. Bur. Min. Res. Geol. Geophy.* **1967**, *87*, 1–187.
30. Murray, P.F.; Megirian, D. Continuity and contrast in middle and late Miocene vertebrate communities from the Northern Territory. *Rec. Mus. Art Gall. Northern Terr.* **1992**, *9*, 195–218.
31. Handley, W.D.; Chinsamy, A.; Yates, A.M.; Worthy, T.H. Sexual dimorphism in the late Miocene mihirung *Dromornis stirtoni* (Aves: Dromornithidae) from the Alcoota Local Fauna of central Australia. *J. Vertebr. Paleontol.* **2016**, *36*, e1180298. [CrossRef]
32. Grellet-Tinner, G.; Spooner, N.A.; Handley, W.D.; Worthy, T.H. The *Genyornis* egg: Response to Miller et al.'s commentary on Grellet-Tinner et al., 2016. *Quarternary Sci. Rev.* **2017**, *161*, 128–133. [CrossRef]
33. Hansford, J.P.; Turvey, S.T. Unexpected diversity within the extinct elephant birds (Aves: Aepyornithidae) and a new identity for the world's largest bird. *Roy. Soc. Open Sci.* **2018**, *5*, 181295. [CrossRef]
34. Archer, M.; Godthelp, H.; Hand, S.J.; Megirian, D. Fossil mammals of Riversleigh, northwestern Queensland: Preliminary overview of biostratigraphy, correlation and environmental change. *Austr. Zool.* **1989**, *25*, 29–65. [CrossRef]
35. Archer, M.; Hand, S.J.; Godthelp, H.; Creaser, P. Correlation of the Cainozoic sediments of the Riversleigh World Heritage fossil property, Queensland, Australia. In Actes du Congrès BiochroM '97. *Mémoires Travaux Inst. Montp.* **1997**, *21*, 131–152.
36. Travouillon, K.J.; Archer, M.; Hand, S.J.; Godthelp, H. Multivariate analyses of Cenozoic mammalian faunas from Riversleigh, northwestern Queensland. *Alcheringa Spec. Iss.* **2006**, *1*, 323–349. [CrossRef]
37. Travouillon, K.J.; Escarguel, G.; Legendre, S.; Archer, M.; Hand, S.J. The use of MSR (Minimum Sample Richness) for sample assemblage comparisons. *Paleobiology* **2011**, *37*, 696–709. [CrossRef]
38. Woodhead, J.; Hand, S.J.; Archer, M.; Graham, I.; Sniderman, K.; Arena, D.A.; Black, K.H.; Godthelp, H.; Creaser, P.; Price, E. Developing a radiometrically-dated chronologic sequence for Neogene biotic change in Australia, from the Riversleigh World Heritage Area of Queensland. *Gondwana Res.* **2016**, *29*, 153–167. [CrossRef]
39. Arena, D.A.; Travouillon, K.J.; Beck, R.M.D.; Black, K.H.; Gillespie, A.K.; Myers, T.J.; Archer, M.; Hand, S.J. Mammalian lineages and the biostratigraphy and biochronology of Cenozoic faunas from the Riversleigh World Heritage Area, Australia. *Lethaia* **2016**, *49*, 43–60. [CrossRef]
40. Megirian, D.; Prideaux, G.J.; Murray, P.F.; Smit, N. An Australian land mammal age biochronological scheme. *Paleobiology* **2010**, *36*, 658–671. [CrossRef]
41. Murray, P.; Megirian, D.; Rich, T.; Plane, M.; Black, K.; Archer, M.; Hand, S.J.; Vickers-Rich, P. Morphology, systematics, and evolution of the marsupial genus *Neohelos* Stirton (Diprotodontidae, Zygomaturinae). *Mus. Gall. Northern Terr. Res. Rep.* **2000**, *6*, 1–141.
42. Yates, A.M. New craniodental remains of *Wakaleo alcootaensis* (Diprotodontia: Thylacoleonidae) a carnivorous marsupial from the late Miocene Alcoota Local Fauna of the Northern Territory, Australia. *PeerJ* **2015**, *3*, e1408. [CrossRef]
43. Yates, A.M. A new species of long-necked turtle (Pleurodira: Chelidae: *Chelodina*) from the late Miocene Alcoota Local Fauna, Northern Territory, Australia. *PeerJ* **2013**, *1*, e170. [CrossRef]
44. Worthy, T.H.; Yates, A. A review of the smaller birds from the late Miocene Alcoota Local Faunas of Australia with a description of a new species. In Proceedings of the 9th International Meeting of the Society of Avian Paleontology and Evolution, Diamante, Argentina, 1–6 August 2016. *Contr. Mus. Arg. Cien. Nat.* **2017**, *7*, 221–252.
45. Yates, A.M.; Worthy, T.H. A diminutive species of emu (Casuariidae: Dromaiinae) from the late Miocene of the Northern Territory, Australia. *J. Vertebr. Paleontol.* **2019**, *39*, e1665057. [CrossRef]
46. Stirton, R.A.; Woodburne, M.O.; Plane, M.D. A phylogeny of the Tertiary Diprotodontidae and its significance in correlation. In Tertiary Diprotodontidae from Australia and New Guinea. *B. Bur. Min. Res. Geol. Geophy.* **1967**, *85*, 149–160.
47. Baumel, J.J.; King, A.S.; Breazile, J.E.; Evans, H.E.; Vanden Berge, J.C. *Handbook of Avian Anatomy: Nomina Anatomica Avium*, 2nd ed.; Nuttall Ornithological Club: Cambridge, MA, USA, 1993; Volume 23, p. 779.
48. Jarvis, E.D.; Güntürkün, O.; Bruce, L.; Csillag, A.; Karten, H.; Kuenzel, W.; Medina, L.; Paxinos, G.; Perkel, D.J.; Shimizu, T. Avian brains and a new understanding of vertebrate brain evolution. *Nat. Rev. Neurosci.* **2005**, *6*, 151–159. [CrossRef] [PubMed]

49. Corfield, J.R.; Wild, J.M.; Parsons, S.; Kubke, M.F. Morphometric analysis of telencephalic structure in a variety of neognath and paleognath bird species reveals regional differences associated with specific behavioral traits. *Brain Behav. Evolut.* **2012**, *80*, 181–195. [CrossRef] [PubMed]

50. Early, C.M.; Iwaniuk, A.N.; Ridgely, R.C.; Witmer, L.M. Endocast structures are reliable proxies for the sizes of corresponding regions of the brain in extant birds. *J. Anat.* **2020**, *237*, 1162–1176. [CrossRef]

51. Rasband, W.S. *ImageJ*; U.S. National Institute of Health: Bethesda, MD, USA, 2018. Available online: https://imagej.nih.gov/ij/ (accessed on 28 January 2021).

52. Cignoni, P.; Callieri, M.; Corsini, M.; Dellepiane, M.; Ganovelli, F.; Ranzuglia, G. Year Meshlab: An open-source mesh processing tool. In *Eurographics Italian Chapter Conference*; Scarano, V., De Chiara, R., Erra, U., Eds.; The Eurographics Association: Geneve, Switzerland, 2008; pp. 129–136. [CrossRef]

53. Wiley, D.F. *Landmark Editor 3.0*; Institute for Data Analysis and Visualization, University of California: Davis, CA, USA, 2006.

54. Bookstein, F.L. *Morphometric Tools for Landmark Data: Geometry and Biology*; Cambridge University Press: Cambridge, UK, 1991; p. 435.

55. Adams, D.C.; Collyer, M.L.; Kaliontzopoulou, A. Geomorph: Software for Geometric Morphometric Analyses. R Package Version 3.1.0. 2019. Available online: https://cran.r-project.org/src/contrib/Archive/geomorph/ (accessed on 25 March 2019).

56. Mosimann, J.E. Size allometry: Size and shape variables with characterizations of the lognormal and generalized gamma distributions. *J. Am. Stat. Assoc.* **1970**, *65*, 930–945. [CrossRef]

57. Claude, J. *Morphometrics with R*; Springer: New York, NY, USA, 2008; p. 316.

58. Klingenberg, C.P. Size, shape, and form: Concepts of allometry in geometric morphometrics. *Dev. Genes Evol.* **2016**, *226*, 113–137. [CrossRef] [PubMed]

59. Sherratt, E.; Vidal-García, M.; Anstis, M.; Keogh, J.S. Adult frogs and tadpoles have different macroevolutionary patterns across the Australian continent. *Nat. Ecol. Evol.* **2017**, *1*, 1385–1391. [CrossRef] [PubMed]

60. R Core Team. *R: A Language and Environment for Statistical Computing*; R Foundation for Statistical Computing: Vienna, Austria, 2019; Available online: https://www.R-project.org/ (accessed on 28 January 2021).

61. RStudio Team. *RStudio: Integrated Development for R*; RStudio, Inc.: Boston, MA, USA, 2019; Available online: http://www.rstudio.com/ (accessed on 28 January 2021).

62. Gower, J.C. Generalized Procrustes analysis. *Psychometrika* **1975**, *40*, 33–51. [CrossRef]

63. Rohlf, F.J.; Slice, D. Extensions of the Procrustes method for the optimal superimposition of landmarks. *Syst. Biol.* **1990**, *39*, 40–59. [CrossRef]

64. Bookstein, F.L. Size and shape spaces for landmark data in two dimensions. *Stat. Sci.* **1986**, *1*, 181–242. [CrossRef]

65. Adams, D.C.; Otárola-Castillo, E. geomorph: An R package for the collection and analysis of geometric morphometric shape data. *Methods Ecol. Evol.* **2013**, *4*, 393–399. [CrossRef]

66. Bookstein, F.L. Landmark methods for forms without landmarks: Morphometrics of group differences in outline shape. *Med. Image Anal.* **1997**, *1*, 225–243. [CrossRef]

67. Bookstein, F.L. Shape and the information in medical images: A decade of the morphometric synthesis. *Comput. Vis. Image Und.* **1997**, *2*, 97–118. [CrossRef]

68. Iwaniuk, A.N.; Nelson, J.E. Can endocranial volume be used as an estimate of brain size in birds? *Can. J. Zool.* **2002**, *80*, 16–23. [CrossRef]

69. Striedter, G.F. *Principles of Brain Evolution*; Sinauer Associates: Sunderland, MA, USA, 2005.

70. Striedter, G.F. Précis of principles of brain evolution. *Behav. Brain Sci.* **2006**, *29*, 1–12. [CrossRef]

71. Witmer, L.M.; Ridgely, R.C.; Dufeau, D.L.; Semones, M.C. Using CT to Peer into the Past: 3D Visualization of the Brain and Ear Regions of Birds, Crocodiles, and Nonavian Dinosaurs. In *Anatomical Imaging, Towards a New Morphology*; Endo, H., Frey, R., Eds.; Springer: Tokyo, Japan, 2008; pp. 67–87.

72. Picasso, M.B.J.; Tambussi, C.; Dozo, M.T. Neurocranial and brain anatomy of a Late Miocene eagle (Aves, Accipitridae) from Patagonia. *J. Vertebr. Paleontol.* **2009**, *29*, 831–836. [CrossRef]

73. Walsh, S.A.; Iwaniuk, A.N.; Knoll, M.A.; Bourdon, E.; Barrett, P.M.; Milner, A.C.; Nudds, R.L.; Abel, R.L.; Sterpaio, P.D. Avian cerebellar floccular fossa size is not a proxy for flying ability in birds. *PLoS ONE* **2013**, *8*, e67176. [CrossRef] [PubMed]

74. Walsh, S.A.; Knoll, F. The Evolution of Avian Intelligence and Sensory Capabilities: The Fossil Evidence. In *Digital Endocasts: From Skulls to Brains*; Bruner, E., Ogihara, N., Tanabe, H.C., Eds.; Springer: Tokyo, Japan, 2018; pp. 59–69.

75. Scofield, R.P.; Ashwell, K.W.S. Rapid somatic expansion causes the brain to lag behind: The case of the brain and behavior of New Zealand's Haast's Eagle (*Harpagornis moorei*). *J. Vertebr. Paleontol.* **2009**, *29*, 637–649. [CrossRef]

76. Lawal, R.A.; Al-Atiyat, R.M.; Aljumaah, R.S.; Silva, P.; Mwacharo, J.M.; Hanotte, O. Whole-genome resequencing of red junglefowl and indigenous village chicken reveal new insights on the genome dynamics of the species. *Front. Genet.* **2018**, *9*, 264. [CrossRef] [PubMed]

77. Ashwell, K.W.S.; Scofield, R.P. Big birds and their brains: Paleoneurology of the New Zealand moa. *Brain Behav. Evolut.* **2008**, *71*, 151–166. [CrossRef] [PubMed]

78. Craigie, E.H. The cerebral cortex of *Rhea americana*. *J. Comp. Neurol.* **1939**, *70*, 331–353. [CrossRef]

79. Martin, G.R.; Wilson, K.-J.; Wild, J.M.; Parsons, S.; Kubke, M.F.; Corfield, J. Kiwi forego vision in the guidance of their nocturnal activities. *PLoS ONE* **2007**, *2*, e198. [CrossRef]

80. Peng, K.; Feng, Y.; Zhang, G.; Liu, H.; Song, H. Anatomical study of the brain of the African ostrich. *Turk. J. Vet. Anim. Sci.* **2010**, *34*, 235–241. [CrossRef]

81. Picasso, M.B.J.; Tambussi, C.P.; Degrange, F.J. Virtual reconstructions of the endocranial cavity of *Rhea americana* (Aves, Palaeognathae): Postnatal anatomical changes. *Brain Behav. Evolut.* **2011**, *76*, 176–184. [CrossRef] [PubMed]

82. Witmer, L.M.; Chatterjee, S.; Franzosa, J.; Rowe, T. Neuroanatomy of flying reptiles and implications for flight, posture and behaviour. *Nature* **2003**, *425*, 950–953. [CrossRef]

83. Milner, A.C.; Walsh, S.A. Avian brain evolution: New data from Palaeogene birds (Lower Eocene) from England. *Zool. J. Linn. Soc. Lond.* **2009**, *155*, 198–219. [CrossRef]

84. Witmer, L.M.; Ridgely, R.C. New insights into the brain, braincase, and ear region of tyrannosaurs (Dinosauria, Theropoda), with implications for sensory organization and behavior. *Anat. Rec.* **2009**, *292*, 1266–1296. [CrossRef]

85. Walsh, S.A.; Luo, Z.-X.; Barrett, P.M. Modern Imaging Techniques as a Window to Prehistoric Auditory Worlds. In *Insights from Comparative Hearing Research*; Koppl, C., Manley, G.A., Popper, A.N., Fay, R.R., Eds.; Springer: New York, NY, USA, 2014; pp. 227–261.

86. Hall, M.I.; Iwaniuk, A.N.; Gutiérrez-Ibáñez, C. Optic foramen morphology and activity pattern in birds. *Anat. Rec.* **2009**, *292*, 1827–1845. [CrossRef]

87. Jerison, H.J. *Evolution of the Brain and Intelligence*; Academic Press: New York, NY, USA, 1973; p. 482.

88. Barton, R.A.; Purvis, A.; Harvey, P.H. Evolutionary radiation of visual and olfactory brain systems in primates, bats and insectivores. *Philos. T. Roy. Soc. B.* **1995**, *348*, 381–392.

89. Barton, R.A.; Aggleton, J.P.; Grenyer, R. Evolutionary coherence of the mammalian amygdala. *P. Roy. Soc. B. Biol. Sci.* **2003**, *270*, 539–543. [CrossRef]

90. Barton, R.A.; Harvey, P.H. Mosaic evolution of brain structure in mammals. *Nature* **2000**, *405*, 1055–1058. [CrossRef] [PubMed]

91. Whiting, B.A.; Barton, R.A. The evolution of the cortico-cerebellar complex in primates: Anatomical connections predict patterns of correlated evolution. *J. Hum. Evolut.* **2003**, *44*, 3–10. [CrossRef]

92. Iwaniuk, A.N.; Dean, K.M.; Nelson, J.E. A mosaic pattern characterizes the evolution of the avian brain. *P. Roy. Soc. B. Biol. Sci.* **2004**, *271*, S148–S151. [CrossRef] [PubMed]

93. Dubbeldam, J.L. Birds. In *The Central Nervous System of Vertebrates*; Nieuwenhuys, R., ten Donkelaar, H.J., Nicholson, C., Eds.; Springer: Berlin, Germany, 1998; pp. 1525–1636.

94. Iwaniuk, A.N.; Heesy, C.P.; Hall, M.I.; Wylie, D.R. Relative wulst volume is correlated with orbit orientation and binocular visual field in birds. *J. Comp. Physiol. A.* **2008**, *194*, 267–282. [CrossRef]

95. Corfield, J.R.; Price, K.; Iwaniuk, A.N.; Gutiérrez-Ibáñez, C.; Birkhead, T.; Wylie, D.R. Diversity in olfactory bulb size in birds reflects allometry, ecology, and phylogeny. *Front. Neuroanat.* **2015**, *9*, 102. [CrossRef] [PubMed]

96. Gutiérrez-Ibáñez, C.; Iwaniuk, A.N.; Wylie, D.R. The independent evolution of the enlargement of the principal sensory nucleus of the trigeminal nerve in three different groups of birds. *Brain Behav. Evolut.* **2009**, *74*, 280–294. [CrossRef] [PubMed]

97. Dubbeldam, J.L. Studies on the somatotopy of the trigeminal system in the mallard, *Anas platyrhynchos* L. II. Morphology of the principal sensory nucleus. *J. Comp. Neurol.* **1980**, *191*, 557–571. [CrossRef]

98. Bubień-Waluszewska, A. The Cranial Nerves. In *Form and Function in Birds*; King, A.S., McLelland, J., Eds.; Academic Press: London, UK, 1981; Volume 2, pp. 385–438.

99. Dubbeldam, J.L.; Brauch, C.S.M.; Don, A. Studies on the somatotopy of the trigeminal system in the mallard, *Anas platyrhynchos* L. III. Afferents and organization of the nucleus basalis. *J. Comp. Neurol.* **1981**, *196*, 391–405. [CrossRef]

100. Wild, J.M.; Zeigler, H.P. Central projections and somatotopic organisation of trigeminal primary afferents in pigeon (*Columba livia*). *J. Comp. Neurol.* **1996**, *368*, 136–152. [CrossRef]

101. Dubbeldam, J.L.; Brus, E.R.; Menken, S.B.J.; Zeilstra, S. The central projections of the glossopharyngeal and vagus ganglia in the mallard, *Anas platyrhynchos* L. *J. Comp. Neurol.* **1979**, *183*, 149–168. [CrossRef]

102. Wild, J.M.; Zeigler, H.P. Central representation and somatotopic organization of the jaw muscles within the facial and trigeminal nuclei of the pigeon (*Columba livia*). *J. Comp. Neurol.* **1980**, *192*, 175–201. [CrossRef]

103. Wild, J.M. Identification and localization of the motor nuclei and sensory projections of the glossopharyngeal, vagus, and hypoglossal nerves of the cockatoo (*Cacatua roseicapilla*), Cacatuidae. *J. Comp. Neurol.* **1981**, *203*, 351–377. [CrossRef]

104. Wild, J.M. Peripheral and central terminations of hypoglossal afferents innervating lingual tactile mechanoreceptor complexes in Fringillidae. *J. Comp. Neurol.* **1990**, *298*, 157–171. [CrossRef] [PubMed]

105. Dubbeldam, J.L. The sensory trigeminal system in birds: Input, organization and effects of peripheral damage. *Arch. Physiol. Biochem.* **1998**, *106*, 338–345. [CrossRef]

106. Wild, J.M. The avian somatosensory system: Connections of regions of body representation in the forebrain of the pigeon. *Brain Res.* **1987**, *412*, 205–223. [CrossRef]

107. Dubbeldam, J.L. Cranial nerves and sensory centres—A matter of definition? Hypoglossal and other afferents of the avian sensory trigeminal system. *Zool. Jahrb.* **1992**, *122*, 179–186.

108. Dubbeldam, J.L. Afferent connections of nervus facialis and nervus glossopharyngeus in the pigeon (*Columba livia*) and their role in feeding behavior. *Brain Behav. Evolut.* **1984**, *24*, 47–57. [CrossRef] [PubMed]

109. Arends, J.J.A.; Dubbeldam, J.L. The subnuclei and primary afferents of the descending trigeminal system in the mallard (*Anas platyrhynchos* L.). *Neuroscience* **1984**, *13*, 781–795. [CrossRef]

110. Davies, S.J.J.F. The food of emus. *Austral Ecol.* **1978**, *3*, 411–422. [CrossRef]
111. Davies, S.J.J.F. *Bird Families of the World: Ratites and Tinamous: Tinamidae, Rheidae, Dromaiidae, Casuariidae, Apterygidae, Struthionidae*; Oxford University Press: Oxford, UK, 2002; p. 310.
112. Wings, O. A review of gastrolith function with implications for fossil vertebrates and a revised classification. *Acta Palaeontol. Pol.* **2007**, *52*, 1–16.
113. Fritz, J.; Hummel, J.; Kienzle, E.; Wings, O.; Streich, W.J.; Clauss, M. Gizzard vs. teeth, it's a tie: Food-processing efficiency in herbivorous birds and mammals and implications for dinosaur feeding strategies. *Paleobiology* **2011**, *37*, 577–586. [CrossRef]
114. Worthy, T.H. Aspects of the biology of two moa species (Aves: Dinornithiformes). *New Zeal. J. Arch.* **1989**, *11*, 77–86.
115. Wood, J.R.; Rawlence, N.J.; Rogers, G.M.; Austin, J.J.; Worthy, T.H.; Cooper, A. Coprolite deposits reveal the diet and ecology of the extinct New Zealand megaherbivore moa (Aves, Dinornithiformes). *Quat. Sci. Rev.* **2008**, *27*, 2593–2602. [CrossRef]
116. Wood, J.R.; Wilmshurst, J.M.; Richardson, S.J.; Rawlence, N.J.; Wagstaff, S.J.; Worthy, T.H.; Cooper, A. Resolving lost herbivore community structure using coprolites of four sympatric moa species (Aves: Dinornithiformes). *Pro. Natl. Acad. Sci. USA* **2013**, *110*, 16910–16915. [CrossRef]
117. Wylie, D.R.; Gutiérrez-Ibáñez, C.; Pakan, J.M.P.; Iwaniuk, A.N. The optic tectum of birds: Mapping our way to understanding visual processing. *Can. J. Exp. Psychol.* **2009**, *63*, 328–338. [CrossRef] [PubMed]
118. Iwaniuk, A.N.; Gutiérrez-Ibáñez, C.; Pakan, J.M.P.; Wylie, D.R. Allometric scaling of the tectofugal pathway in birds. *Brain Behav. Evolut.* **2010**, *75*, 122–137. [CrossRef]
119. Wylie, D.R.; Iwaniuk, A.N. Neural Mechanisms Underlying Visual Motion Detection in Birds. In *How Animals See the World: Comparative Behavior, Biology, and Evolution of Vision*; Lazareva, O.F., Shimizu, T., Wasserman, E.A., Eds.; Oxford University Press: New York, NY, USA, 2012; pp. 289–318.
120. Iwaniuk, A.N.; Wylie, D.R. The evolution of stereopsis and the wulst in caprimulgiform birds: A comparative analysis. *J. Comp. Physiol. A* **2006**, *192*, 1313–1326. [CrossRef] [PubMed]
121. Wild, J.M.; Williams, M.N. Rostral wulst in passerine birds. I. Origin, course, and terminations of an avian pyramidal tract. *J. Comp. Neurol.* **2000**, *416*, 429–450. [CrossRef]
122. Miceli, D.; Marchand, L.; Repérant, J.; Rio, J.-P. Projections of the dorsolateral anterior complex and adjacent thalamic nuclei upon the visual wulst in the pigeon. *Brain Res.* **1990**, *518*, 317–323. [CrossRef]
123. Deng, C.; Wang, B. Overlap of somatic and visual response areas in the wulst of pigeon. *Brain Res.* **1992**, *582*, 320–322.
124. Stingelin, W. *Vergleichende Morphologische Untersuchungen am Vorderhirn der Vögel auf Cytologischer und Cytoarchitektonischer Grundlage*; Verlag Helbing & Lichtenhahn: Basel, Switzerland, 1957; p. 123.
125. Pettigrew, J.D. Evolution of Binocular Vision. In *Visual Neuroscience*; Pettigrew, J.D., Sanderson, K.J., Levick, W.R., Eds.; Springer: New York, NY, USA, 1986; pp. 208–222.
126. Rogers, L. Behavioral, structural and neurochemical asymmetries in the avian brain: A model system for studying visual development and processing. *Neurosci. Biobehav. Rev.* **1996**, *20*, 487–503. [CrossRef]
127. Wild, J.M.; Kubke, M.F.; Peña, J.L. A pathway for predation in the brain of the barn owl (*Tyto alba*): Projections of the gracile nucleus to the "claw area" of the rostral wulst via the dorsal thalamus. *J. Comp. Neurol.* **2008**, *509*, 156–166. [CrossRef]
128. Wild, J.M. The avian somatosensory system: The pathway from wing to wulst in a passerine (*Chloris chloris*). *Brain Res.* **1997**, *759*, 122–134. [CrossRef]
129. Manger, P.R.; Elston, G.N.; Pettigrew, J.D. Multiple maps and activity-dependent representational plasticity in the anterior wulst of the adult barn owl (*Tyto alba*). *Eur. J. Neurosci.* **2002**, *16*, 743–750. [CrossRef] [PubMed]
130. Pettigrew, J.D.; Frost, B.J. A tactile fovea in the Scolopacidae? *Brain Behav. Evolut.* **1985**, *26*, 185–195. [CrossRef]
131. Martin, G.R. What is binocular vision for? A birds' eye view. *J. Vision.* **2009**, *9*, 1–19. [CrossRef]
132. Wylie, D.R.; Gutiérrez-Ibáñez, C.; Iwaniuk, A.N. Integrating brain, behavior, and phylogeny to understand the evolution of sensory systems in birds. *Front. Neurosci. Switz.* **2015**, *9*, 281. [CrossRef] [PubMed]
133. Pettigrew, J.D.; Konishi, M. Neurons selective for orientation and binocular disparity in the visual wulst of the barn owl (*Tyto alba*). *Science* **1976**, *193*, 675–678. [CrossRef]
134. Pettigrew, J.D. Binocular visual processing in the owl's telencephalon. *Proc. Roy. Soc. B Biol. Sci.* **1979**, *204*, 435–454. [CrossRef]
135. Van der Willigen, R.F.; Frost, B.J.; Wagner, H. Stereoscopic depth perception in the owl. *NeuroReport* **1998**, *9*, 1233–1237. [CrossRef]
136. Orlowski, J.; Harmening, W.; Wagner, H. Night vision in barn owls: Visual acuity and contrast sensitivity under dark adaptation. *J. Vision.* **2012**, *12*, 1–8. [CrossRef]
137. Gutiérrez-Ibáñez, C.; Iwaniuk, A.N.; Lisney, T.J.; Wylie, D.R. Comparative study of visual pathways in owls (Aves: Strigiformes). *Brain Behav. Evolut.* **2013**, *81*, 27–39. [CrossRef] [PubMed]
138. Kulemeyer, C.; Asbahr, K.; Gunz, P.; Frahnert, S.; Bairlein, F. Functional morphology and integration of corvid skulls–a 3D geometric morphometric approach. *Front. Zool.* **2009**, *6*, 1–14. [CrossRef] [PubMed]
139. Northcutt, R.G. Evolution of the telencephalon in nonmammals. *Annu. Rev. Neurosci.* **1981**, *4*, 301–350. [CrossRef] [PubMed]
140. Wild, J.M.; Arends, J.J.A.; Zeigler, H.P. Telencephalic connections of the trigeminal system in the pigeon (*Columba livia*): A trigeminal sensorimotor circuit. *J. Comp. Neurol.* **1985**, *234*, 441–464. [CrossRef]
141. Hall, M.I.; Ross, C.F. Eye shape and activity pattern in birds. *J. Zool.* **2007**, *271*, 437–444. [CrossRef]
142. Hall, M.I. The anatomical relationships between the avian eye, orbit and sclerotic ring: Implications for inferring activity patterns in extinct birds. *J. Anat.* **2008**, *212*, 781–794. [CrossRef] [PubMed]

143. Iwaniuk, A.N.; Heesy, C.P.; Hall, M.I. Morphometrics of the eyes and orbits of the nocturnal swallow-tailed gull (*Creagrus furcatus*). *Can. J. Zool.* **2010**, *88*, 855–865. [CrossRef]

144. Corfield, J.R.; Gsell, A.C.; Brunton, D.; Heesy, C.P.; Hall, M.I.; Acosta, M.L.; Iwaniuk, A.N. Anatomical specializations for nocturnality in a critically endangered parrot, the Kakapo (*Strigops habroptilus*). *PLoS ONE* **2011**, *6*, e22945. [CrossRef]

145. Corfield, J.R.; Wild, J.M.; Hauber, M.E.; Parsons, S.; Kubke, M.F. Evolution of brain size in the palaeognath lineage, with an emphasis on New Zealand ratites. *Brain Behav. Evolut.* **2008**, *71*, 87–99. [CrossRef]

146. Garamszegi, L.Z.; Møller, A.P.; Erritzøe, J. Coevolving avian eye size and brain size in relation to prey capture and nocturnality. *Proc. Roy. Soc. B Biol. Sci.* **2002**, *269*, 961–967. [CrossRef]

147. Dubbeldam, J.L.; Visser, A.M. The organization of the nucleus basalis—neostriatum complex of the mallard (*Anas platyrhynchos* L.) and its connections with the archistriatum and the paleostriatum complex. *Neuroscience* **1987**, *21*, 487–517. [CrossRef]

148. Reiner, A.; Davis, B.M.; Brecha, N.C.; Karten, H.J. The distribution of enkephalinlike immunoreactivity in the telencephalon of the adult and developing domestic chicken. *J. Comp. Neurol.* **1984**, *228*, 245–262. [CrossRef]

149. Salzen, E.A.; Parker, D.M.; Williamson, A.J. A forebrain lesion preventing imprinting in domestic chicks. *Exp. Brain Res.* **1975**, *24*, 145–157. [CrossRef] [PubMed]

150. Iwaniuk, A.N.; Wylie, D.R. Sensory systems in birds: What we have learned from studying sensory specialists. *J. Comp. Neurol.* **2020**, *528*, 2902–2918. [CrossRef] [PubMed]

151. Berkhoudt, H.; Dubbeldam, J.L.; Zeilstra, S. Studies on the somatotopy of the trigeminal system in the mallard, *Anas platyrhynchos* L. IV. Tactile representation in the nucleus basalis. *J. Comp. Neurol.* **1981**, *196*, 407–420. [CrossRef]

152. Schneider, E.R.; Mastrotto, M.; Laursen, W.J.; Schulz, V.P.; Goodman, J.B.; Funk, O.H.; Gallagher, P.G.; Gracheva, E.O.; Bagriantsev, S.N. Neuronal mechanism for acute mechanosensitivity in tactile-foraging waterfowl. *Pro. Natl. Acad. Sci. USA* **2014**, *111*, 14941–14946. [CrossRef] [PubMed]

153. Schneider, E.R.; Anderson, E.O.; Mastrotto, M.; Matson, J.D.; Schulz, V.P.; Gallagher, P.G.; LaMotte, R.H.; Gracheva, E.O.; Bagriantsev, S.N. Molecular basis of tactile specialization in the duck bill. *Pro. Natl. Acad. Sci. USA* **2017**, *114*, 13036–13041. [CrossRef]

154. Schneider, E.R.; Anderson, E.O.; Feketa, V.V.; Mastrotto, M.; Nikolaev, Y.A.; Gracheva, E.O.; Bagriantsev, S.N. A cross-species analysis reveals a general role for Piezo2 in mechanosensory specialization of trigeminal ganglia from tactile specialist birds. *Cell Rep.* **2019**, *26*, 1979–1987. [CrossRef]

155. Wild, J.M.; Arends, J.J.A.; Zeigler, H.P. A trigeminal sensorimotor circuit for pecking, grasping and feeding in the pigeon (*Columba livia*). *Brain Res.* **1984**, *300*, 146–151. [CrossRef]

156. Wild, J.M.; Farabaugh, S.M. Organization of afferent and efferent projections of the nucleus basalis prosencephali in a passerine, *Taeniopygia guttata*. *J. Comp. Neurol.* **1996**, *365*, 306–328. [CrossRef]

157. Von Eugen, K.; Tabrik, S.; Güntürkün, O.; Ströckens, F. A comparative analysis of the dopaminergic innervation of the executive caudal nidopallium in pigeon, chicken, zebra finch, and carrion crow. *J. Comp. Neurol.* **2020**, *528*, 2929–2955. [CrossRef]

158. Hellmann, B.; Güntürkün, O.; Manns, M. Tectal mosaic: Organization of the descending tectal projections in comparison to the ascending tectofugal pathway in the pigeon. *J. Comp. Neurol.* **2004**, *472*, 395–410. [CrossRef]

159. Gibson, J.J. The visual perception of objective motion and subjective movement. *Psychol. Rev.* **1954**, *61*, 304–314. [CrossRef] [PubMed]

160. Wylie, D.R.; Gutiérrez-Ibáñez, C.; Gaede, A.H.; Altshuler, D.L.; Iwaniuk, A.N. Visual-cerebellar pathways and their roles in the control of avian flight. *Front. Neurosci. Switz.* **2018**, *12*, 223. [CrossRef] [PubMed]

161. Simpson, J.I. The accessory optic system. *Annu. Rev. Neurosci.* **1984**, *7*, 13–41. [CrossRef]

162. Simpson, J.I.; Leonard, C.S.; Soodak, R.E. The accessory optic system. Analyzer of self-motion. *Ann. N. Y. Acad. Sci.* **1988**, *545*, 170–179. [CrossRef] [PubMed]

163. Giolli, R.A.; Blanks, R.H.; Lui, F. The accessory optic system: Basic organization with an update on connectivity, neurochemistry, and function. *Prog. Brain Res.* **2006**, *151*, 407–440.

164. Gaede, A.H.; Gutiérrez-Ibáñez, C.; Armstrong, M.S.; Altshuler, D.L.; Wylie, D.R. Pretectal projections to the oculomotor cerebellum in hummingbirds (*Calypte anna*), zebra finches (*Taeniopygia guttata*), and pigeons (*Columba livia*). *J. Comp. Neurol.* **2019**, *527*, 2644–2658. [CrossRef]

165. Pakan, J.M.P.; Wylie, D.R. Two optic flow pathways from the pretectal nucleus lentiformis mesencephali to the cerebellum in pigeons (*Columba livia*). *J. Comp. Neurol.* **2006**, *499*, 732–744. [CrossRef] [PubMed]

166. Iwaniuk, A.N.; Keirnan, A.R.; Janetzki, H.; Mardon, K.; Murphy, S.; Leseberg, N.P.; Weisbecker, V. The endocast of the Night Parrot (*Pezoporus occidentalis*) reveals insights into its sensory ecology and the evolution of nocturnality in birds. *Sci. Rep.* **2020**, *10*, 9258. [CrossRef] [PubMed]

167. Kay, R.F.; Kirk, E.C. Osteological evidence for the evolution of activity pattern and visual acuity in primates. *Am. J. Phys. Anthropol.* **2000**, *113*, 235–262. [CrossRef]

168. Bennett, P.M.; Harvey, P.H. Relative brain size and ecology in birds. *J. Zool.* **1985**, *207*, 151–169. [CrossRef]

169. Lau, K.L.; Glover, R.G.; Linkenhoker, B.; Wylie, D.R. Topographical organization of inferior olive cells projecting to translation and rotation zones in the vestibulocerebellum of pigeons. *Neuroscience* **1998**, *85*, 605–614. [CrossRef]

170. Wylie, D.R. Projections from the nucleus of the basal optic root and nucleus lentiformis mesencephali to the inferior olive in pigeons (*Columba livia*). *J. Comp. Neurol.* **2001**, *429*, 502–513. [CrossRef]

171. Wylie, D.R. Processing of visual signals related to self-motion in the cerebellum of pigeons. *Front. Behav. Neurosci.* **2013**, *7*, 1–15. [CrossRef] [PubMed]
172. Iwaniuk, A.N.; Hurd, P.L.; Wylie, D.R. Comparative morphology of the avian cerebellum: II. Size of folia. *Brain Behav. Evolut.* **2007**, *69*, 196–219. [CrossRef]
173. Walsh, S.A.; Knoll, M.A. Directions in palaeoneurology. *Spec. Pap. Palaeontol.* **2011**, *86*, 263–279. [CrossRef]
174. Martin, H.A. Cenozoic climatic change and the development of the arid vegetation in Australia. *J. Arid Environ.* **2006**, *66*, 533–563. [CrossRef]
175. Macphail, M.K. Late Neogene climates in Australia: Fossil pollen-and spore-based estimates in retrospect and prospect. *Aust. J. Bot.* **1997**, *45*, 425–464. [CrossRef]
176. Stirling, E.C. The recent discovery of fossil remains at Lake Callabonna, South Australia II. *Nature* **1894**, *50*, 206–211.
177. Hamm, G.; Mitchell, P.; Arnold, L.J.; Prideaux, G.J.; Questiaux, D.; Spooner, N.A.; Levchenko, V.A.; Foley, E.C.; Worthy, T.H.; Stephenson, B. Cultural innovation and megafauna interaction in the early settlement of arid Australia. *Nature* **2016**, *539*, 280–283. [CrossRef]
178. Szabo, M.J. Stout-legged moa. *New Zeal. Birds Online*. 2013. Available online: http://www.nzbirdsonline.org.nz (accessed on 17 April 2020).
179. Alexander, R.M. Allometry of the leg bones of moas (Dinornithes) and other birds. *J. Zool. Lond.* **1983**, *200*, 215–231. [CrossRef]
180. Campbell, K.E., Jr.; Marcus, L. The relationship of hindlimb bone dimensions to body weight in birds. *Nat. Hist. Mus. Los Angeles County Sci. Ser.* **1992**, *36*, 395–412.

Article

Histovariability and Palaeobiological Implications of the Bone Histology of the Dromornithid, *Genyornis newtoni*

Anusuya Chinsamy [1,*] **and Trevor H. Worthy** [2]

[1] Department of Biological Sciences, University of Cape Town, Private Bag X3, Rhodes Gift, Cape Town 7700, South Africa

[2] College of Science and Engineering, Flinders University, GPO 2100, Adelaide 5001, Australia; trevor.worthy@flinders.edu.au

* Correspondence: anusuya.chinsamy-turan@uct.ac.za; Tel.: +27-21-650-3604

Abstract: The bone microstructure of extinct animals provides a host of information about their biology. Although the giant flightless dromornithid, *Genyornis newtoni*, is reasonably well known from the Pleistocene of Australia (until its extinction about 50–40 Ka), aside from various aspects of its skeletal anatomy and taxonomy, not much is known about its biology. The current study investigated the histology of fifteen long bones of *Genyornis* (tibiotarsi, tarsometatarsi and femora) to deduce information about its growth dynamics and life history. Thin sections of the bones were prepared using standard methods, and the histology of the bones was studied under normal and polarised light microscopy. Our histological analyses showed that *Genyornis* took more than a single year to reach sexual maturity, and that it continued to deposit bone within the OCL for several years thereafter until skeletal maturity was attained. Thus, sexual maturity and skeletal maturity were asynchronous, with the former preceding the latter. Our results further indicated that *Genyornis* responded to prevailing environmental conditions, which suggests that it retained a plesiomorphic, flexible growth strategy. Additionally, our analyses of the three long bones showed that the tibiotarsus preserved the best record of growth for *Genyornis*.

Keywords: Australia; Pleistocene fossil bird; dromornithid; *Genyornis*; bone histology; osteohistology

Citation: Chinsamy, A.; Worthy, T.H. Histovariability and Palaeobiological Implications of the Bone Histology of the Dromornithid, *Genyornis newtoni*. *Diversity* **2021**, *13*, 219. https://doi.org/10.3390/d13050219

Academic Editor: Eric Buffetaut

Received: 3 May 2021
Accepted: 17 May 2021
Published: 20 May 2021

Publisher's Note: MDPI stays neutral with regard to jurisdictional claims in published maps and institutional affiliations.

1. Introduction

It is well recognised that, during life, the bones of vertebrates are living tissues that record various aspects of their life history and biology, e.g., [1–3]. A host of studies have shown that even after millions of years of fossilisation, the bone microstructure is often well preserved, e.g., [1,2,4–6]. Thus, by studying the bone histology of extinct vertebrates, various inferences into their biology can be made.

Genyornis newtoni was a giant flightless galloansere bird that belonged to the Dromornithidae [7,8]. The dromornithids first appeared in the fossil record in the Eocene and reached the height of their diversity in the middle Miocene. By the late Oligocene, the dromornithids had already attained a large size and were flightless with reduced wings, a morphology they retained for the next 25 ma [7]. However, by the Pleistocene, *Genyornis newtoni* [9,10] was the only surviving member of the family.

Genyornis newtoni was widespread, though usually rare, in south eastern Australia in the mid-to-late Pleistocene and remains the only species of dromornithid for which individual skeletal assemblages are known. *Genyornis* went extinct in the late Pleistocene along with many other megafaunal animals about 50–40 Ka [7,11]. Aside from some general reconstructions of *Genyornis* as a medium-sized dromornithid (180–250 kg) that stood about 2–2.5 m tall, little is known about its biology. However, the well-defined bimodality in skeletal measurements is attributed to marked sexual dimorphism [12], where the males are assumed to be the larger sex, as was demonstrated for its larger relative *Dromornis stirtoni* for which medullary tissue was found in examples of the smaller morph [13]. Here, we investigated its bone histology to deduce various aspects of its biology, particularly to infer how their growth dynamics allowed them to reach their large body size.

Modern birds generally, grow to adult body size within a single year [14,15]. Even large modern birds such as *Sagittarius serpentarius*, the secretary bird [16,17], and the largest modern bird, *Struthio camelus*, the ostrich [16,17], reach adult body size within a single year. Thus, the bones of modern birds generally have uninterrupted growth until an adult body size is reached, which is usually coincident with the attainment of sexual maturity [4]. However, once their growth rate slows down (usually upon or close to sexual maturity), their rate of bone deposition (osteogenesis) slows down, and a different type of bone tissue develops in their compacta. This outer band of tissue, called the outer circumferential layer (OCL) [18], often shows lines of arrested growth (LAGs) therein, indicating that these birds experience periodic arrests in growth as they slowly accrete bone for a few more years until skeletal maturity is reached. Although most modern birds grow like this, there are exceptions to such rapid growth rates among birds. This is particularly the case among insular birds such as the Apterygiformes (kiwi) [19], the Dinornithiformes (moa) [20] and the Aepyornithidae (elephant birds) from Madagascar [5]. Among the aepyornithids, *Vorombe titan* is largest, and like the large moa (Emeidae and Dinornithidae) [20], it takes several years to reach adult body size. It has been suggested that these birds grow much slower than other birds, because they are island birds without the pressure of mammalian predators [15]. Long-lived birds possibly also experience slower growth rates, but this needs to be verified—a single bone of a parrot (*Amazona amazonica*) had a growth mark therein [21], but there are no details of the approximate age of the individual.

The aim of our study was to assess the histology of *Genyornis* to decipher information pertaining to their growth dynamics and life history. As different bones would be studied, we also assessed the histovariability of the long bones studied, and we ascertained which elements were more reliable for growth assessment.

2. Materials

The *Genyornis* bones we studied were recovered from late Pleistocene lacustrine deposits at Lake Callabonna (48–45,000 years ago) and from mid-to-late Pleistocene fluvial deposits along Cooper Creek and Billeroo Creek [22] Figure 1. These deposits are assumed to have sampled palaeoenvironments of arid grassland/shrubland with some trees along watercourses, although current palaeoenvironmental studies are underway and should shed more light on the palaeoecology of these sites.

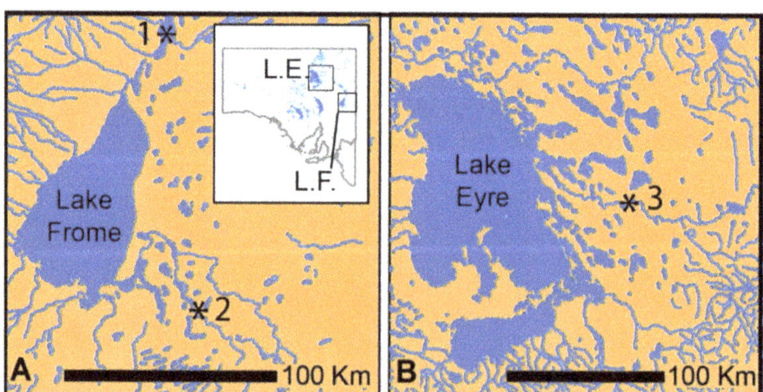

Figure 1. Sites in South Australia (inset) where the sampled bones of *Genyornis* were collected. (**A**) Lake Frome area (L.F.) and (**B**) Lake Eyre area (L.E.). 1, Lake Callabonna; 2, Billeroo Creek sites; 3, Cooper Creek sites (Malkuni Waterhole and Site 73).

Fifteen specimens of *Genyornis* were studied: eight tibiotarsi, three tarsometatarsi and four femora. As far as possible, standard measurements were taken of the limb bones studied, and the least shaft circumference of the tibiotarsus was measured in all specimens where it was preserved. The latter ranged from 134 to 146 mm (Table 1). For specimen SAM P.53833, we were able to sample different bones of the skeleton: a femur and tibiotarsus. All the other successfully sampled bones were from different individuals. In some specimens, parts of the shaft were opportunistically sampled (through natural breakages, etc.), but the majority of the specimens were core-sampled in the midshaft region to obtain the best possible track record of growth (see details in Table 1) [1].

Table 1. Details of the *Genyornis* specimens sampled. All sites are in South Australia. Unsampled elements of individuals are listed for their ability to reveal the size of the individual. Shading is used to show associated bones of individuals.

Catalogue No.	Locality	Field ID	Element	Histology Sample	Comment	Sex
SAM P.25017	Cooper Creek		Distal left tibiotarsus	y	distal width, 85 mm; minimum shaft circumference, 140 mm; sampled section of shaft 1/3 length from distal end.	Female?
SAM P.53826	Callabonna	Geny 1A	Distal right tibiotarsus	y	distal width, 88 mm, shaft inflated by salt degradation; sampled caudal facies.	Female?
SAM P.53826	Callabonna	Geny 1A	Right tarsometatarsus		max proximal width, 105 mm; min shaft width, 45 mm; distal width, 99 mm	
SAM P.53831	Callabonna	Geny 9A	Right tarsometatarsus	y	Length trochlea III-cotyla lateralis, 358 mm; min shaft diameter, 50 mm; distal width, 110 mm; sampled medioplantar facies.	Male
SAM P.53832	Callabonna	Geny 9B	Left tarsometatarsus	y	min shaft diameter, 44 mm; distal width, 95 mm; sampled medioplantar facies.	Female
SAM P.53833	Callabonna	Geny 10	Right femur	y	midshaft diameter, 76 mm; max distal width, 154 mm, sampled mid-caudal facies	
SAM P.53833	Callabonna	Geny 10	Left tibiotarsus	y	distal width, 86 mm; total length, 610 mm; minimum shaft circumference, 144 mm; sampled mid-medial facies	Female?
SAM P.53833	Callabonna	Geny 10	Right tarsometatarsus		length, 347 mm; TL2, 355 mm, distal width medial-lateral, 102; min shaft width, 39 mm; max distal width, 92 mm	
SAM P.54333	Cooper Creek	Geny C	Distal left tibiotarsus	y	distal width, 91 mm; sampled anterolateral facies	Female?
SAM P.54334	Cooper Creek	Cooper Creek 73-B	Distal left tibiotarsus	y	distal width, 92 mm; minimum shaft circumference, ±140 mm; sampled section midshaft	Female?
FU2750	Callabonna	CB2018-98	part left tarsometatarsus	y	min shaft diameter, 38 mm; sampled medial facies	Female?
FU2755	Callabonna	CB2018-75 Ind 1	Right femur	y	shaft width, 87 mm; small indiv; sampled mid-caudal facies	
FU2755	Callabonna	CB2018-75 Ind 1	Right tibiotarsus	y	distal width, 88 mm; minimum shaft circumference, 144 mm; small indiv; 2 samples distomedial facies.	Female?
FU2755	Callabonna	CB2018-75 Ind 1	Left tarsometatarsus		distal width, 90 mm; min shaft width, 40 mm; max proximal width, 102 mm; small indiv.	
FU2756	Callabonna	CB2018-75 Ind 2	Crushed left and right femora			
FU2756	Callabonna	CB2018-75 Ind 2	Right tibiotarsus	y	distal width, 96 mm; minimum shaft circumference 146 mm; big indiv; sampled midshaft medial facies	boundary-male/female?
FU2756	Callabonna	CB2018-75 Ind 2	Left tarsometatarsus		min shaft diameter, 46 mm; max proximal width, 109 mm; max distal width, ±105 mm	
FU2760	Callabonna	CB2010-75 Ind 3	Right femur	y	midshaft width, ~66 mm; sampled mid-caudal facies	Male
FU2758	Billeroo Creek	NA	Left femur	y	midshaft width, 70 mm; surface texture is porous; crista trochanteris is not fully formed; sampled mid-caudal facies.	young indiv?
FU2759	Billeroo Creek	NA	Distal left tibiotarsus	y	Sampled medial facies	Small-female?

The sampled bones were embedded in resin and were thin-sectioned according to standard petrographic methods [23]. They were then sectioned along the midline, resulting in two blocks labelled as A and B (which permitted the investigation of the histology closest to the neutral region, i.e., the area least affected by remodelling changes [1]). Four thin sections were prepared from these blocks (AI, AII; BI, BII). The sections were studied under petrographic microscopes (Nikon Eclipse E200 with a Nikon DS-Fi1 camera or a Zeiss Ax10 Lab.A1 with an Axiocam camera). All thin section preparations and photomicroscopies were performed at the University of Cape Town, South Africa.

Given that the dromornithid *Dromornis stirtoni* has a marked bimodality with females shown to be the smaller morph through the presence of medullary bone [13], its size has the potential to establish the sex of *Genyornis* bones. *Genyornis newtoni* shows a nonoverlapping bimodal size distribution [12], where tibiotarsi with a least shaft circumference of 137–150 mm are presumed females and those with values greater than 150 mm are males (assuming that dimorphism is the same in species of *Dromornis* and *Genyornis*, which is reasonable as males are uniformly larger birds in all extant Galloanseres). However, we used the totality of measurements from associated bones of an individual to infer sex, as for several individuals, the minimum shaft circumference was not measurable, but values for the tibiotarsal distal width or width for the femora and tarsometatarsi of the same individual were available. These revealed two individuals that were much larger than those biggest based on tibiotarsal shaft circumference and, thus, were inferred as males (Table 1).

Here, we followed the traditional histological terminology sensu [1,18,24]. Although we used the orientation of the canals in the bones as a proxy for the orientation of the vascular canals, we are aware that this does not accurately reflect the orientation of the blood vessels therein [15].

3. Results

The histology of several examples of hind limb elements are described and summarized as follows: eight tibiotarsi, four femora and three tarsometatarsi.

3.1. Tibiotarsi

3.1.1. FU2759, Billeroo Creek, F

Thin sections of this specimen showed that the bone tissue had experienced some taphonomic damage by infiltration of the surrounding minerals and sediment into the bone, concordant with its fluvial deposition. Thus, most of the specimen showed more extensive damage to the peripheral parts of the compacta, whereas the internal bone tissues were better preserved. Figure 2 shows the alteration in the bone microstructure caused by this infiltration damage. Nevertheless, it is evident that the bone has vascular canals right up to the margin, and there appears to be no distinctive slowing down in the rate of bone deposition (Figure 2A). Deeper in the compacta, the tissue continues to be primary periosteal bone with abundant vascular canals in a woven bone matrix. The vascular canals tend to be a mixture of longitudinal and short circumferentially oriented canals in a laminar arrangement (Figure 2B). In the innermost region of the compacta, closest to the medullary cavity, the bone tissue has a different appearance; here, the bone is still fibrolamellar tissue, but the vascular canals have a much more disarrayed arrangement with a predominantly longitudinal-to-radial arrangement in the woven bone matrix (Figure 2C. Throughout the compacta, secondary osteons are rare, sparse erosion cavities occur and no growth marks (i.e., LAGs or annuli) are apparent in the compacta (Figure 2).

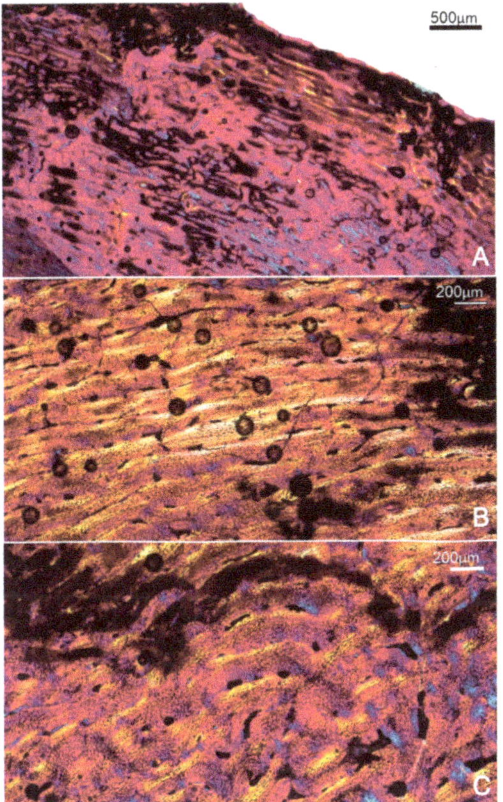

Figure 2. Specimen FU2759; tibiotarsus. (**A**) Outermost region of the compacta showing some diagenetic features but no change in the rate of bone formation, and vascular canals right up to the edge of the bone wall. (**B**) Mid-cortical region showing the circumferential organisation of the vascular canals. (**C**) Deep cortex of the compacta showing the more haphazard arrangement of the vascular canals within a woven bone matrix.

3.1.2. Specimen SAM P.54333, Cooper Creek (Geny C), ?F

The histology of this specimen is reasonably well preserved with the entire compacta from the periphery to the endosteal region preserved (Figure 3). Localised differences in the orientation of the vascular canals, and the nature of the bone tissue are evident; the outermost bone tissue, which is the most recently formed bone, comprises a narrow (~250 μm) band of avascular lamellar tissue, the OCL (Figure 3A,B). Preceding the OCL, there is a band of reticular bone tissue (Figure 3A,B), whereas deeper into the compacta, the tissue changes to a mix of plexiform to circumferential laminar bone tissue (Figure 3A,B). In the perimedullary region, several enlarged cavities are evident, and there are several secondary osteons that occur right up to the margin of the medullary cavity (Figure 3A), and in some areas, cancellous tissue extends into the medullary cavity. A narrow layer of lamellar bone, the inner circumferential layer (ICL), lines the medullary cavity in localised areas (Figure 3C). Except for this region, the rest of the cortex comprises mostly primary bone tissue, although there are a few scattered secondary osteons. In the thickest part of the bone wall, the innermost bone tissue comprises FBL with more longitudinal and circumferentially arranged primary osteons (Figure 3C). It is likely that this tissue is bone formed during the early stages of ontogeny.

3.1.3. Specimen SAM P.54334, Cooper Creek site 73B, F

The bone wall is incompletely preserved but there is enough of the compacta visible to make a histological description. The outermost part of the bone wall has a layer of lamellar bone, the OCL, that varies in thickness around the bone wall. Several lines of arrested growth (LAGs or rest lines) are visible in the OCL (Figure 4A). Below the OCL is a layer of richly vascularised bone tissue that varies from a reticular to a plexiform type of bone tissue with several growth marks in the form of annuli (Figure 4A,B). Note that the innermost and outermost annuli appear to be quite wide (Figure 4A). In the perimedullary region, there occurs a region of well-vascularised bone tissue that appears to be bone formed during the early stages of ontogeny (Figure 4A,C). Closer to the medullary cavity, there is a large amount of secondary reconstruction (Figure 4A) where numerous secondary osteons are visible, but these do not reach a high density. In the endosteal region, extensive remodelling of the compacta is evident, and there are large excavations into the compacta (Figure 4A). Small patches of ICL are visible in localised areas (Figure 4A).

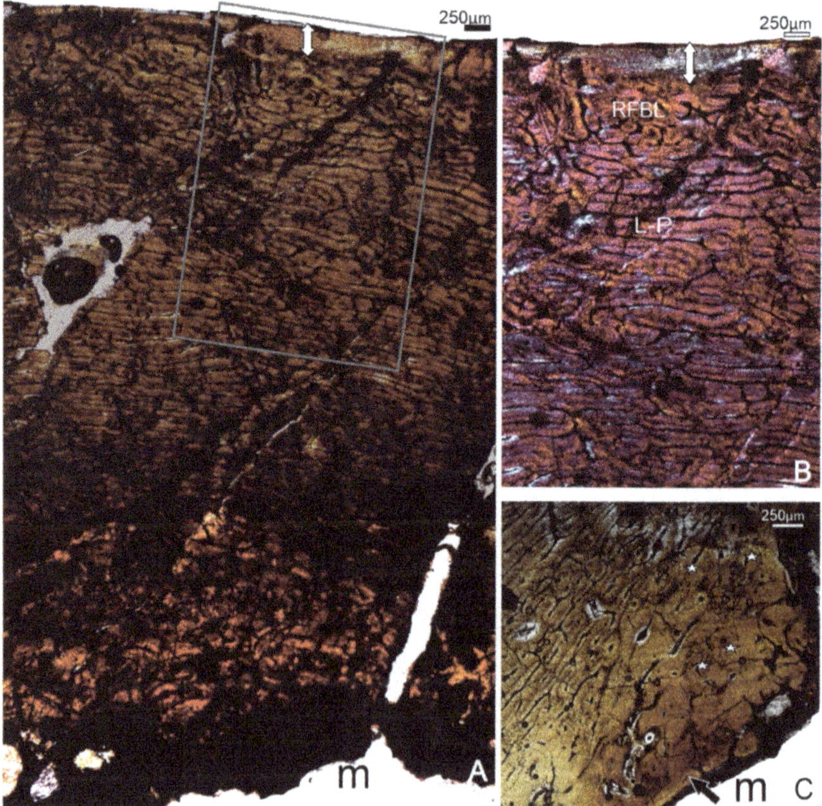

Figure 3. Specimen SAM P.54333; tibiotarsus. (**A**) Overview of the histology of the bone wall. (**B**) Higher magnification of the framed region in (**A**), showing the OCL (double-headed arrow) below which is a band of reticular organised FBL(RFLB), and the more laminar-plexiform (L-P) organised bone tissue deeper in the compacta. (**C**) A view of the perimedullary region showing a narrow ICL (arrow), and some remnants of the early formed reticular bone tissue, secondarily enlarged vascular canals and a few secondary osteons (stars). m, medullary cavity.

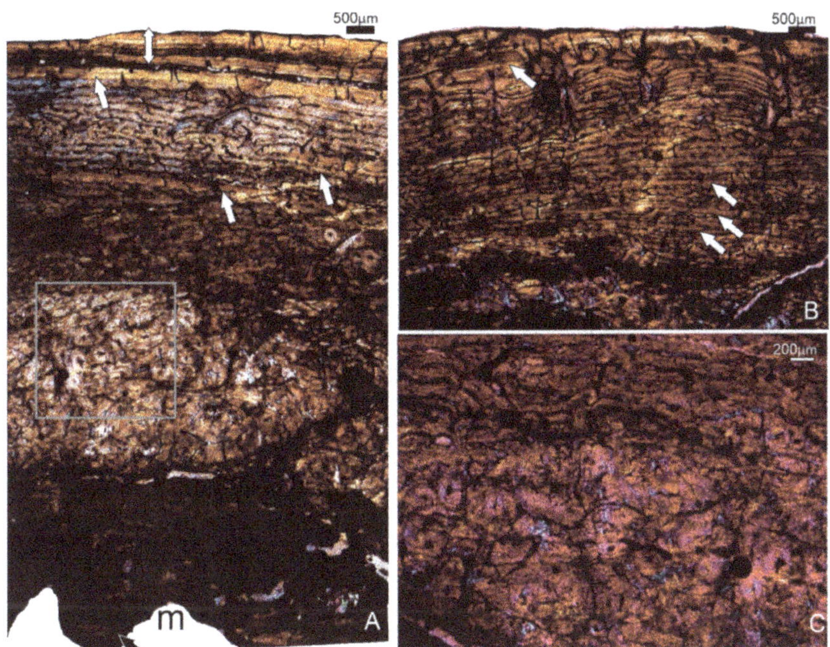

Figure 4. Specimen SAM P.54334; tibiotarsus. (**A**) Overview of the compacta showing an OCL in the peripheral region (double-headed arrow), several growth marks (annuli) (arrows) and a resorptive medullary (m) margin. The black arrow points to remnants of an ICL. (**B**) A different part of the compacta showing four growth marks (arrows) in the compacta. (**C**) Higher magnification of the framed region in (**A**) showing the reticular FBL bone tissue formed during early ontogenetic stages. m, medullary cavity.

3.1.4. Specimen SAM P.53826, Callabonna Geny 1A, F

Unfortunately, we were unable to retrieve a full core from this specimen, and only the outer part of the bone wall was preserved. Microscopical examination revealed that the bone is badly fractured, concordant with salt damage, but a distinct OCL is visible with at least three (perhaps four) growth marks in the form of narrow annuli present. Below the OCL, several circumferentially organised vascular canals are visible.

3.1.5. Specimen FU2756 Callabonna CB2018-75, Indiv 2, ?M/F

The outermost compacta consist of a wide layer of poorly vascularised lamellar bone tissue that forms the OCL, which is interrupted by at least 10 growth marks (LAGs) (Figure 5A). Below this outer band of tissue, the cortex consists of a more richly vascularised, more laminarly organised FLB tissue (Figure 5B). A narrow annulus with a LAG interrupts the deposition of this tissue (Figure 5B). Deeper in the compacta, the bone formed during the early stages of ontogeny is visible and appears to be FLB with mainly longitudinally and reticular organised vascular canals (Figure 5C). Some of these vascular canals have been enlarged by secondary reconstruction, and in some, there are secondary deposits of centripetally formed lamellar bone which form secondary osteons (Figure 5C).

3.1.6. Specimen FU2755, Callabonna CB2018-75, Indiv 1, F

The bone tissue of this tibiotarsus was not well preserved. The outer cortex was not sampled, so the most recently formed bone tissue cannot be described (and we cannot assess whether or not an OCL is present). The part of the compacta that was preserved shows a richly vascularised primary compacta with predominantly longitudinal and reticular-oriented vascular canals.

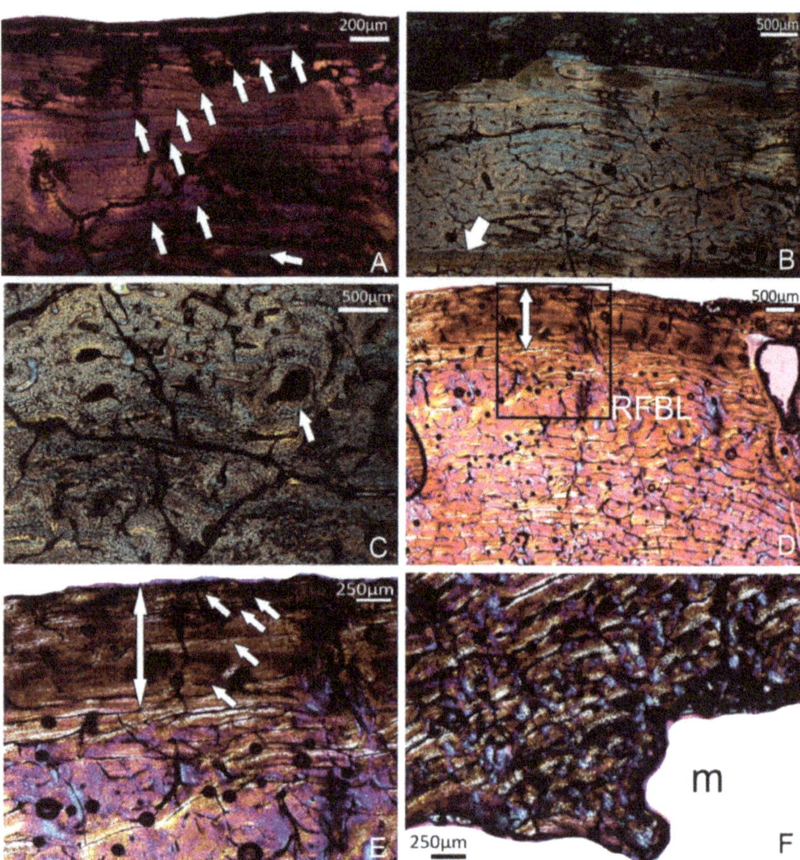

Figure 5. (**A–C**) Specimen FU2756; tibiotarsus. (**A**) The OCL with at least 10 LAGs (arrows). (**B**) A deeper part of the compacta showing the laminar organised FBL bone tissue, and a single annulus with a LAG (arrow). (**C**) FBL bone tissue with more longitudinal and reticular organised vascular canals. Arrows point to the lamellar bone tissue in the process of being deposited around an enlarged cavity. (**D–F**) Specimen SAM P.25017, tibiotarsus. (**D**) A general view of the compacta. The double-headed arrow indicates the OCL. RFBL indicates the band of reticular organised FBL bone tissue that precedes the OCL. (**E**) Higher magnification of the framed region of (D). The OCL with 5 LAGs (arrows). (**F**) The innermost part of the compacta being actively resorbed. m, medullary cavity.

3.1.7. Specimen SAM P.25017, Cooper Creek, Malkuni WH, F

The well-preserved compacta of this tibiotarsus shows that it is heavily vascularised with a distinctly wide OCL comprised of lamellar bone tissue with at least five LAGs (Figure 5D,E). A few blood vessels occur in the OCL, but overall, it is much more sparsely vascularised than the underlying bone tissue. Preceding the OCL is a region comprising reticular primary bone tissue, whereas deeper parts of the compacta have a more plexiform arrangement. Overall, the bone wall appears to be primary in nature, but there are some scattered secondary osteons visible. In parts of the compacta, at least three narrow lamellar deposits interrupt the rapid deposition of bone. The perimedullary region of the cortex is uneven due to extensive resorption which cuts into the original early formed primary compacta of the bone wall (Figure 5E).

3.1.8. Specimen SAM P.53833, Callabonna, Geny 10, F?

A striking feature of the compacta is the presence of two distinct bands of lamellar bone towards the outer cortex (Figure 6A,B). These wide bands of more slowly deposited tissue are separated by an almost equally thick region of fibrolamellar bone tissue. Note that there are several incursions of blood vessels through both the lamellar layers, but overall, these layers are not as well vascularised as the tissue below. The outer band of lamellar tissue has about three LAGs, and this appears to be the OCL (Figure 6A,B). Below the more inner wide band of lamellar tissue, a narrow annulus is visible, and perhaps another, but the latter cannot be followed around the compacta (Figure 6A). Deeper in the cortex (the inner 40% of bone wall) is extensively reconstructed and, in places, reaches dense Haversian bone levels where even interstitial bone is secondary (Figure 6A). Many erosion cavities are visible, and there are many examples of these connecting with one another to form even larger cavities (Figure 6C). No ICL is present.

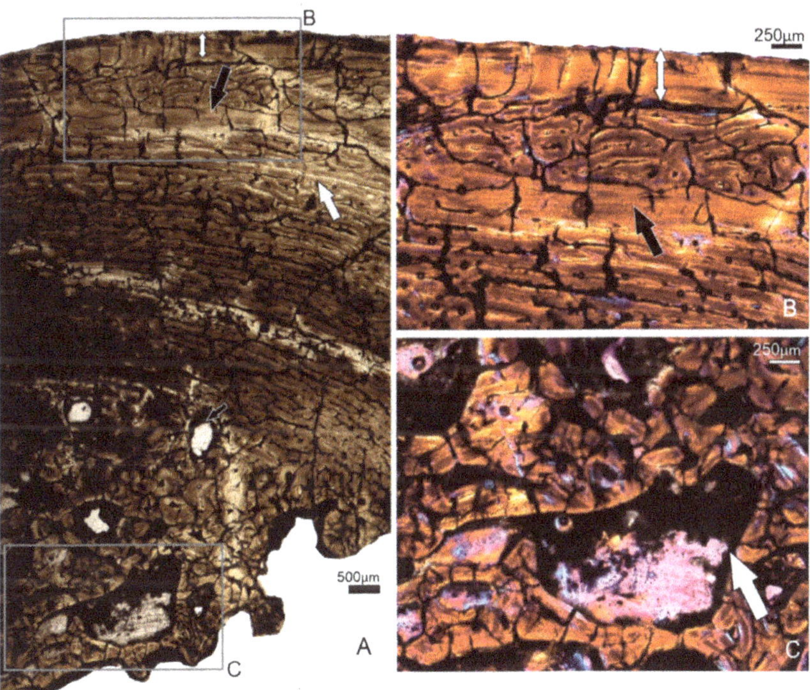

Figure 6. Specimen SAM P.53833, tibiotarsus. (**A**) Overview of the compacta. Subperiosteally, there is a wide OCL present (double-headed arrow), and further in the compacta, another wide band of lamellar tissue occurs (black arrow), below which is a narrow annulus (white arrow). Deep parts of the compacta show secondary remodelling, and several large erosion cavities are visible (small black arrow). (**B**) Higher magnification of the framed region of (A), showing the OCL and the wide inner band of lamellar tissue. (**C**) Higher magnification of the framed region in (A), showing the extensively secondary remodelled perimedullary region.

3.1.9. Summary of the Tibiotarsus Histology

The eight tibiotarsi studied here provide information about the overall growth of this skeletal element. There are obvious changes that are related to ontogenetic age, such as the development of the OCL, which clearly indicates that specimen FU2759 is the youngest of all the tibiotarsi, whereas specimen SAM P.54333 appears to be a slightly older individual that has already begun slowing down its overall rate of growth and only has a narrow OCL present. Besides the onset of OCL formation, the latter specimen also shows much more secondary remodelling as compared to specimen FU2759, but less than the other tibiotarsi studied. This suggests that secondary remodelling increases with age. All the other specimens have a relatively wide OCL with varying numbers of LAGs present—specimen FU2756 has the widest OCL with at least 10 LAGs present (although it is uncertain whether any of these are double or triple LAGs). Specimen SAM P.54334 clearly shows at least four annuli that precede the deposition of the OCL, and interestingly, the deepest annulus and the last one appear quite wide.

3.2. Femora

3.2.1. Specimen FU2758 Left Femur, Billeroo Creek

Overall, the bone wall is not very well preserved, but histological details are discernible. The compacta is richly vascularised, and there is a thin band of lamellar bone tissue visible along the outermost peripheral part of the bone wall (Figure 7A). Below this, there are mainly longitudinally arranged primary osteons. In the mid-cortex, a few secondarily enlarged erosion cavities are evident, and a few completely formed secondary osteons can be seen (Figure 7B). A narrow ICL is present in places, suggesting that medullary expansion has been completed. No growth marks are visible anywhere in the compacta.

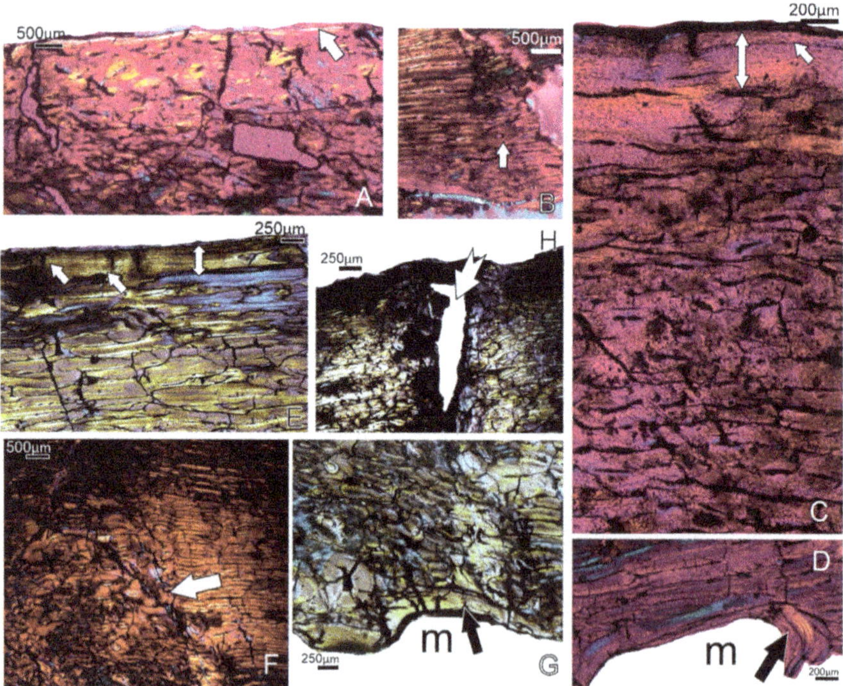

Figure 7. (**A,B**) Specimen FU2758; femur. (**A**) The white arrow points to the narrow deposit of lamellar bone tissue subperiosteally. (**B**) Some secondary osteons (arrow) are visible in the compacta. (**C,D**) Specimen FU2760; femur. (**C**) The well-vascularised compacta, and an OCL (double-headed arrow) with one LAG (arrow). (**D**) The deposition of an ICL (black arrow). (**E–H**) Specimen SAM P.53833; femur. (**E**) The OCL (double-headed arrow) and the laminar bone tissue that precedes it. (**F**) The tract of secondary osteons (arrow) that are coincident with the linea. (**G**) The uneven perimedullary margin, and the narrow ICL (arrow) in places. (**H**) The white arrow shows the entry of the nutrient foramen into the femur. Notice the changes in the orientation of the bone tissue along the margins of the canal. m, medullary cavity.

3.2.2. Specimen FU2760, Callabonna CB2018-75, Indiv 3, M

The compacta comprise predominantly short circumferentially organised vascular canals in a laminar arrangement (Figure 7C). An OCL is present, and a distinct LAG is visible therein (Figure 7C). In the perimedullary region, a well-developed ICL is visible, which continues along a bony strut that projects into the medullary cavity (Figure 7D).

3.2.3. Specimen FU2755, Callabonna CB2018-75, Indiv 1, F

In this specimen, the core does not penetrate the bone wall completely and, as a result, only the outer part of the cortex is preserved. In this region, an OCL with at least ~6–7 LAGs is preserved. Below this band of tissue, the cortex consists of laminar FBL bone tissue with circumferentially organised vascular canals.

3.2.4. Specimen SAM P.53833, Callabonna Geny10, ?F

This specimen preserves a fairly thick layer of cortical bone tissue. A distinct OCL is evident along the peripheral margin of the bone wall, and at least two LAGs are evident therein (Figure 7E). Below the OCL, the bone is mainly primary tissue consisting of laminar-plexiform organised FBL bone tissue (Figure 7E,F). In the outer part of the compacta, the tissue appears to be more laminarly textured with a predominance of circumferentially oriented canals (Figure 7E). Several scattered secondary osteons are present throughout the compacta, but in localised parts of the cortex, a tract of secondary osteons extends from the peripheral margin to the endosteal region (Figure 7F); this is likely related to the linea intermuscularis caudalis. In places, an ICL is present, but overall, the endosteal margin is highly resorptive (Figure 7G). The entry of the nutrient foramen into the bone cortex is evident in Section 10FB, and there are distinctive changes in the orientation of the bone tissue around the foramen (Figure 7H).

3.2.5. Summary of the Femoral Histology

The four femora studied showed features related to ontogenetic status. Specimen FU2758 appears to be from a young individual, which has just begun to deposit lamellar bone tissue, but an OCL is not yet present in the compacta. In this bone, there are sparse secondary osteons visible. Specimen FU2756 appears to be a slightly older individual—a well-developed OCL is present, and a LAG occurs therein. Specimen SAM P.53833 has the most mature compacta, with an OCL with multiple LAGs, and compacta with evidence of much more secondary reconstruction. It should be noted, however, that this section intersects the linea intermuscularis caudalis and the nutrient foramen, and therefore, it is expected to show localised changes as a consequence. Specimen FU2755 appears to be the most mature of the four femora—it has five or six LAGs in the OCL, but we cannot decipher any more of the nature of the bone tissue, because of core failure.

3.3. Tarsometatarsi

3.3.1. Specimen FU2750, Callabonna CB2018-98

Overall, the bone tissue appears richly vascularised, although it is apparent that the deeper cortex is much more vascularised than the outer compacta (Figure 8). Towards the outer part of the bone wall, there are two distinct growth marks in the compacta (Figure 8). Following the outer LAG, there is a distinct OCL, and near the margin of the bone, there appears to be a LAG present (Figure 8).

3.3.2. Specimen SAM P.53832, Callabonna, Geny 9B, M

This section of the TMT shows a distinct OCL in the outermost part of the cortex (Figure 9A). Except for the OCL, the rest of the compacta appears to be intensely secondarily remodelled (Figure 9A–C). Although there is a lot of secondary reconstruction, dense Haversian bone proportions are not reached in the mid-cortical regions, as there is still primary bone between neighbouring secondary osteons. However, towards the medullary cavity, the secondary reconstruction is more extensively developed, and it appears to reach dense Haversian levels, but there are also several large unfilled erosion cavities visible in this area (Figure 9C). In places, a narrow layer of lamellar bone (ICL) is visible (Figure 9C).

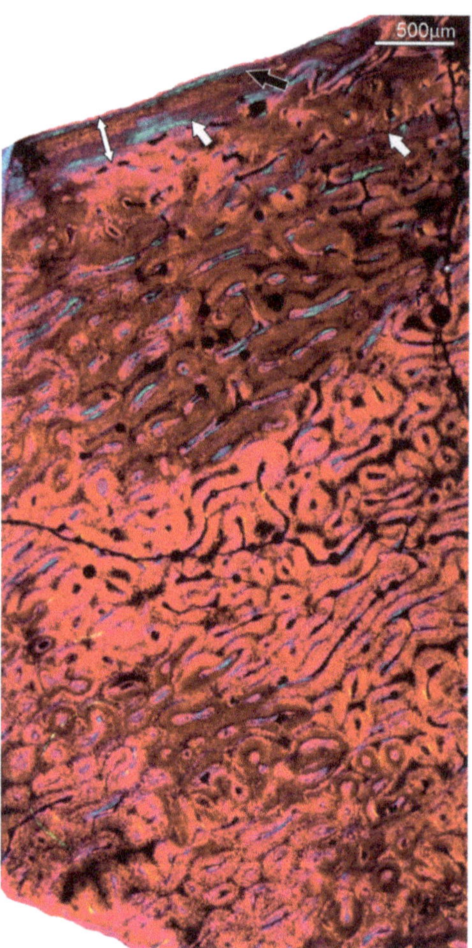

Figure 8. Specimen FU2750; Tarsometatarsus. Overview of the well-vascularised, primary compacta of the bone. In the peripheral region, a narrow OCL (double-headed arrow) and 2 narrow annuli (arrows) that precede it are visible. Subperiosteally, a LAG is present (black arrow).

3.3.3. Specimen SAM P.53831, Callabonna, Geny 9A, M

Overall, the bone tissue is as described for Specimen SAM P.53832, i.e., the compacta is extensively remodelled right up to the OCL, and the perimedullary region consists of dense Haversian bone.

3.3.4. Summary of Tarsometatarsi Histology

Similar to the femora and the tibiotarsi, the tarsometatarsi studied here show ontogenetic changes in the nature of the bone tissue. In this sample, the tarsometatarsus from the youngest individual (FU2750) has an OCL, but its compacta are still predominantly primary in nature, whereas in the tarsometatarsi from more mature individuals, except for the OCL, the compacta are intensively reconstructed and reaches dense Haversian characteristics in the deep cortex.

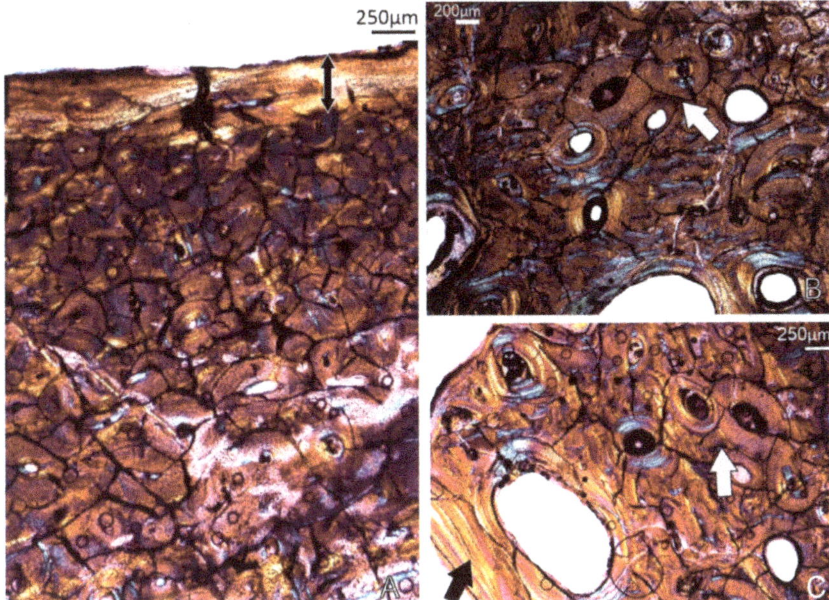

Figure 9. (**A–C**) Specimen SAM P.53832 TMT. (**A**) Except for the OCL (double-headed arrow), the compacta have been extensively reconstructed. (**B**) The dense development of secondary osteons (white arrow). (**C**) The perimedullary margin lined by an ICL (black arrow).

4. Discussion

4.1. Growth Pattern

In the *Genyornis* bones sampled, it is evident that during the earliest stages of growth, FLB was deposited. This tissue is typically formed in young fast-growing birds, e.g., *Struthio camelus* [16], *Sagittarius serpentarius* [16], *Aptenodytes patagonicus* [25] and *Calonectris leucomelas* [26]. This early bone has a number of longitudinally and reticular arranged primary osteons, and a large number of globular osteocytes in the woven bone matrix (see Figure 2C) [1]. In some of the bones, none of this early bone remains, whereas in a few bones where remodelling changes have not been extensive, the bone formed during early stages of ontogeny is preserved (e.g., tibiotarsus SAM P.54333), and we therefore have a continuous record of all the bone tissues formed during development. Often, this early bone is overlain by a laminar–plexiform bone, which tends to dominate the cortices of the tibiotarsus and the femora. In the tarsometatarsus, the predominant bone tissue is Haversian bone. Localised differences in the bone tissue were observed in response to the anatomy of the bones—for example, in the femur, in the region of muscle insertions, there is a radial tract of secondary osteons present (Figure 7F), and in the area where the nutrient foramen penetrates the bone, the bone tissue is organised so as to accommodate the foramen (Figure 7H).

In contrast to the bone rapidly forming during early ontogeny, in late stages of ontogeny, a distinct layer of lamellar bone tissue forms subperiosteally. This band of tissue, the OCL [1,18], marks the change to a slower rate of bone formation. Among many modern vertebrates, such a change is linked to the attainment of sexual maturity and the subsequent slow-down in growth, e.g., [1,27]. Thus, the occurrence of the OCL directly suggests that a slow-down in growth has been reached, which means that, thereafter, only slow accretionary growth will occur from this stage onwards. This seems to be the case in the extant kiwi [19], but in ducks, sexually immature ducks have been reported to show an OCL [27], and Watanabe [26] found the same in three species of water birds (*Calonectris leucomelas*,

Ardea cinerea and *Phalacrocorax capillatus*). Whether the OCL in *Genyornis* is linked to the attainment of sexual maturity is uncertain, but in the current study, it is apparent that *Genyornis* bones that morphologically appear to be from juveniles do not have an OCL (see later).

In several of the bones, narrow bands of lamellar bone tissue (annuli) were observed to have been deposited prior to the deposition of the OCL. The deposition of these periodic annuli interrupts the rapid phase of growth and reflect a slow-down in the overall rate of bone deposition [1,24], and they are generally thought to be formed annually in vertebrates, e.g., [17,27–29]. As most of our samples were cores, we cannot be certain that these annuli extend around the complete bone wall. In one specimen (tibiotarsus SAM P.54334), at least four such narrow annuli are observed, suggesting that this individual took at least 4 years before an OCL formed. Interestingly, in specimen SAM P.54334, the width of the annuli appears to be quite variable—even around the section (see Figure 4A,B). In parts of the compacta, some annuli appear as relatively wide bands of lamellar tissue. In most of the other individuals, there appears to only be about 1–2 growth marks evident in the compacta. Thus, it is evident that, unlike the largest of modern birds, ostriches (*Struthio camelus*), which weigh in at about 150 kg, *Genyornis* (estimated to have weighed about 250 kg) took more than a single year to reach skeletal maturity. *Vorombe titan*, the largest aepyornithid (and possibly the largest of all birds) [30] also took several years to reach skeletal maturity [5], and the extinct Dinornithiformes, such as the emeids, *Euryapteryx* and *Anomalopteryx*, as well as *Megalapteryx*, all experienced extended periods of cyclical growth to somatic maturity [20]. The extinct Mesozoic bird, *Gargantuavis*, was also found to have taken at least a decade to reach skeletal maturity [31]. Thus, it is evident that several large terrestrial birds experienced protracted growth to adult body size. It appears that other island birds which are not as large, such as the kiwi, *Apteryx* species [19], the dodo, *Raphus cucullatus* [28] and the solitaire, *Pezophaps solitaria* [32], also adopted slow, extended growth rates in response to reduced predation on the islands.

We were unable to identify any sex differences between the *Genyornis* femora studied. In *Dromornis stirtoni*, medullary bone [1], a tissue formed in female birds during ovulation was identified in several tibiotarsi [13], which verified their assignment as females. However, in the case of our *Genyornis* sample, medullary bone was not observed in any of the bones examined. It is likely that the birds were mired during a protracted drought, which may explain the lack of evidence of breeding (both in the form of hatchlings, and females with medullary bone), but it is also possible that our small sample size precluded the observation of sex-specific tissues.

4.2. Histological Variations Evident in Different Skeletal Elements

4.2.1. Histological Differences among Bones without OCL

The tibiotarsus, specimen FU2759, and the femur, FU2758, were recovered from different individuals and, as such, we cannot directly compare their growth dynamics. However, on the basis of the histology evident, they appear to be from young individuals as they both do not have an OCL present. Furthermore, of these two bones, the tibiotarsus appears to be younger than the femur as rapidly formed FLB occurs subperiosteally in the tibiotarsus, whereas in the femur, a narrow band of lamellar bone tissue is present, indicating that the rate of bone deposition had begun to slow down. Interestingly, the surface texture of specimen FU2758 is clearly porous, and the crista trochanteris is also not fully formed, which further indicate that this is a young individual. The overall small diameter of the tibiotarsus specimen FU2759 also agrees with its young ontogenetic age assessment.

4.2.2. Histological Variations Evident among Bones with OCL

Except for the bones mentioned above, all the other bones have a distinctive OCL, which means that the appositional growth had passed its most rapid phase of growth [16]. In some of the specimens, it is apparent that within the OCL, there are LAGs—assuming that each LAG is formed annually, this indicates that some individuals are older than others, e.g., [31] (Table 2). This is further supported by an external morphology typical of fully adult birds; for example, as defined for dinornithiforms by Turvey and Holdaway [33], where femora all have the adult form of the condyles, all tibiotarsi have a fully ossified pons supratendineus and fully developed condyles with no sign of the synostosis between the tibia and proximal tarsal, and tarsometatarsi have no distinction of the fused metatarsi or distal tarsal.

Table 2. Summary of histology data for specimens studied. Spec. no., specimen number; OCL, outer circumferential layer; Lags, lines of arrested growth; ICL, inner circumferential layer; hav, Haversian; Resorpt. Perimedull., resorption cavities in the perimedullary region; Ontog., ontogenetic. Y denotes presence, N denotes absence and ? denotes uncertain.

Specimen	Element	OCL	Lags in OCL	Annuli	ICL	Dense Hav Bone	Early Bone	Resorpt. Perimedull.	Ontog. State
SAM P.54334	Tibiotarsus	Y	4–5	3–4	Y	Y	Y	Y	adult
SAM P.54333	Tibiotarsus	Y	N	1?	Y	N	Y	Y	young adult
SAM P.53833	Femur	Y	2	Y	Y	Y	N	Y	adult
SAM P.53833	Tibiotarsus	Y	3	2?	N	Y	N	Y	adult
SAM P.53832	Tarsometatarsus	Y	2?	N	Y	Y	N	Y	adult
SAM P.53831	Tarsometatarsus	Y	2?	N	?	Y	N	Y	adult
SAM P.53826	Tibiotarsus	Y	3–4	2?	?	?	?	?	adult
SAM P.25017	Tibiotarsus	Y	5	3	Y	N	Y	Y	adult
FU2760	Femur	Y	1	?	Y	N	N	Y	adult
FU2759	Tibiotarsus	N	N	N	N	N	Y	?	immature
FU2758	Femur	Y	N	N	Y	N	Y	Y	young adult
FU2756	Tibiotarsus	Y	~10	1	N	N	Y	Y	mature
FU2750	Tarsometatarsus	Y	1	2	?	N	Y	Y	young adult
FU2755	Tibiotarsus	?	?	?	?	N	Y?	?	?
FU2755	Femur	Y	6–7	?	?	N	Y?	?	mature

Our sample of *Genyornis* bones provides evidence for different growth dynamics between individuals; although most individuals showed a periodic slow-down in growth in the form of narrow annuli, some individuals (e.g., tibiotarsus, SAM P.53833 from the Callabonna locality) had a wide band of lamellar tissue, suggesting that it experienced a particularly stressful period that was lengthy in duration. One specimen, SAM P.54334 (from Cooper Creek), showed four narrow annuli in the tibiotarsus, indicating that it had at least four periods of slowed growth, and two of these were longer in duration. The facts that, in some specimens, we find no annuli interrupting growth, and up to four in one individual, as well as the widely varying thickness of the annuli, suggest that *Genyornis* experienced variable growth dynamics, which may have been correlated with the particular environment during which the individuals were growing up. Such plasticity in growth appears to be a plesiomorphic trait inherited from their dinosaurian ancestors [1,34].

One of the main reasons for this discrepancy could be the fact that the specimens studied come from different localities, which, although they are not greatly separated, i.e., Billeroo Creek is perhaps 100 km from Lake Callabonna, which is 500 km at most from

Cooper Creek (Figure 1), they may have had different local ecologies. It is also possible that the specimens derive from slightly different time periods of the late Pleistocene, which makes it likely that they experienced different environmental conditions during their lives, i.e., they do not reflect a single contemporaneous population. The strikingly wide annulus present in the tibiotarsus specimen SAM P.53833 indicates that this individual experienced a prolonged stressed period when osteogenesis slowed down (Figure 6) [1,6]. However, once the conditions changed to a more favourable situation, osteogenesis recovered to a rapid rate, resulting in FLB tissue being formed. This is directly contrasted with the tarsometatarsus specimen FU2750, which had no annuli prior to the OCL formation. In the tibiotarsus specimen SAM P.54334 from Cooper Creek, two distinct wider-than-usual annuli are also observed (Figure 4A).

The Cooper Creek specimens (SAM P.25017, 54,333 and 54334) and Billeroo Creek specimens (FU2758, 2759) were deposited in fluvial sediments in a riverine situation, which does mean they had abundant water at their death. The Callabonna specimens were all trapped in the dried-out bed of a lake during a protracted drought.

4.2.3. Histological Differences among Specimens Recovered from the Same Site

Specimen SAM P.53832 and SAM P.53831 were both recovered from Lake Callabonna within 1 m of the other and facing the same direction, which suggests that they may have been trapped together. Interestingly, both these TMT show that they are mature adult individuals with a well-developed OCL and heavily reconstructed compacta. The overall dimension of these bones suggests that SAM P.53832 was the larger bird and likely to be a male, while SAM P.53831 was a probable female and was similar in size to SAM P.53833.

The tibiotarsi specimens, FU2756 and FU2755, were sampled from individuals that were recovered from Callabonna CB2018-75. The tibiotarsus from FU2756 has about 10 closely associated LAGs in the OCL, although we cannot be sure if any of them are part of double or triple LAGs which are known to occur in some vertebrates when conditions are recurrently unfavourable [1,27]. It must be noted that these lines do not interrupt the rapid phase of growth but are located in the outermost part of the compacta and are bone deposits that are responsible for the thickening or robustness of the bones (i.e., they do not contribute to the lengthening of the bone) [6]. As this tibiotarsus occurs in the boundary size range of male/female, these histology findings suggest that it is a mature female (rather than a young male). Unfortunately, the outer compacta of the tibiotarsus of FU2755 is not as well preserved, and we cannot determine whether or not it had passed its most rapid phase of growth. Its relatively small size suggests it is also a female individual.

4.3. Secondary Reconstruction

Secondary reconstruction was observed in all the skeletal elements, but compared to the femur and the tibiotarsi studied, the tarsometatarsus was the most extensively reconstructed element, with dense Haversian bone tissue present. This finding agrees with [35] that there is a proximodistal gradient in terms of secondary reconstruction with more distal elements being more extensively remodelled. Note that in the aepyornithids, the fibula was the most reconstructed element [5], but in the current study, fibulae were not sampled. It is possible that the high incidence of Haversian bone in the tarsometatarsus suggests that this element bears more weight and is subjected to more biomechanical forces than the tibiotarsi [36].

It is also apparent that the extent of secondary reconstruction in the compacta is age-related—the tarsometatarsus specimen FU2750 has an OCL but shows hardly any secondary reconstruction, whereas the other two tarsometatarsus studied have compacta that are completely remodelled right up to the OCL. These findings suggest that secondary reconstruction increases with age.

As in the aepyornithids [5], the tibiotarsi in *Genyornis* appears to provide the best record of growth and preserves most of the primary compacta. The tarsometatarsus is useful in young adults, but older individuals show extensive secondary remodelling that removes the primary bone tissues, and hence the growth record.

5. Conclusions

Genyornis took more than a single year to reach sexual maturity, whereupon an OCL develops, indicating a change in the rate of osteogenesis.

Of the three skeletal elements studied, the tibiotarsus preserves the best record of growth for *Genyornis*.

The occurrence of several LAGs in the OCL indicates that it continued to accrete bone for several years to reach skeletal maturity. This further indicates that, in *Genyornis*, sexual and skeletal maturity were asynchronous, with sexual maturity preceding skeletal maturity.

Genyornis retained a plesiomorphic flexible growth strategy [1,15] and responded to prevailing environmental conditions at the time.

Author Contributions: A.C. and T.H.W. conceived the project and sampled the bones together; A.C. carried out the histological descriptions, took the photomicrographs and wrote the first draft of the MS. All authors have read and agreed to the published version of the manuscript.

Funding: This research was funded by Australian Research Grant, ARC Discovery Project DP180101913 "Extricating extinction histories at Lake Callabonna's megafauna necropolis", to T. Worthy, L. Arnold, and A. Chinsamy-Turan. AC is also supported by the National Research Foundation (NRF), grant number 117716. The APC was funded by the journal.

Informed Consent Statement: Not applicable.

Data Availability Statement: All data is within the manuscript.

Acknowledgments: Joshua van der Blerk is acknowledged for technical assistance with some of the thin section preparation. We thank Warren Handley for assistance with sampling the specimens. Three anonymous reviewers are thanked for their constructive comments.

Conflicts of Interest: The authors declare no conflict of interest.

References

1. Chinsamy-Turan, A. *The Microstructure of Dinosaur bones: Deciphering Biology through Fine Scale Techniques*; John Hopkins University Press: Baltimore, MD, USA, 2005.
2. Chinsamy-Turan, A. *The Forerunners of Mammals: Radiation, Histology, Biology*; Indiana University Press: Bloomington, IN, USA, 2012.
3. Erickson, G.M. Assessing dinosaur growth patterns: A microscopic revolution. *Trends Ecol. Evol.* **2005**, *20*, 677–684. [CrossRef]
4. Erickson, G.M. On dinosaur growth. *Ann. Rev. Earth Planetary Sci.* **2014**, *42*, 675–697. [CrossRef]
5. Chinsamy, A.; Angst, D.; Canoville, A.; Göhlich, U.B. Bone histology yields insights into the biology of the extinct elephant birds (*Aepyornithidae*) from Madagascar. *Biol. J. Linn. Soc.* **2020**, *130*, 268–295. [CrossRef]
6. Chinsamy, A.; Warburton, N.M. Ontogenetic growth and the development of a unique fibrocartilage entheses in *Macropus fuliginosus*. *Zoology* **2020**, *144*, 125860. [CrossRef] [PubMed]
7. Murray, P.F.; Vickers-Rich, P. *Magnificent Mihirungs: The Colossal Flightless Birds of the Australian Dreamtime*; Indiana University Press: Bloomington, IN, USA, 2004.
8. Worthy, T.H.; Degrange, F.J.; Handley, W.D.; Lee, M.S. The evolution of giant flightless birds and novel phylogenetic relationships for extinct fowl (Aves, Galloanseres). *Roy. Soc. Open Sci.* **2017**, *4*, 170975. [CrossRef] [PubMed]
9. Stirling, E.C.; Zietz, A.H.C. Fossil remains of Lake Callabonna. Part III. Description of the vertebrae of *Genyornis newtoni*. *Mem. Roy. Soc. S. Aust.* **1905**, *1*, 81–110.
10. Stirling, E.C.; Zietz, A.H.C. Preliminary notes on *Genyornis newtoni*: A new genus and species of fossil struthious bird found at Lake Callabonna, South Australia. *Trans. Proc. Rep. Roy. Soc. S. Aust.* **1896**, *20*, 171–190.
11. Saltré, F.; Rodríguez-Rey, M.; Brook, B.W.; Johnson, C.N.; Turney, C.S.; Alroy, J.; Cooper, A.; Beeton, N.; Bird, M.I.; Fordham, D.A. Climate change not to blame for late Quaternary megafauna extinctions in Australia. *Nat. Commun.* **2016**, *7*, 1–7. [CrossRef]
12. Grellet-Tinner, G.; Spooner, N.; Handley, W.D.; Worthy, T.H. The *Genyornis* egg: Response to Miller et al.'s commentary on Grellet-Tinner et al., 2016. *Quat. Sci. Rev.* **2017**, *61*, 128–133. [CrossRef]
13. Handley, W.D.; Chinsamy, A.; Yates, A.M.; Worthy, T.H. Sexual dimorphism in the late Miocene mihirung *Dromornis stirtoni* (Aves: Dromornithidae) from the Alcoota Local Fauna of central Australia. *J. Vert. Paleontol.* **2016**, *36*, e1180298. [CrossRef]
14. Chinsamy, A.; Chiappe, L.M.; Dodson, P. Growth rings in Mesozoic birds. *Nature* **1994**, *368*, 196–197. [CrossRef]

15. Starck, J.M.; Chinsamy, A. Bone microstructure and developmental plasticity in birds and other dinosaurs. *J. Morphol.* **2002**, *254*, 232–246. [CrossRef]
16. Chinsamy, A. Histological perspectives on growth in the birds *Struthio camelus* and *Sagittarius serpentarius*. *Acta Palaeornithol.* **1995**, *181*, 317–323.
17. Chinsamy, A. The Osteohistology of Femoral Growth within a Clade: A Comparison of a Crocodile, *Crocodylus niloticus*, the Dinosaurs, *Massospondylus* and *Syntarsus* and the Birds, *Struthio* and *Sagittarius*. Ph.D. Thesis, University of the Witwatersrand, Johannesburg, South Africa, 1990.
18. Ponton, F.; Elżanowski, A.; Castanet, J.; Chinsamy, A.; Margerie, E.D.; de Ricqlès, A.; Cubo, J. Variation of the outer circumferential layer in the limb bones of birds. *Acta Ornithol.* **2004**, *39*, 137–140. [CrossRef]
19. Bourdon, E.; Castanet, J.; de Ricqlès, A.; Scofield, P.; Tennyson, A.; Lamrous, H.; Cubo, J. Bone growth marks reveal protracted growth in New Zealand kiwi (Aves, Apterygidae). *Biol. Lett.* **2009**, *5*, 639–642. [CrossRef]
20. Turvey, S.T.; Green, O.R.; Holdaway, R.N. Cortical growth marks reveal extended juvenile development in New Zealand moa. *Nature* **2005**, *435*, 940–943. [CrossRef]
21. De Ricqlès, A.; Padian, K.; Horner, J.R. The bone histology of basal birds in phylogenetic and ontogenetic perspectives. In *New Perspectives on the Origin and Early Evolution of Birds: Proceedings of the International Symposium in Honor of John, H. Ostrom*; Gauthier, J., Gall, L.F., Eds.; Allen Press: Lawrence, KS, USA, 2001; pp. 411–426.
22. Nanson, G.C.; Price, D.M.; Jones, B.G.; Maroulis, J.C.; Coleman, M.; Bowman, H.; Cohen, T.J.; Pietsch, T.J.; Larsen, J.R. Alluvial evidence for major climate and flow regime changes during the middle and late Quaternary in eastern central Australia. *Geomorphology* **2008**, *101*, 109–129. [CrossRef]
23. Chinsamy, A.; Raath, M.A. Preparation of fossil bone for histological examination. *Palaeontol. Afr.* **1992**, *29*, 39–44.
24. Francillon-Vieillot, H.; De Buffrénil, V.; Castanet, J.; Géraudie, J.; Meunier, F.J.; Sire, J.Y.; Zylberberg, L.; De Ricqlès, A. Microstructure and mineralisation of vertebrate skeletal tissues. In *Skeletal Biomineralisation: Patterns, Processes and Evolutionary Trends*; Carter, J.G., Ed.; Van Nostrand Reinhold: New York, NY, USA, 1990; pp. 471–530.
25. De Margerie, E.; Robin, J.P.; Verrier, D.; Cubo, J.; Groscolas, R.; Castanet, J. Assessing a relationship between bone microstructure and growth rate: A fluorescent labelling study in the king penguin chick (*Aptenodytes patagonicus*). *J. Exp. Biol.* **2004**, *207*, 869–879. [CrossRef]
26. Watanabe, J. Ontogeny of surface texture of limb bones in modern aquatic birds and applicability of textural ageing. *Anat. Rec.* **2018**, *301*, 1026–1045. [CrossRef]
27. Castanet, J.; Vieillot, H.F.; Meunier, F.J.; De Ricqlès, A. Bone and individual aging. In *Bone, Vol. 7, Bone Growth*; Hall, B.K., Ed.; CRC Press: Boca Raton, FL, USA, 1993; pp. 245–283.
28. Angst, D.; Chinsamy, A.; Steel, L.; Hume, J.P. Bone histology sheds new light on the ecology of the dodo (*Raphus cucullatus*, Aves, Columbiformes). *Sci. Rep.* **2017**, *7*, 7993. [CrossRef] [PubMed]
29. Castanet, J.; Grandin, A.; Arbourachid, A.; de Ricqles, A. Expression of growth dynamic in the structure of the periosteal bone in the mallard, *Anas platyrhynchos*. *C. R. Acad. Sci. Paris Ser. 3 Sci. Vie* **1996**, *319*, 301–308.
30. Hansford, J.P.; Turvey, S.T. Unexpected diversity within the extinct elephant birds (Aves: Aepyornithidae) and a new identity for the world's largest bird. *R. Soc. Open Sci.* **2018**, *5*, 181295. [CrossRef] [PubMed]
31. Chinsamy, A.; Buffetaut, E.; Angst, D.; Canoville, A. Insight into the growth dynamics and systematic affinities of the Late Cretaceous *Gargantuavis* from bone microstructure. *Naturwissenschaften* **2014**, *101*, 447–452. [CrossRef]
32. Hume, J.P.; Steel, L. Fight club: A unique weapon in the wing of the solitaire, *Pezophaps solitaria* (Aves: Columbidae), an extinct flightless bird from Rodrigues, Mascarene Islands. *Biol. J. Linn. Soc.* **2013**, *110*, 32–44. [CrossRef]
33. Turvey, S.T.; Holdaway, R.N. Postnatal ontogeny, population structure, and extinction of the giant moa *Dinornis*. *J. Morphol.* **2005**, *265*, 70–86. [CrossRef]
34. Chinsamy, A.; Marugán-Lobón, J.; Serrano, F.J.; Chiappe, L. Osteohistology and life history of the basal pygostylian, *Confuciusornis sanctus*. *Anat. Rec.* **2019**, *303*, 949–996. [CrossRef]
35. De Ricqlès, A.; Bourdon, E.; Legendre, L.J.; Cubo, J. Preliminary assessment of bone histology in the extinct elephant bird *Aepyornis* (Aves, Palaeognathae) from Madagascar. *Comptes Rendus Palevol.* **2016**, *15*, 197–208. [CrossRef]
36. Martin, R.B.; Burr, D.B. *Structure, Function, and Adaptation of Compact Bone*; Raven Pr: New York, NY, USA, 1989.

 *diversity*

MDPI

Review

A Giant Ostrich from the Lower Pleistocene Nihewan Formation of North China, with a Review of the Fossil Ostriches of China

Eric Buffetaut [1,2,*] **and Delphine Angst** [3]

[1]   Centre National de la Recherche Scientifique—CNRS (UMR 8538), Laboratoire de Géologie de l'Ecole Normale Supérieure, PSL Research University, 24 rue Lhomond, CEDEX 05, 75231 Paris, France
[2]   Palaeontological Research and Education Centre, Maha Sarakham University, Maha Sarakham 44150, Thailand
[3]   School of Earth Sciences, University of Bristol, Life Sciences Building, 24 Tyndall Avenue, Bristol BS8 1TQ, UK; angst.delphine@gmail.com
*   Correspondence: eric.buffetaut@sfr.fr

Abstract: A large incomplete ostrich femur from the Lower Pleistocene of North China, kept at the Muséum National d'Histoire Naturelle (Paris), is described. It was found by Father Emile Licent in 1925 in the Nihewan Formation (dated at about 1.8 Ma) of Hebei Province. On the basis of the minimum circumference of the shaft, a mass of 300 kg, twice that of a modern ostrich, was obtained. The bone is remarkably robust, more so than the femur of the more recent, Late Pleistocene, *Struthio anderssoni* from China, and resembles in that regard *Pachystruthio* Kretzoi, 1954, a genus known from the Lower Pleistocene of Hungary, Georgia and the Crimea, to which the Nihewan specimen is referred, as *Pachystruthio* indet. This find testifies to the wide geographical distribution of very massive ostriches in the Early Pleistocene of Eurasia. The giant ostrich from Nihewan was contemporaneous with the early hominins who inhabited that region in the Early Pleistocene.

Keywords: ostrich; China; Nihewan; Pleistocene; femur

Citation: Buffetaut, E.; Angst, D. A Giant Ostrich from the Lower Pleistocene Nihewan Formation of North China, with a Review of the Fossil Ostriches of China. *Diversity* **2021**, *13*, 1085. https://doi.org/10.3390/d13020047

Academic Editor: Michael Wink
Received: 30 December 2020
Accepted: 22 January 2021
Published: 26 January 2021

Publisher's Note: MDPI stays neutral with regard to jurisdictional claims in published maps and institutional affiliations.

1. Introduction

The Lower Pleistocene fossiliferous beds of the Nihewan Basin (Figure 1) in northern Hebei Province (North China) have been known for their vertebrate remains since the 1920s. More recently, abundant evidence of early human occupation has also come to light ([1], and references therein). The fossil mammals from the various formations of the Nihewan Basin have attracted considerable attention, starting with the pioneering paper by Teilhard de Chardin and Piveteau [2]. However, although bird bones have been mentioned, few of them have been described in detail, with the notable exception of a metatarsus belonging to a crow (*Corvus*) [3].

Here we describe an ostrich femur, collected in the 1920s and kept in the paleontology collection of the Muséum National d'Histoire Naturelle (Paris, France). Although this bone is poorly preserved, a body mass estimate based on its circumference shows that it belonged to a giant ostrich, significantly larger than the living *Struthio camelus*. It provides new evidence of the wide geographical distribution of giant ostriches in the Early Pleistocene of Eurasia.

A note on spelling: In this paper we have used the modern pīnyīn spelling for place names. In the 1920s, a different transliteration was used by paleontologists working in China: "Nihowan" instead of Nihewan, "Sangkan Ho" instead of Sanggan He, etc.

Institutional abbreviations: MNHN: Muséum National d'Histoire Naturelle, Paris, France. IVPP: Institute of Vertebrate Paleontology and Paleoanthropology, Beijing, China.

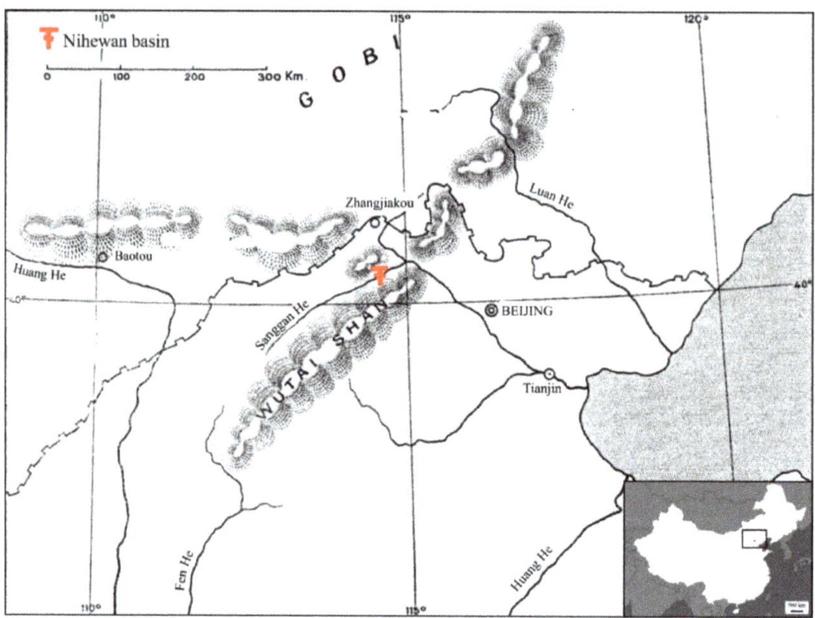

Figure 1. Map of part of northern China showing location of the Nihewan Basin, WNW of Beijing, with general location in China (small map at lower right corner). Modified after Teilhard de Chardin and Piveteau [2].

2. Discovery and Geological Setting of the Specimen

Following discoveries of fossil bones by Father Vincent, a missionary based in Nihewan village (some 150 km NW of Beijing; Figure 1), in 1920 [4–6], the area was visited independently but almost simultaneously by the British geomorphologist George B. Barbour [7] and the French Jesuit and naturalist Emile Licent in 1924 [4]. Abundant vertebrate (mainly mammal) remains were subsequently collected from the Nihewan Basin in the course of field trips led by Licent in 1925 and Licent and Teilhard de Chardin in 1926 [6].

The first mention of an ostrich bone from the Nihewan Basin is in a section on the antiquity of the ostrich in eastern Asia in the monograph by Boule et al. [8] on the Paleolithic in China. The authors note (p. 92) that Licent has found in the "Sanmenian" [Lower Pleistocene] beds of the Sanggan He (the river which flows through the Nihewan Basin) an ostrich femur more than 340 mm in length, indicating a bird larger than the living ostrich. This brief mention seems to have attracted little attention, although it was noted by Lowe [9] and Lambrecht [10]. Later, in their study of the Nihewan mammals, Teilhard de Chardin and Piveteau [2] briefly mentioned in a footnote (p. 126) the few bird remains in their collection, viz. a humerus of a large vulture and an ostrich femur. These two bones are kept together at the MNHN, the ostrich bone bearing number NIH008. A second number, 1927–13, refers to a catalog entry briefly listing a collection of vertebrate fossils from Nihewan brought back from his second mission to China by Teilhard de Chardin on 20 November 1927. The words "*Struthio*" and "Femur" are written in pencil on the bone. There is therefore no doubt that the femur described below is that mentioned by Teilhard de Chardin and Piveteau in 1930. Whether it is the same bone as that mentioned by Boule et al. [8] is not so clear, because specimen MNHN–NIH008, in its present condition, is 247 mm in length, while the length provided by Boule et al. [8] is more than 340 mm. This may suggest that two distinct ostrich femora were found in the Nihewan beds in the 1920s and that one of them may have remained in China while the other was sent to Paris. However, no ostrich femur is currently kept at the Hoang Ho Pai Ho Museum in Tianjin, where Licent's collections are kept, and there is no evidence that such a bone was part of the fossils that were transferred from the Hoang Ho Pai Ho Museum to Beijing in 1940 by Teilhard de Chardin and Leroy (see Leroy [11] about this transfer) and are now kept at the IVPP in Beijing. We therefore suppose that MNHN–NIH008 is indeed the bone mentioned by Boule et al. [8], which is no longer as complete as it was when Licent found it (probably during his 1925 collecting trip, when Teilhard de Chardin was not with him, since Licent alone is credited with the discovery), having lost a good part of the distal end.

Many vertebrate localities are currently known in the Nihewan Basin, in formations of different geological ages (see Cai et al. [12], for a recent review), and the exact place where the ostrich femur was found is unclear, all the more so given that Licent does not mention this find in his publications about his collecting trips in the Nihewan Basin ([4,13]). However, the early collections made by Licent and Teilhard de Chardin in the region were restricted to a relatively small area around the villages of Nihewan and Xiashagou (see map in Teilhard de Chardin and Piveteau ([2], p. 8). This corresponds to what Cai et al. [12] call the "classic Nihewan fauna" from the middle part of the Nihewan Formation. This fauna is about 1.8 Ma in age, according to Cai et al. [12].

3. Description

Specimen MNHN–NIH008 is a right femur (Figure 2) lacking the distal end (at least one-third of the bone seems to be missing) and the proximal articular head (caput femoris). Some craniocaudal compression seems to have occurred. Some areas in the proximal region have been roughly repaired with plaster. The bone is poorly preserved, its cortex being broken into many pieces on the cranial surface, whereas the caudal surface has been less affected, except in its proximal part. At the level of the distal break, it can be seen that the shaft is hollow and filled with brownish clay. Its bony walls are up to 6 mm in thickness. Cancellous bone can be seen at the proximal end where the cortex is broken. The bony structure of the specimen is thus clearly avian.

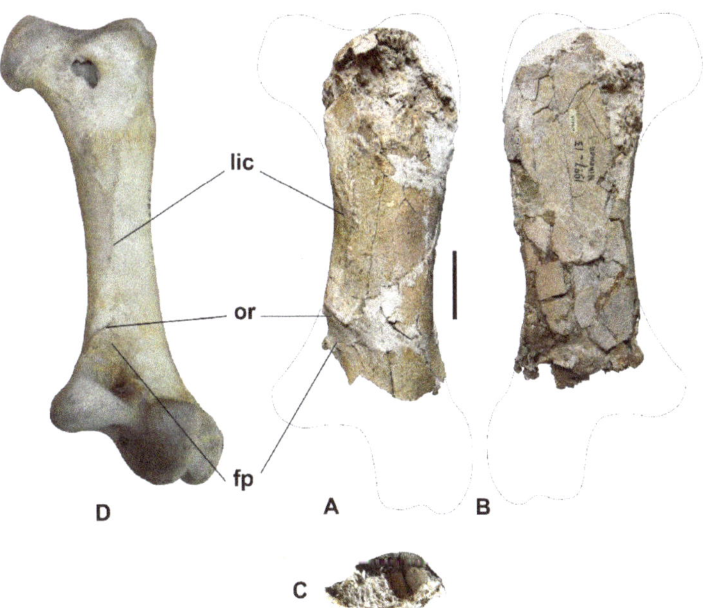

Figure 2. Right femur of giant ostrich (*Pachystruthio* indet.), MNHN–NIH008, from the Lower Pleistocene of the Nihewan Basin, northern China, in caudal (**A**), cranial (**B**) and distal (**C**) views, compared with a femur of the living ostrich *Struthio camelus* in caudal view (**D**). Abbreviations: fp: fossa poplitea; lic: linea intermuscularis caudalis; or: oblique ridge. Scale bar: 50 mm.

Very little is preserved of the proximal articular area, both the caput femoris and the trochanter femoris being destroyed. The cranial face of the bone is relatively flat and featureless. On the caudal face, a long, well-marked longitudinal ridge, the linea intermuscularis caudalis (Figure 2, lic), is visible in the medial half of the bone. It is broad and strongly rugose in its proximal part and becomes sharper distally. At its distal end, it meets a shorter, oblique ridge that extends mediodistally from the medial margin of the bone. Beyond this oblique ridge, the surface of the bone is somewhat depressed, indicating the beginning of the fossa poplitea (Figure 2, fp). At roughly this level, the width of the shaft begins to increase, but nothing is preserved of the distal articular area.

The minimum width of the bone is 74 mm. Its length as preserved is 247 mm. Considering that at least one-third of the bone is missing distally and that the trochanter femoris is broken, its original length must have been very close to the 340 mm mentioned by Boule et al. [8].

Although the specimen is very incomplete, the characters that can be observed, in particular the extent and development of the linea intermuscularis caudalis (Figure 2, lic), are in agreement with an attribution to an ostrich, confirming Teilhard de Chardin and Piveteau's identification.

4. Body Mass Estimate, Comparison with Other Giant Ostriches and Identification

Using the equation published by Campbell and Marcus [14] [LogM = 2.411 × LogLCF − 0.065], we used the minimum circumference of the shaft (LCF = 199 mm) of MNHN–NIH008 to estimate the body mass of the Nihewan ostrich. The estimated mass, 300 kg, is twice that of a large male *Struthio camelus* (Figure 3; Table 1). This Early Pleistocene ostrich was clearly a very large bird, in the mass range of some of the largest known birds, such as the giant moa, *Dinornis robustus* [15]. This suggests that the Nihewan ostrich was even larger than the giant ostrich from the Late Pleistocene of China, *Struthio anderssoni*. According to a mass estimate based on the minimum circumference of the shaft of a femur from the Upper Cave at Zhoukoudian (IVPP V6943), *S. anderssoni* reached a weight of 269 kg, a result in good agreement with estimates based on the dimensions of various Pleistocene eggs from the loess of North China referred to that large ostrich [16]. Morphologically, MNHN–NIH008 differs from the Zhoukoudian femur (described by Shaw [17] and Hou [18]) in having a more robust shaft (Figure 4A). A complete femur from Zhoukoudian Upper Cave (present whereabouts unknown) measured by Shaw [17] was 355 mm in length and 69 mm in diameter at midlength, which indicates a more slender bone than MNHN–NIH008.

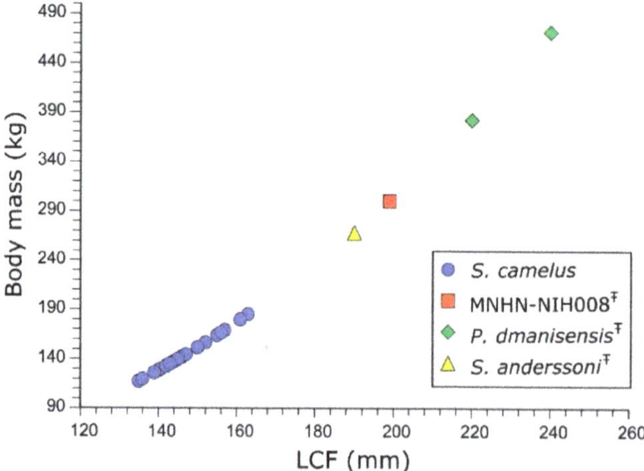

Figure 3. Body mass estimated from the minimum circumference of the shaft (LCF) for several fossil ostriches found in the Pleistocene of Eurasia, compared with the living *Struthio camelus*.

Table 1. Summary table of the measurements of the fossil Struthionidae from the Pleistocene of Eurasia and Africa compared with the modern *Struthio camelus*. **Bold** total length values are estimated. LCF: minimum circumference of the shaft of the femora.

Specimen Number	Species	Age	Geological Location	LCF (mm)	Body Mass Estimated (kg)	Total Length (mm)	Minimum Shaft Width (mm)	Stoutness Index (%)	Source
NIH008	*Pachystruthio* indet.	Early Pleistocene	Nihewan Formation (China)	199	300	**340.0**	74.0	21.76	Boule et al. 1928; this paper
D70	*Pachystruthio dmanisensis*	Early Pleistocene	Dmanisi (Georgia)	220	382	**380.0**	76.0	20.00	Burchak–Abramovich & Vekua 1990; Vekua 2013
D5768	*Pachystruthio dmanisensis*	Early Pleistocene	Dmanisi (Georgia)	220	382	**385.0**	76.0	19.74	Burchak–Abramovich & Vekua 1990; Vekua 2013
PIN 5644/56	*Pachystruthio dmanisensis*	Early Pleistocene	Taurida Cave (Crimea)	240	472	**390.0**	74.7	19.15	Zelenkov et al. 2019
IVPP V6943	*Struthio anderssoni*	Late Pleistocene	Zhoukoudian (China)	190	269	/	/	/	this paper
/	*Struthio anderssoni*	Late Pleistocene	Zhoukoudian (China)	/	/	355.0	69.0	19.44	Shaw 1937
/	*Struthio camelus*	Modern	/	140	129	301.0	38.4	12.75	this paper
/	*Struthio camelus*	Modern	/	140.5	130	327.0	36.9	11.29	this paper
1888.377	*Struthio camelus*	Modern	/	146	142	300.0	49.0	16.33	this paper
1888.377	*Struthio camelus*	Modern	/	147	145	300.0	50.0	16.67	this paper
1889.171	*Struthio camelus*	Modern	/	144	138	330.0	45.0	13.64	this paper
1889.171	*Struthio camelus*	Modern	/	145	140	307.0	40.0	13.03	this paper
1389.31	*Struthio camelus*	Modern	/	135	118	310.0	49.0	15.81	this paper
1389.31	*Struthio camelus*	Modern	/	136	120	330.0	45.0	13.64	this paper
1912.49	*Struthio camelus*	Modern	/	155	164	315.0	52.0	16.51	this paper
1922.105	*Struthio camelus*	Modern	/	142	133	286.0	32.0	11.19	this paper
1922.105	*Struthio camelus*	Modern	/	157	170	320.0	43.0	13.44	this paper
1923.2163	*Struthio camelus*	Modern	/	152	157	320.0	42.0	13.13	this paper
1923.2163	*Struthio camelus*	Modern	/	150	152	315.0	46.0	14.60	this paper
1937.114	*Struthio camelus*	Modern	/	139	126	325.0	45.0	13.85	this paper
/	*Struthio camelus*	Modern	Baku	156	167	**310.0**	51.0	16.45	Burchak–Abramovich and Vekua 1990
/	*Struthio camelus*	Modern	Moscow	143	135	314.0	46.0	14.65	Burchak–Abramovich and Vekua 1990
/	*Struthio camelus*	Modern	Krakow	163	186	325.0	51.0	15.69	Burchak–Abramovich and Vekua 1990
/	*Struthio camelus*	Modern	Krakow	161	180	318.0	49.0	15.41	Burchak–Abramovich and Vekua 1990
/	*Struthio oldawayi*	Early Pleistocene	Olduvai Gorge, Tanzania	/	/	400.0	64.0	16.00	Leakey 1967; Vekua 2013

Other Early Pleistocene large ostriches for which the femur is known include *Struthio oldawayi* Lowe, 1933 from Olduvai Gorge, Tanzania [19]. The femur of *S. oldawayi* figured by Leakey [20] has a more slender shaft than that of the Nihewan bone (minimum width about 64 mm) but is longer (total length about 400 mm). Ostrich femora more closely resembling the Nihewan bone are known from the Early Pleistocene of Georgia (Dmanisi [21,22]) and the Crimea (Taurida Cave [23]) (Figures 4 and 5). The giant ostrich from Dmanisi was originally described as *Struthio dmanisensis* by Burchak–Abramovich and Vekua [21]. More recently, Zelenkov et al. [23] have placed both the ostrich from Dmanisi and that from the Taurida Cave in the genus *Pachystruthio* Kretzoi, originally erected as a subgenus of *Struthio* by Kretzoi [24] for a large phalanx and eggshell remains from the Early Pleistocene of Hungary, described as *Struthio (Pachystruthio) pannonicus*. Femora of *P. dmanisensis* described by Vekua [22] are ca. 380 mm and 385 mm in length, with smallest (mediolateral) shaft widths of 76 mm. The femur of the giant bird from Taurida Cave is ca. 390 mm in length, with a smallest shaft width of 74.7 mm [23]. In terms of shaft width, the Nihewan ostrich thus seems more reminiscent of the giant ostriches from Georgia and Crimea than of *S. oldawayi*.

Burchak–Abramovich and Vekua [22] and Vekua [23] used what they called the stoutness (or massiveness) index, i.e., the minimum shaft width/total length ratio, expressed in percent, to compare the femora of various ostriches (Figure 6; Table 1). In the living *Struthio camelus* specimens measured by Burchak–Abramovich and Vekua [22], the index ranges from 13.8 to 16.4; it is ca. 16 in *Struthio oldawayi* and 20.0 in *P. dmanisensis*, which has an exceptionally massive femur, as noted by Vekua [23]. The stoutness index for the femur from Taurida Cave, calculated on the basis of the measurements provided by Zelenkov et al. [23], is 19.15. Based on the measurements provided by Shaw [17], the stoutness index for the femur of *Struthio anderssoni* is 19.44. Calculating the stoutness index for MNHN–NIH008 is of course difficult because the bone in its present state is incomplete. If we accept that MNHN–NIH008 is the femur mentioned by Boule et al. [8], which was more than 340 mm in length, we obtain a stoutness index of 21.76, using 340 mm as the total length; this is higher than the index for *Pachystruthio dmanisensis* (Figure 6; Table 1). However, the index calculated for the Nihewan specimen is probably slightly exaggerated because, according to Boule et al. [8], the total length of the bone was more than 340 mm. Nevertheless, it can be assumed that MNHN–NIH008 had a high stoutness index, comparable to that of *Pachystruthio dmanisensis*, which it resembles in the robustness of the shaft and the development of the linea intermuscularis caudalis (although the latter is in a more central position on the shaft in the specimen from Taurida Cave than in those from Nihewan and Dmanisi). Although a precise identification of MNHN–NIH008 is difficult because of the incompleteness of the specimen, these similarities with the more or less coeval species from Georgia and Crimea are notable and, pending the discovery of more material from the Nihewan Formation, we refer to the specimen as *Pachystruthio* indet. Although the eastern European localities and Nihewan are some 6000 km apart (Figure 5), the occurrence of the same taxon of ostrich at both localities cannot be ruled out, because there were considerable similarities between the vertebrate faunas of various parts of Eurasia, from China to western Europe, in the Early Pleistocene, possibly linked to the development of extensive grasslands [25]—an idea already put forward to explain the Pleistocene distribution of the ostrich in Eurasia by Andersson [26].

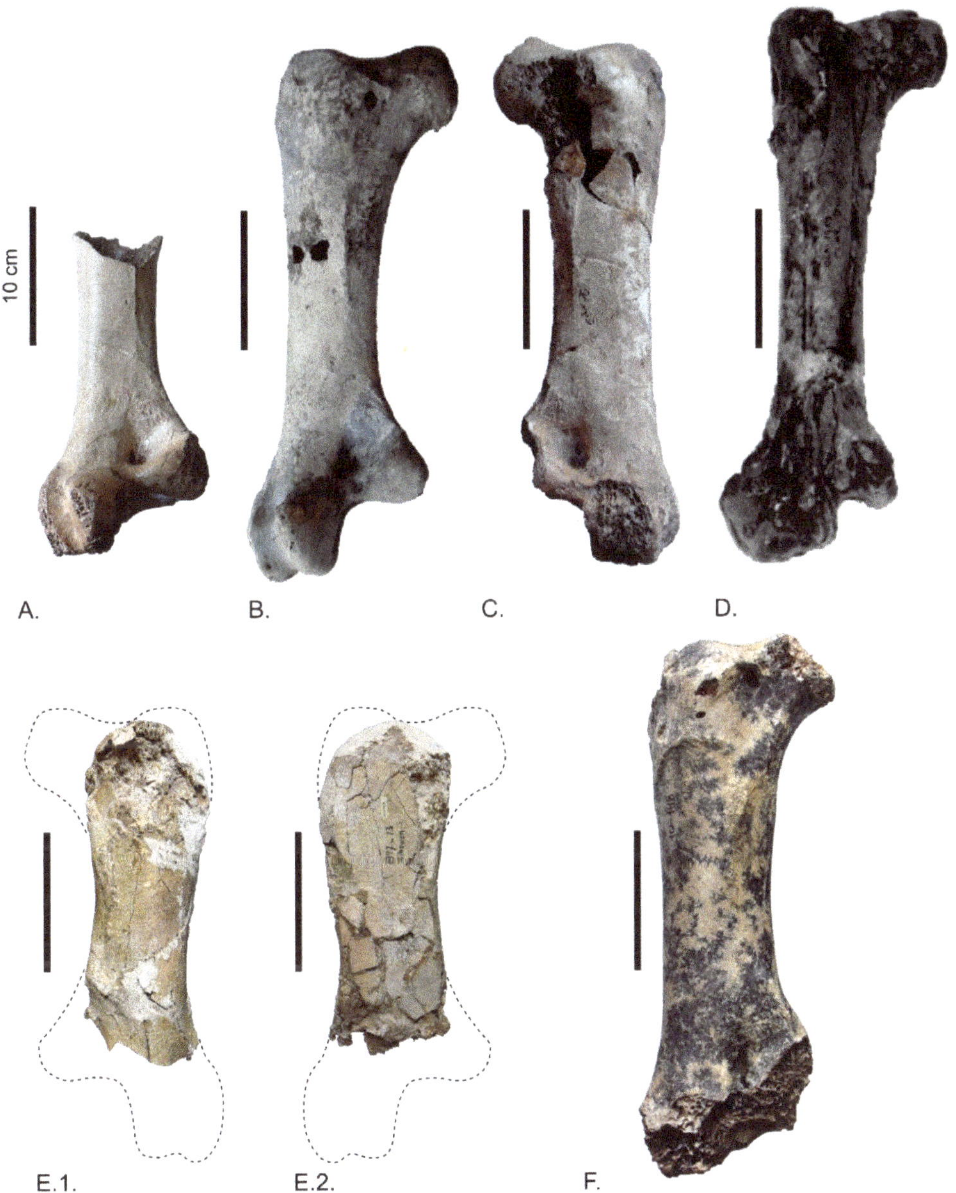

Figure 4. Comparison of the femur MNHN–NIH008 with other femora of Pleistocene ostriches. (**A**) Femur of *Struthio anderssoni*, from Zhoukoudian (China), specimen number IVPP V6943, in caudal view. (**B**) Femur of *Pachystruthio* cf. *dmanisensis*), from Dmanisi (Georgia), specimen number D5768, in cranial view (from [22]). (**C**) Femur of *Pachystruthio* cf. *dmanisensis*), from Dmanisi (Georgia), specimen number D70, in caudal view (from [22]). (**D**) Femur of *Struthio oldawayi*, from Olduvai (Tanzania), in cranial view (from [20]); E: femur of giant ostrich (*Pachystruthio indet.*), specimen number MNHN–NIH008, from the Lower Pleistocene of the Nihewan Basin, northern China, in caudal (**E.1**) and cranial (**E.2**) views. (**F**) Femur of *Pachystruthio* cf. *dmanisensis*, from Taurida Cave (Crimea), specimen number PIN 5644/56, in caudal view (from [23]).

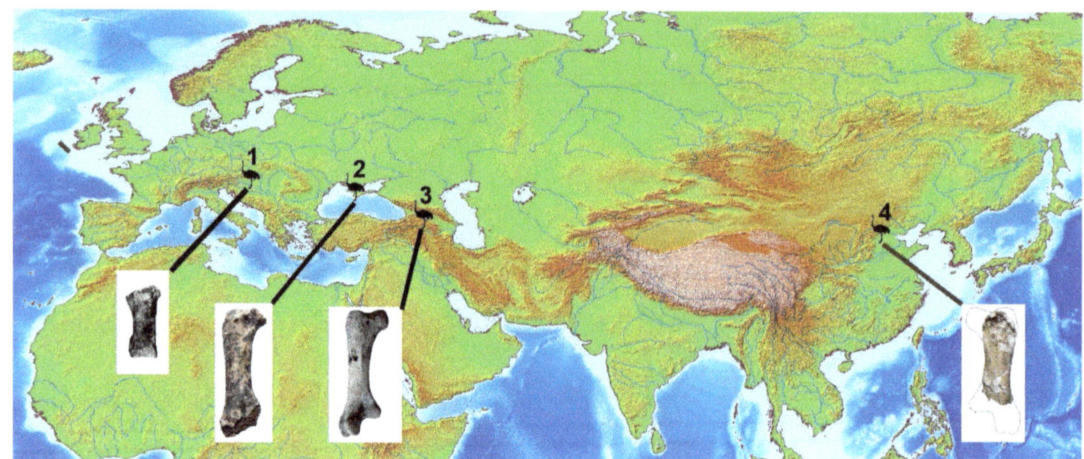

Figure 5. Map showing the distribution of skeletal remains referrable to *Pachystruthio* in the Lower Pleistocene of Eurasia. 1: Kisláng, Hungary (phalanx, after [24]); 2: Taurida Cave, Crimea (femur, after [23]); 3: Dmanisi, Georgia (femur, after [22]); 4: Nihewan, China (femur, this paper). The bones are not to scale.

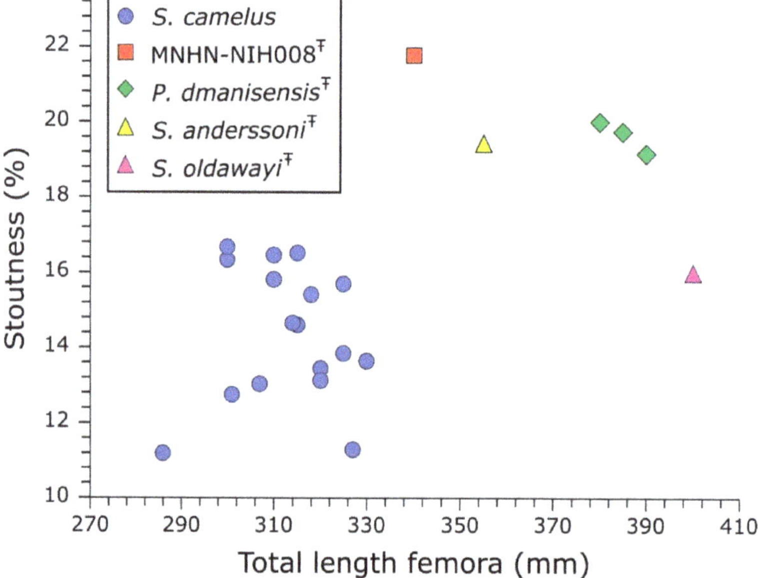

Figure 6. Stoutness (minimum shaft width/total length ratio) vs. total length of the femora of various Pleistocene fossil ostriches, compared with the living *Struthio camelus*.

5. A Brief Review of the Fossil Record of Ostriches in China

The giant ostrich from the Nihewan Formation adds an Early Pleistocene link to a succession of large to very large ostriches known from Neogene and Quaternary formations in China. The earliest record of ostriches or ostrich-like birds from China seems to be eggshell fragments from two Lower Miocene localities in Inner Mongolia [27]. However, Mikhailov and Zelenkov [28] have suggested that the eggshell fragments may have been derived from more recent sediments. This suggestion is based on the fact that the eggshell fragments belong to a type which supposedly was not present in Asia at such an early date, and clearly this claim must be checked, possibly by further field observations, before the hypothesis of an erroneous dating is confirmed. The earliest skeletal remains have been referred to two more or less coeval Late Miocene species, *Struthio wimani* Lowe, 1931 [9] and *S. linxiaensis* Hou et al., 2005 [29]. The type specimen of *Struthio wimani*, a pelvis, was found in the so-called *Hipparion* red clay of the Baode area of NW Shanxi [9,30]. These highly fossiliferous deposits were traditionally referred to the Pontian, a stage once placed in the Pliocene, but are now referred to the Late Miocene [31]. *Struthio linxiaensis* was erected on the basis of a pelvis from the Liushu Formation (Late Miocene) of the Linxia Basin in Gansu Province [29]. Since both taxa appear to be of roughly the same geological age, it may be wondered whether they should really be considered as separate species, a point already made by Mikhailov and Zelenkov [28]. However that may be, both have been described as being larger than the living *Struthio camelus*. Ostrich eggshell remains from the Late Miocene ("*Hipparion* fauna") of Shanxi and Gansu, many of them collected by Emile Licent, were reported by Andersson [30]. Lowe [9] referred these Miocene eggshell fragments to *Struthio wimani*.

Next in age is the *Pachystruthio* femur from the Lower Pleistocene Nihewan Formation described in the present paper. Eggshell remains have been reported from various anthropic sites in the Nihewan Basin (see below).

The Late Pleistocene species *Struthio anderssoni* was originally described by Lowe [9] on the basis of large eggs from many localities in the loess of northern China [30]. Many more eggs from the loess referrable to *S. anderssoni* have subsequently been reported (e.g., [32,33]). The name was later applied to skeletal remains (femora) from the Upper Cave at Zhoukoudian [17,18,34,35], for which dates ranging from 35.1 to 33.5 ky are available [36]. Eggshell fragments are known from several of the karstic localities at Zhoukoudian, of various geological ages, some being significantly older than the Upper Cave [32,35,37]. A discussion of the stratigraphic distribution of *Struthiolithus* eggshells in the loess of China and of the validity of applying the egg-based taxon *Struthio anderssoni* to skeletal remains is beyond the scope of this paper. It may be mentioned that mass estimates based on eggs referred to *Struthio anderssoni* and on a femur from the Upper Cave yield very similar results, viz. about 270 kg [16]. Both the eggs and the few skeletal remains thus indicate an ostrich significantly larger than the living one.

On the basis of C14 dates, Janz et al. [38] have suggested that ostriches survived in north-eastern Asia, including China, until the Holocene. This is in agreement with the suggestion by Kurochkin et al. [39], based on C14 dates from eggshell fragments, that they may have become extinct in the Holocene in Mongolia and Siberia. However, Khatsenovich et al. [40] have urged caution about dates obtained from ostrich eggshell, a material that poses special problems (including different ages for the outside and the inside of the shell) and which in some instances provides ages that are significantly younger than those obtained from bones from the same sites.

The available fossil record thus suggests that the ostrich has been present in China possibly from the Early Miocene to the Late Pleistocene, a time span covering some 20 My. Because of the scarcity of skeletal material (as opposed to the abundance of eggshell remains), it is difficult to reconstruct the evolution of ostriches in that part of the world and the relationships between the several species that have been described are unclear. The fact that these fossil ostriches were larger than the living species seems to be well established, and the form from the Nihewan Formation may have been the most massive of them all.

6. Conclusions: The Giant Ostrich from Nihewan in Its Environment

Although the presence of ostrich remains among the Early Pleistocene vertebrate assemblages of the Nihewan Basin was recorded as early as 1928, they have received little attention. *Struthio* is mentioned in various recent papers about the Nihewan Basin [1,41–45], but often no details are given about the nature of the material, an exception being the paper by Pei et al. [46], which lists eggshells and a coracoid fragment from the Feiliang site. In this context, the ostrich is sometimes considered as an environmental indicator. Dennell [44] considers the presence of the ostrich at the Majuangou III site as in agreement with a warm, moist climate, but also lists *Struthio dmanisensis* at Dmanisi as a steppe indicator. Pei et al. [46] list the ostrich, together with equids, as suggesting large open temperate grasslands, although other evidence (cervids) indicates that forest environments were also present. However, the ecology of the giant Nihewan ostrich may not have been completely similar to that of modern ostriches, which are clearly adapted to open environments. As noted by Vekua [22], the very robustly built femur of *Pachystruthio dmanisensis* differs from the more slender one of other large Pleistocene ostriches such as *S. oldawayi*, and may indicate a bird less well adapted to rapid running. Similarly, Zelenkov et al. [23] suppose that *Pachystruthio* may not have been as good a runner as modern ostriches because of its great body mass. The same probably applies to the very robust Nihewan ostrich. Only the discovery of more complete material will allow a more accurate assessment of the locomotion of these giant ostriches. More generally, the scantiness of the available material makes it difficult to reconstruct the general proportions of these birds and to draw conclusions about their paleobiology. It should be noted that the widespread occurrence of ostrich eggshell fragments (complete eggs being much less common) in the loess of China [8,30], which was deposited under a cold climate during glacial episodes, shows that the presence of ostrich remains cannot be used as evidence for a warm climate.

Largely on the basis of eggshell microstructure and ornamentation, Mikhailov and Zelenkov [46] have proposed a reconstruction of ostrich evolutionary history that is rather complex and will not be discussed here in detail. Suffice it to say that their hypothesis of a Late Pliocene/Early Pleistocene dispersal of giant ostriches belonging to the genus *Pachystruthio*, from eastern Europe to Central Asia, seems convincing. The occurrence of a giant ostrich referrable to *Pachystruthio* in the Lower Pleistocene beds of the Nihewan Basin indicates that this dispersal reached much farther eastward than previously realized. The easternmost occurrences of *Pachystruthio* mentioned by Mikhailov and Zelenkov [28] were finds of Late Pliocene/earliest Pleistocene eggshell remains from eastern Kazakhstan. The Nihewan Basin is located roughly 3000 km farther east, and the occurrence there of *Pachystruthio* shows that this giant ostrich in fact inhabited a very large part of central and north-eastern Eurasia in the Early Pleistocene. This in agreement with the idea of ostrich dispersal along the Eurasian steppes, all the way to North China, already put forward by Andersson in 1929 [26].

Zelenkov et al. [23] have suggested that the very large size of *Pachystruthio* may have been an adaptation to low-nutrition food linked to increased aridity, a hypothesis already put forward by Murray and Vickers–Riche [47] to explain the increasingly large size of Australian dromornithids. This is a possible explanation, but it should be remarked that in China *Pachystruthio* lived in an environment that was not especially arid (see above). Moreover, the later *Struthio anderssoni*, although large, was smaller than *Pachystruthio* despite the fact that it apparently lived under a more arid climate (under which loess was deposited). Moreover, the living ostrich *Struthio camelus*, which is significantly smaller than *Pachystruthio*, lives (or used to live) in arid environments such as the margins of the Sahara or the Syrian desert. One could also invoke Bergmann's rule, which states that within a zoological group forms living at higher latitudes under colder climates tend to be larger than those from warmer climates at lower latitudes. However, it can hardly be used to explain the large size of Eurasian ostriches, and especially *Pachystruthio*, since very large ostriches, such as *Struthio oldawayi*, are also known from the Pleistocene of tropical Africa.

A final point worth noting is that, like at Dmanisi, Taurida Cave and Olduvai, giant ostriches cohabited with early humans in the Nihewan Basin. Their remains are sometimes found at anthropic sites, such as Goudi [43] and Feiliang [46]. Although there is factual evidence of Paleolithic humans at least butchering ostriches [48], whether the early hominins of the Nihewan Basin hunted the giant ostrich is uncertain: a bird twice the weight of the living ostrich cannot have been an easy prey—although eggshell collecting may have been less hazardous.

Author Contributions: Both E.B. and D.A. conceived and carried out the study. All authors have read and agreed to the published version of the manuscript.

Funding: This research received no external funding.

Data Availability Statement: Data is contained within the article.

Acknowledgments: We are grateful to Christine Argot and Vincent Pernègre at the Muséum National d'Histoire Naturelle (Paris) for locating the femur from Nihewan and making it available for study. Access to the ostrich femur from Zhoukoudian at the IVPP was made possible by Zhou Zhonghe and useful information about Chinese collections was provided by Deng Tao (IVPP, Beijing) and the staff of the Hoang Ho Pai Ho Museum (Tianjin), especially Zhang Xiaoxiao.

Conflicts of Interest: The authors declare no conflict of interest.

References

1. Ao, H.; Deng, C.; Dekkers, M.J.; Sun, Y.; Liu, Q.; Zhu, R. New evidence for early presence of hominids in North China. *Sci. Rep.* **2013**, *3*, 2403. [CrossRef] [PubMed]
2. Teilhard de Chardin, P.; Piveteau, J. Les mammifères fossiles de Nihowan (Chine). *Ann. Paléont.* **1930**, *19*, 1–134.
3. Wang, M.; O'Connor, J.K.; Zhou, Z. The first fossil crow (*Corvus* sp. indet.) from the Early Pleistocene Nihewan Paleolithic sites in North China. *J. Archaeol. Sc.* **2013**, *40*, 1623–1628. [CrossRef]
4. Licent, E. Voyage aux terrasses du Sang Kan Ho, à l'entrée de la plaine de Sining Hien. *Pub. Mus. Hoang Ho Pai Ho* **1924**, *4*, 1–14.
5. Barbour, G.B. Preliminary observations in the Kalgan area. *Bull. Geol. Soc. China* **1924**, *3*, 153–168. [CrossRef]
6. Barbour, G.B.; Licent, E.; Teilhard de Chardin, P. Geological study of the deposits of the Sangkanho basin. *Bull. Geol. Soc. China* **1926**, *5*, 263–278. [CrossRef]
7. Barbour, G.B. The deposits of the Sang Kan Ho valley. *Bull. Geol. Soc. China* **1925**, *4*, 53–55. [CrossRef]
8. Boule, M.; Breuil, H.; Licent, E.; Teilhard, P. Le Paléolithique de la Chine. *Arch. Inst. Paléont. Hum.* **1928**, *4*, 1–138.
9. Lowe, P.R. Struthious remains from Northern China and Mongolia, with descriptions of *Struthio wimani*, *Struthio anderssoni* and *Struthio mongolicus*, spp. nov. *Palaeont. Sin. C* **1931**, *6*, 1–47.
10. Lambrecht, K. *Handbuch der Palaeornithologie*; Borntrager: Berlin, Germany, 1933.
11. Leroy, P. L'Institut de Géobiologie à Pékin, 1940–1946. Les dernières années du P. Teilhard de Chardin en Chine. *L Anthropologie* **1965**, *69*, 360–367.
12. Cai, B.Q.; Zheng, S.H.; Liddicoat, J.; Li, Q. Review of the litho-, bio-, and chronostratigraphy in the Nihewan Basin, Hebei, China. In *Fossil Mammals of Asia*; Wang, X., Flynn, L.J., Fortelius, M., Eds.; Columbia University Press: New York, NY, USA, 2013; pp. 218–242.
13. Licent, E. Comptes-rendus de onze années (1923–1933) de séjour et d'exploration dans le bassin du Fleuve Jaune, du Pai Ho et des autres tributaires du golfe du Pai Tcheu ly. *Pub. Mus. Hoang Ho Pai Ho* **1935**, *38*, 1–1131.
14. Campbell, K.E.; Marcus, L. The relationship of hindlimb bone dimensions to body weight in birds. *Nat. Hist. Mus. Los Angeles County Sc. Ser.* **1992**, *36*, 395–412.
15. Angst, D.; Buffetaut, E. *Paleobiology of Giant Flightless Birds*; Elsevier & ISTE Press: Oxford/London, UK, 2017.
16. Buffetaut, E.; Angst, D. How large was the giant ostrich of China? *Evoluçao* **2018**, *2*, 6–8.
17. Shaw, T.H. Einige Bemerkungen zum Oberschenkelknochen des fossilen Strausses *Struthio anderssoni* Lowe von Chou Kou Tien in Nord-China. *Ornithol. Monatsber.* **1937**, *45*, 201–202.
18. Hou, L. Avian fossils of Pleistocene from Zhoukoudian. *Mem. Inst. Vert. Palaeont. Palaeoanthrop. Acad. Sin.* **1993**, *19*, 165–297.
19. Lowe, P.R. On some struthious remains: -1. Description of some pelvic remains of a large fossil ostrich, *Struthio oldawayi*, sp. n., from the Lower Pleistocene of of Oldaway (Tanganyika Territory); 2. Egg-shell fragments referable to *Psammornis* and other Struthiones collected by H. St. John Philby in southern Arabia. *Ibis* **1933**, *75*, 652–657.
20. Leakey, L.S.B. *Olduvai Gorge 1951–1961*; Cambridge University Press: Cambridge, UK, 1967; Volume 1.
21. Burchak-Abramovich, N.I.; Vekua, A. The fossil ostrich *Struthio dmanisensis* sp. n. from the Lower Pleistocene of eastern Georgia. *Acta Zool. Cracov.* **1990**, *33*, 121–132.
22. Vekua, A. Giant ostrich in Dmanisi fauna. *Bull. Georg. Nat. Acad. Sc.* **2013**, *7*, 143–148.
23. Zelenkov, N.V.; Lavrov, A.V.; Startsev, D.B.; Vislobokova, I.A.; Lopatin, A.V. A giant early Pleistocene bird from eastern Europe: Unexpected component of terrestrial faunas at the time of early *Homo* arrival. *J. Vert. Paleont.* **2019**. [CrossRef]
24. Kretzoi, M. Ostrich and camel remains from the Central Danube basin. *Acta Geologica* **1954**, *2*, 231–242.
25. Dennell, R.W. The colonization of "Savannahstan": Issues of timing(s) and patterns of dispersal across Asia in the Late Pliocene and Early Pleistocene. In *Asian Paleoanthropology: From Africa to China and Beyond*; Norton, C.J., Braun, D.R., Eds.; Springer: Dordrecht, The Netherlands, 2010; pp. 7–30.
26. Andersson, J.G. Der Weg über die Steppen. *Bull. Mus. Far East. Antiquit.* **1929**, *1*, 143–165.
27. Wang, S.; Hu, Y.; Wang, L. New ratite eggshell material from the Miocene of Inner Mongolia, China. *Chinese Birds* **2011**, *2*, 18–26. [CrossRef]

28. Mikhailov, K.E.; Zelenkov, N. The late Cenozoic history of the ostriches (Aves: Struthionidae), as revealed by fossil eggshell and bone remains. *Earth Sc. Rev.* **2020**, *208*, 103270. [CrossRef]
29. Hou, L.; Zhou, Z.; Zhang, F.; Wang, Z. A Miocene ostrich fossil from Gansu Province, northwest China. *Chin. Sci. Bull.* **2005**, *50*, 1808–1810. [CrossRef]
30. Andersson, J.G. Essays on the Cenozoic of northern China. *Mem. Geol. Surv. China Ser. A* **1923**, *3*, 1–152.
31. Qiu, Z.X.; Qiu, Z.D.; Deng, T.; Li, C.K.; Zhang, Z.Q.; Wang, B.Y.; Wang, X. Neogene Land Mammal Ages/Stages of China. In *Fossil Mammals of Asia*; Wang, X., Flynn, L.J., Fortelius, M., Eds.; Columbia University Press: New York, NY, USA, 2013; pp. 29–90.
32. Young, C.C. On the new finds of fossil eggs of *Struthio anderssoni* Lowe in North China, with remarks on the egg remains found in Shansi, Shensi and in Choukoutien. *Bull. Geol. Soc. China* **1933**, *12*, 145–152. [CrossRef]
33. Young, C.C.; Sun, A.L. New material of ostrich eggs and its stratigraphic significance. *Paleovert. Paleoanthrop.* **1960**, *2*, 115–119. (In Chinese)
34. Shaw, T.H. Preliminary observations on the fossil birds from Chou-Kou-Tien. *Bull. Geol. Soc. China* **1935**, *14*, 77–81. [CrossRef]
35. Qi, T. *Choukoutienology*; Zhejiang University Press and Archaeopress: Hangzhou, China, 2018.
36. Li, F.; Bae, C.J.; Ramsey, C.B.; Chen, F.; Gao, X. Re-dating Zhoukoudian Upper Cave, northern China and its regional significance. *J. Hum. Evol.* **2018**, *121*, 170–177. [CrossRef]
37. Zhao, Z.; Yuan, G.; Wang, J.; Zhong, Y. On the amino acid composition and microstructure of fossil ostrich eggshell from Sinanthropus site, Choukoutien. *Vert. Palasiat.* **1981**, *19*, 327–337.
38. Janz, L.; Elston, R.G.; Burr, G.S. Dating North Asian surface assemblages with ostrich eggshell: Implications for palaeoecology and extirpation. *J. Archaeol. Sci.* **2009**, *36*, 1982–1989. [CrossRef]
39. Kurochkin, E.N.; Kuzmin, Y.V.; Antoshchenko-Olenev, I.V.; Zabelin, V.I.; Krivonogov, S.K.; Nohrina, T.I.; Lbova, L.V.; Burr, G.S.; Cruz, R.J. The timing of ostrich existence in Central Asia: AMS 14C age of eggshells from Mongolia and southernSiberia (a pilot study). *Nucl. Instr. Meth. Phys. Res. B* **2010**, *268*, 1091–1093. [CrossRef]
40. Khatsenovich, A.M.; Rybin, E.P.; Gunchinsuren, B.; Bolorbat, T.; Odsuren, D.; Angaragdulguun, G.; Margad-Erdene, G. Human and *Struthio asiaticus*: One page of Paleolithic art in the eastern part of Central Asia. *Izv. Irkutsk. gosudarst. Univers. Ser. Geoarkheol. Etnol. Antrop.* **2017**, *21*, 80–106.
41. Zhu, R.X.; Potts, H.; Xie, F.; Hoffman, K.A.; Deng, C.L.; Shi, C.D.; Pan, Y.X.; Wang, H.Q.; Shi, G.H.; Wu, N.Q. New evidence on the earliest human presence at high northern latitudes in northeast Asia. *Nature* **2004**, *431*, 559–562. [CrossRef] [PubMed]
42. Deng, C.; Zhu, R.; Zhang, R.; Ao, H.; Pan, Y. Timing of the Nihewan formation and faunas. *Quat. Res.* **2008**, *69*, 77–90. [CrossRef]
43. Shen, C.; Gao, X.; Wei, Q. The earliest hominin occupations in the Nihewan Basin of Northern China: Recent progress in field investigations. In *Asian Paleoanthropology: From Africa to China and Beyond*; Norton, C.J., Braun, D.R., Eds.; Springer: Dordrecht, The Netherlands, 2010; pp. 169–180.
44. Dennell, R.W. The Nihewan Basin of North China in the Early Pleistocene: Continuous and flourishing, or discontinuous, infrequent and ephemeral occupation? *Quat. Intern.* **2012**, *295*, 223–236. [CrossRef]
45. Liu, P.; Deng, C.; Li, S.; Cai, S.; Cheng, H.; Yuan, B.; Wei, Q.; Zhu, R. Magnetostratigraphic dating of the Xiashagou Fauna and implication for sequencing the mammalian faunas in the Nihewan Basin, North China. *Palaeogeogr. Palaeoclimat. Palaeoecol.* **2012**, *315*, 75–85. [CrossRef]
46. Pei, S.; Gao, X.; Wang, H.; Kuman, K.; Bae, C.J.; Chen, F.; Guan, Y.; Zhang, Y.; Zhang, X.; Peng, F.; et al. Early Pleistocene archaeological occurrences at the Feiliang site, and the archaeology of human origins in the Nihewan Basin, North China. *PLoS ONE* **2017**, *12*, e0187251. [CrossRef]
47. Murray, P.F.; Vickers-Rich, P. *Magnificent Mihirungs: The Colossal Flightless Birds of the Australian Dreamtime*; Indiana University Press: Bloomington/Indianapolis, IN, USA, 2004.
48. Bonilauri, S.; Boëda, E.; Griggo, C.; Al-Sakhel, H.; Muhesen, S. Un éclat de silex moustérien coincé dans un bassin d'autruche (*Struthio camelus*) à Umm el Tlel (Syrie centrale). *Paléorient* **2007**, *33*, 39–46. [CrossRef]

Review

The Enigmatic Avian Oogenus *Psammornis*: A Review of Stratigraphic Evidence

Eric Buffetaut [1,2]

[1] Centre National de la Recherche Scientifique (UMR 8538), Laboratoire de Géologie de l'Ecole Normale Supérieure, PSL Research University, 24 rue Lhomond, CEDEX 05, 75231 Paris, France; eric.buffetaut@sfr.fr

[2] Palaeontological Research and Education Centre, Maha Sarakham University, Maha Sarakham 44150, Thailand

Abstract: *Psammornis rothschildi* is an avian taxon established by Andrews in 1911 on the basis of eggshell fragments surface-collected near the city of Touggourt, in the north-eastern part of the Algerian Sahara. Since the initial discovery, a number of *Psammornis* specimens have been reported from various localities in North Africa (Algeria, Tunisia, Libya, Mauritania) and the Middle East (Saudi Arabia, Iran). Most of the finds lack a stratigraphic context, which has resulted in considerable confusion about the geological age of *Psammornis*, with attributions ranging from the Eocene to the Holocene. A review of the available evidence shows that only two groups of localities provide reasonably reliable stratigraphic evidence: the Segui Formation of SW Tunisia, apparently of latest Miocene age, and the Aguerguerian (Middle Pleistocene) of NW Mauritania. This suggests a fairly long time range for *Psammornis*. *Psammornis* eggs are, in all likelihood, those of giant ostriches, although the lack of associated skeletal material makes it difficult to interpret the eggshell fragments in evolutionary terms. However, the oological record suggests that giant ostriches have been present in Africa since the late Miocene, which leads to the reconsideration of some hypotheses about the palaeobiogeographical history of the Struthionidae. The lack of *Psammornis* eggs transformed by humans suggests that this giant ostrich did not survive until Epipalaeolthic or Neolithic times.

Keywords: *Psammornis*; ostrich; eggs; Africa; Miocene; Pleistocene

Citation: Buffetaut, E. The Enigmatic Avian Oogenus *Psammornis*: A Review of Stratigraphic Evidence. *Diversity* **2022**, *14*, 123. https://doi.org/10.3390/d14020123

Academic Editors: Juan Manuel López-García and Michael Wink

Received: 31 December 2021
Accepted: 2 February 2022
Published: 8 February 2022

Publisher's Note: MDPI stays neutral with regard to jurisdictional claims in published maps and institutional affiliations.

1. Introduction

The genus *Psammornis* was established in 1911 by Andrews [1], with *P. rothschildi* as the type species, on the basis of eggshell fragments collected in southern Algeria in the course of one of Lord Rothschild's expeditions to North Africa [2]. Since then, more eggshell material collected at a number of localities in Africa and the Middle East has been attributed to *Psammornis*; this rather enigmatic taxon has been the subject of much speculation, both because it is represented solely by fragmentary eggshell material not clearly associated with any skeletal remains, and because the stratigraphic provenance of the specimens is often very uncertain due to many of them being surface-collected rather than found in situ in sedimentary formations. Despite these uncertainties, *Psammornis* is still often mentioned in works on ostrich evolution [3] as well as in papers on fossilisation [4], with different geological age estimates—Pleistocene, according to Mikhailov and Zelenkov [3], and Holocene, according to Wiemann et al. [4].

The systematic position of *Psammornis* has been the subject of much debate [5]. Although aepyornithid affinities have been suggested [6,7] and Dughi and Sirugue [8,9] thought that the microstructure of *Psammornis* eggshells was closer to that of rheas and moas, it is now widely accepted that the *Psammornis* eggs are probably those of giant ostriches [1,3,5,10]; according to Mikhailov and Zelenkov [3], they belong to their "non-specialised type S", and are similar to both the thin-shelled eggs of the modern subspecies *Struthio camelus camelus* and the thick-shelled eggs of the Pliocene *S. chersonensis* from

Europe. However, the purpose of the present paper is not to discuss systematic issues or morphological and microstructural characteristics, but to review whatever solid evidence may be available about the stratigraphic provenance of *Psammornis* eggshells, on the basis of a survey of the available literature; this includes papers, mainly in French, which have been overlooked by many authors dealing with the question. Bearing in mind that the geological age of the type material of *Psammornis rothschildi* can rightly be considered as uncertain [11], widely different opinions have been expressed about the geological age of *Psammornis* specimens from various localities; this, in turn, has resulted in divergent interpretations of ratite evolutionary history in Africa and other continents.

2. The Discovery of *Psammornis* and the Beginning of the Stratigraphic Conundrum

In 1909, during a visit to southern Algeria, Lord Rothschild and Ernst Hartert explored the area around the city of Touggourt [2], in the north-eastern part of the Algerian Sahara. They found abundant fragments of ostrich eggshell on the ground in various places. While picking up some of them about 22 miles east of Touggourt, Hartert found "three pieces of a very much thicker egg-shell of a much browner colour" ([2], p. 550). Rothschild immediately thought that they must belong to "an extinct large Struthionid bird". The fragments were handed over for study to C.W. Andrews, a palaeontologist at the British Museum (Natural History), who described them as *Psammornis rothschildi* [1]. Andrews gave a detailed, but not illustrated, description of the microstructure and surface features of the eggshell, noting its considerable thickness (3.2 to 3.4 mm), much greater than that of eggs of the living ostrich (up to 2.10 mm) and second only to that of the eggshell of *Aepyornis titan*, from Madagascar. His conclusion was that the fragments were evidence of a "hitherto unknown bird which laid an egg considerably larger than the largest produced by any modern ostrich" and that, although there were some similarities with *Aepyornis*, "with *Struthio* the relationship was probably very close" ([1], p. 172).

Besides the lack of associated skeletal material, which made precise identification difficult, Andrews also noted uncertainties about the geological age of the specimens. Although they had been surface-collected and had been abraded by drifting sand, they were highly mineralised and had been found in the vicinity of a well. Andrews suggested that they might have been brought up from a considerable depth when digging the well. This gave no clue as to the exact antiquity of the eggshell fragments. Hartert [12], discussing *Psammornis*-like eggshell fragments found in the western Algerian Sahara, questioned Andrews' suggestion, pointing out that the *Psammornis* fragments often had the same preservation as *Struthio camelus* eggshell fragments and, everywhere, were found together with the latter.

Since the initial discovery in Algeria, many eggshell fragments (but no complete eggs) attributed to *Psammornis* have been reported from various parts of North Africa (including Algeria, Tunisia and Mauritania; Figure 1) and the Middle East (Saudi Arabia, Iran), but many of them were surface-collected and lack details about their stratigraphic origin. In addition, some of these purported *Psammornis* specimens appear to be significantly different from the fragments described by Andrews [1], and their attribution to the genus seems dubious.

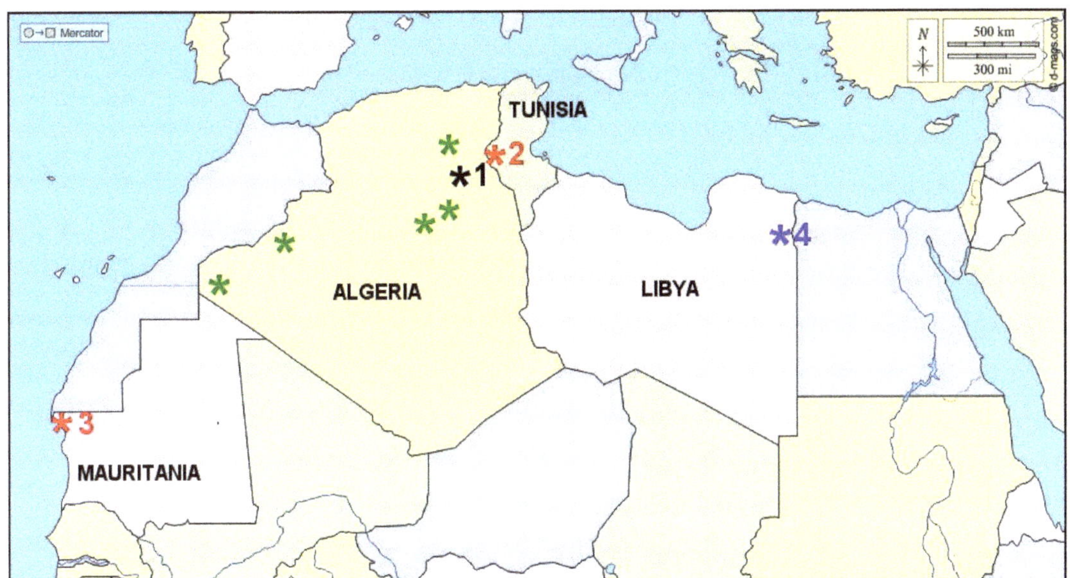

Figure 1. Distribution map of *Psammornis* localities in North Africa: black star (1)—type locality of *Psammornis rothschildi* Andrews, near Touggourt, Algeria; red stars—localities with *Psammornis* eggshells in situ in a stratigraphic context: 2—Segui Formation (latest Miocene), Chebket Safra near Moulares, south-western Tunisia, 3—Aguerguerian (Middle Pleistocene), Lévrier Bay area, north-western Mauritania; blue star (4)—type locality of *Psammornis libycus* Moltoni (a probable synonym of *Struthio camelus*), near Giarabub, Libya; unnumbered green stars—*Psammornis* finds without a stratigraphic context in the Algerian Sahara.

3. An Eocene Age for *Psammornis*?

Rothschild [13] (1911) listed *Psammornis* among what he called the Heterornithes, a loosely defined group of large birds which he considered as "fore-runners or ancestral forms of the Ratite section of the Palaeognathae" ([13], p. 148). This highly heterogenous group contained birds as different as gastornithids, phorusrhacoids and even the Jurassic *Laopteryx* (now considered a pterosaur). *Psammornis* was listed together with *Eremopezus eocaneus* as an Eocene representative of the Heterornithes from North Africa. *Eremopezus eocaneus* was described by Andrews [14] on the basis of the distal end of a tibiotarsus from the Upper Eocene Jebel Qatrani Formation of the Fayum, in Egypt. It was long believed to be an early ratite, but Rasmussen et al. [15] considered it a representative of an endemic group of large African birds.

Why Rothschild chose to associate *Psammornis* and *Eremopezus* is obscure. In his description of *Psammornis*, Andrews [1] mentioned *Eremopezus*, mainly to remark that it was only the size of a small ostrich (and, therefore, unlikely to have produced eggs the size of *Psammonis* eggs). Moreover, there was no solid reason to believe that the type material of *Psammornis* from Algeria came from an Eocene deposit, despite the fact that Andrews had suggested that the fragments might have been brought up to the surface from a considerable depth by a well. However, as mentioned above, as early as 1913, Hartert [12] had doubted that the Algerian *Psammornis* fragments could have been brought up to the surface by a well, and in 1918, Geyr von Schweppenburg [16] had noted that it was geologically unlikely that they were Eocene in age. The hypothesis that the Algerian specimens may have originated from Eocene deposits was dismissed on geological evidence by Dughi and Sirugue [8,9]. Sauer [5] reviewed, at some length, the question of the possible relationships between *Psammornis* and *Eremopezus* and concluded that there was no reason to believe they were related.

Nonetheless, Rothschild's idea of an Eocene age for *Psammornis* was taken up first by Abel [17] and then by Lambrecht [6], who placed it together with *Eremopezus* in the family Eremopezinae within the Aepyornithiformes. In 1933, Lambrecht [7] still placed *Psammornis* close to *Eremopezus* but as a genus *incertae sedis*, and gave it an Eocene age with a question mark. More recently, Piveteau [18] noted that *Psammornis* had been discovered in the Eocene of Touggourt. Brodkorb [19] still considered *Psammornis* as a possible synonym of *Eremopezus*, and also attributed it to the Eocene, with a question mark. Dementiev [20] used question marks concerning the placement of *Psammornis* among the "Eremopezidae" and its attribution to the Eocene. As late as 1965, Swinton [21] wrote that *Psammornis* had been found in the Eocene of southern Algeria. Obviously, Rothschild's ill-founded assertion that *Psammornis* was an Eocene taxon has had long-lasting consequences. There is, in fact, no convincing stratigraphic evidence anywhere suggesting that *Psammornis*-type eggshells have been found in Eocene deposits.

4. *Psammornis* Outside Africa

4.1. Saudi Arabia

In 1933, Lowe [22] attributed to *Psammornis* a thick eggshell fragment collected by the explorer and secret agent St John Philby at Shuqqat al Khalfat in Saudi Arabia. He compared it with Andrews' original material from Algeria and found close similarities in thickness and microstructure. The specimen had been surface-collected and lacked accurate information about its geological origin.

In this connection it should be noted that Bibi et al. [23] attributed eggshell fragments from the Upper Miocene Baynunah Formation of the United Arab Emirates not to *Psammornis*, but to *Diamantornis*, previously described from Namibia and Kenya. The pore complexes of the specimens from the United Arab Emirates are clearly different from what has been described in *Psammornis*, and Philby's specimen from Saudi Arabia, whatever its age, may indicate a bird different from that from the Baynunah Formation.

4.2. Iran

Dughi and Sirugue [8] reported on eggshell fragments collected at several localities in the Lut desert of eastern Iran by the geographer Jean Dresch and the naturalist Théodore Monod in 1969 and 1970. Although these fragments were significantly thinner (1.8 to 2.3 mm) than the *Psammornis* material described by Andrews, Dughi and Sirugue attributed them to *Psammornis* on microstructural evidence, considering that they belonged to an early, less advanced form than that from North Africa. However, the Iranian specimens seem to have been collected in a sand dune environment without a well-defined stratigraphic context. Caution should be urged when trying to use the putative *Psammornis* specimens from Iran for evolutionary interpretations. In connection with these specimens, Dughi and Sirugue mentioned the large ostrich eggs known from the Neogene and Quaternary periods of China [3,24–26], which may indeed be more relevant than the African *Psammornis*.

5. African Records of *Psammornis* of Doubtful Identification or Stratigraphic Position

5.1. Algeria

In addition to the type material of *Psammornis rothschildi* from the Touggourt region in the north-eastern part of the Algerian Sahara, Rothschild and Hartert [2] reported that in 1911, Hilgert had surface-collected a number of large eggshell fragments, rather different from the type of *P. rothschildi*, near Biskra, north of Touggourt. Schönwetter [27] discussed the fragments found by Hilgert and found them similar to the type material from Touggourt. He also mentioned eggshell fragments collected at Ouargla and El Golea (SW of Touggourt) by Hilgert and Hartert and at Temassin (S of Touggourt) by Fromholz and found them different from the type material. In 1960, Schönwetter [28] came back to the topic of

Psammornis and mentioned additional material from the Iguidi region, in the southern part of the Algerian Sahara, collected by Colonel Le Pivain.

Le Pivain also collected thick eggshell fragments at various points during a trip across the western Algerian Sahara in 1930; they were referred to *Psammornis* by Heim de Balsac [29].

Dughi and Sirugue [8] mentioned additional *Psammornis* eggshell fragments collected at various localities in the north-eastern Algerian Sahara (Souf region). No indications about the geological context were provided.

Apparently, all of the above-mentioned Algerian specimens were surface-collected and do not provide any reliable stratigraphic evidence about the age of *Psammornis*.

5.2. Libya

Psammornis libycus was described by Moltoni [30] on the basis of brownish-red eggshell fragments found in the dunes south of Giarabub (al-Jaghbub), a town in the eastern Libyan desert. No stratigraphic information was available. Moltoni established a new species because the fragments were significantly thinner (2.1 mm) than those of *Psammornis rothschildi* eggs. Schönwetter [28] concluded that these fragments had nothing to do with *Psammornis* and were reminiscent of specimens from Algeria possibly belonging to *Struthio camelus*. Sauer [5] described the type specimens of *Psammornis libycus* in detail and found similarities with *Struthio camelus*, wondering whether they could be the earliest *S. camelus* (or a link between *Psammornis* and *Struthio*). However, nothing is known about the geological age of *P. libycus*. Although this species was mentioned by some authors (e.g., [7,19]), there seems to be no reason to attribute it to *Psammornis*. The fragments described by Moltoni are more likely to belong to *Struthio*.

Psammornis has also been reported from the important early Miocene Jebel Zelten vertebrate locality in north-central Libya. However, this record is not based on oological evidence, and therefore, is completely unreliable. In a preliminary report, Arambourg and Magnier [31] initially mentioned a giant bird belonging to the Aepyornithidae, without specifying what kind of material this identification was based on. In 1961, this was clarified by Arambourg [32], who briefly attributed to an aepyornithid an incomplete large tibia (more properly, a tibiotarsus) lacking both ends, which in terms of size, shape and proportions was supposed to be reminiscent of *Aepyornis*. Arambourg then mentioned that large eggshell fragments from "Mio-Pliocene" deposits in the northern Sahara had been described as *Psammornis rothschildi*, but refrained from clearly attributing the Jebel Zelten giant bird to that genus. In 1967, however, in a list of the Jebel Zelten fauna, he mentioned: "Aepyornithide [sic] (*Psammornis*)", without specifying what kind of material this was based on [33]. This mention obviously caused some confusion among subsequent authors who appear to have been unaware of Arambourg's 1961 paper, in which it is clearly stated that the evidence for the presence of a giant bird at Jebel Zelten is based on a fossil bone. Curiously enough, Savage and Hamilton [34], who cite Arambourg's 1961 paper, list "? *Eremopezus*" among the fossil vertebrates from Jebel Zelten; this is probably a consequence of the above-mentioned confusion, initiated by Lord Rothschild, between *Eremopezus* and *Psammornis*. Vickers-Rich [35], not being aware of Arambourg's 1961 paper, was not sure on what material the mention of *Psammornis* was based. Mlíkovský [36], who apparently was not aware of Arambourg's 1961 paper either, misquoted Arambourg and Magnier (his quote clearly refers to Arambourg's 1967 paper, not cited in his list of references) and noted that the material on which a giant bird was identified at Jebel Zelten was "unknown" but probably consisted of eggshell fragments, since it was attributed to *Psammornis*. This, of course, is erroneous, since Arambourg's mention of a giant bird was based on a tibiotarsus.

To sum up, there is no evidence of *Psammornis*-type eggshell material from Jebel Zelten. The large tibiotarsus from Jebel Zelten mentioned by Arambourg [32] was never described in detail or illustrated. It is of potentially considerable importance for our understanding of the evolution of giant birds in Africa, but it cannot provide much reliable evidence concerning the question of the geological age of *Psammornis*, and cannot be used as evidence for the presence of this egg-based taxon in the early Miocene of Libya.

6. *Psammornis* from Tunisia: A Late Miocene Record?

As early as 1911, Rothschild and Hartert [2] reported that the German naturalists Erlanger and Hilgert had found many fragments of large eggshells in the South Tunisian desert. Bédé [37] noted that he had found eggshell fragments of a brownish colour and thicker than those of the "ordinary ostrich" at Mezzouna, 100 km SW of the city of Sfax; he thought they belonged to *Psammornis*.

A stratigraphically more significant discovery of *Psammornis* specimens was reported by Choumowitch (not 'Choumwitch' as erroneously printed in his paper) in 1951 [38]. The eggshell fragments mostly came from outcrops SW of the city of Moularès, in south-western Tunisia. They were found in abundance in red beds deeply dissected by erosion (Figure 2) in an area known as Chebket Safra ("yellow network"). The brown-coloured eggshell fragments were attributed to *Psammornis* because of their thickness (3 mm). Choumowitch also noted that they were less convex than ostrich eggshells, thus indicating larger eggs.

Figure 2. The *Psammornis*-bearing red clays on the southern flank of the Chebket Safra, SW of the city of Moulares in south-western Tunisia. The level containing remains of broken eggs preserved in situ is shown by the yellow asterisk. Photo by E.G. Gobert, modified after Chomowitch (1951).

As emphasised by Choumowitch, contrary to previous *Psammornis* finds, the eggshell fragments from Chebket Safra came from well-dated sediments. They were found in palaeosols containing calcareous concretions and fossil helicid gastropods, notably *Leucochroa tissoti*, considered as indicating a Pontian age. Choumowitch's investigations showed that the eggshell fragments did not come from Quaternary pebble deposits topping the hills. Digging into the red clays to a depth of one metre below the surface, he found an accumulation of eggshell fragments (weighing altogether more than 11 kg) arranged into small groups, each of which apparently corresponded to an egg. The whole accumulation was interpreted as a nest.

According to the geological map, the egg-bearing red beds were Pontian in age, overlying fluvial sands. Choumowitch noted that similar *Psammornis* eggshell fragments occurred at other localities in the area. Other records from Tunisia (Metlaoui, Gabès) were listed by Dughi and Sirugue [8] on the basis of information provided by Gobert.

More details about the geology of the localities were sent to Dughi and Sirugue by Gobert and published by them [8,9]. He mentioned that within the Pontian red marls, the eggshell fragments were concentrated in a well-defined zone about 50 cm in thickness, containing abundant helicid gastropods and calcareous concretions, which were interpreted as having formed around grass roots. The egg-bearing deposit was considered a paleosol formed in a dry steppe environment.

The observations made by Choumowitch and Gobert thus indicate that at Chebket Safra, the *Psammornis* eggshells were found in situ, and not reworked.

Dughi and Sirugue [8,9] gave detailed descriptions of the morphology and microstructure of the eggshells from Chebket Safra (Figure 3A). They had no hesitation about attributing the Tunisian specimens to *Psammornis rothschildi*. However, they noted that Andrews' description of the pores and canals of the *Psammornis* eggshell had to be significantly modified. This raises the question of whether the Tunisian specimens can really be attributed to *Psammornis rothschildi*. The comparison between Andrews' material from Algeria and the Tunisian fragments is made difficult by the fact that the former have suffered aeolian abrasion so that their surface features are somewhat obscured [1]. On the basis of the type material from Touggourt, Andrews [1], and especially Sauer [5], emphasised the similarities between the eggshells of *Struthio* and *Psammornis*. Dughi and Sirugue [8], mainly on the basis of the Tunisian specimens, came to a somewhat different conclusion, viz. that *Psammornis* was more closely related to Rheiformes and Dinornithiformes. In a later paper [9], they found similarities with Aepyornithiformes, Rheiformes, Casuariformes and Dinornithiformes, rather than with ostriches. A detailed comparison between eggshell fragments from Chebket Safra and the type material of *Psammornis rothschildi* from Touggourt would be useful to check whether they can really be attributed to the same ootaxon.

By contrast with the various specimens from Algeria, which lack a stratigraphic context, the eggshells from Chebket Safra are of considerable importance concerning the geological age of *Psammornis*. However, the age attributions proposed at the time of the discovery need to be discussed in light of current knowledge.

On the basis of previous work in the area, notably that of Solignac [39], Choumowitch attributed the fossil-bearing red beds to the Pontian. The Pontian is a local stage for the Peri-Tethyan regions around the Black Sea corresponding to deposits close to the Miocene–Pliocene boundary. According to Rybkina and Rostovtseva [40], the Pontian of the Black Sea area can be correlated with the Messinian Salinity Crisis of the upper part of the Messinian stage (latest Miocene). However, correlations of the so-called Pontian of south-western Tunisia with the standard stratigraphic scale is not obvious. Moreover, the red beds of the Chebket Safra lack useful biostratigraphic markers. The abundant helicid gastropods they contain, in particular *Helix* (*Leucochroa*) *tissoti*, were once considered valuable stratigraphic indicators for a Pontian age in North Africa [38,41]. However, this has been doubted [42].

The lithostratigraphic characteristics and the abundant helicid gastropods lead to the consideration that the *Psammornis*-bearing red beds belong to the Segui Formation. This is in agreement with the Metlaoui sheet of the geological map of Tunisia at 1/100,000 [43], which shows a great development of this formation (attributed to the Pliocene) SW of Moularès, in the area of the Chebket Safra. Although Robinson and Wiman [42] described the Segui Formation as a temporal enigma, its stratigraphic position is relatively clear. As early as 1910, Roux and Douvillé [41] placed the red clays containing *Helix tissoti* in the upper part of the Miocene series and observed that they overlie white and yellow sands yielding mammal remains. This sandy formation is now called the Beglia Formation and two fossiliferous levels within it have yielded a rich vertebrate assemblage indicative of an age close to the middle–late Miocene boundary ([44] and references therein). The Beglia Formation has yielded skeletal remains of an ostrich [45], *Struthio* sp., which is not larger than the living *Struthio camelus*, and therefore, unlikely to have produced the very large *Psammornis* eggs. The Segui Formation, which in the Moularès region directly overlies the Beglia Formation, was considered Messinian to Pliocene [46], or Mio-Pliocene to Villafranchian [47]. In her reviews of Miocene formations in Tunisia, Mannai Tayech [18,19] suggested a late Tortonian–early Messinian age for the Segui Formation. Mannai-Tayech's lithological description of the Segui Formation suggests that the level with *Psammornis* eggshells is in the upper part of the formation. It is worth noting that Robinson and Wiman [42] mentioned the occurrence of gypsum deposits in the Segui Formation as possible evidence for contemporaneity with the Messinian salinity crisis.

Although some uncertainties remain about the age of the upper boundary of the Segui Formation, the above-mentioned evidence strongly suggests a late Miocene, possibly Messinian age, for the *Psammornis* eggshells from the Chebket Safra.

It is worth noting that in 1952, Arambourg [50] mentioned that, according to Gobert, *Psammornis* eggshell fragments were relatively frequent in continental deposits, probably Pontian in age, in southern Tunisia. However, he did not cite Choumowitch's paper and seems to have been unaware of it (or possibly Choumowitch's paper was published after Arambourg wrote his paper). Vickers-Rich [10] noted Arambourg's mention of *Psammornis* in southern Tunisia, but she was unaware of Choumowitch's and Dughi and Sirugue's papers, so she could not appreciate the real significance of the Tunisian finds.

7. *Psammornis* from Mauritania: A Pleistocene Record

In 1939, Monod [51] was the first to report the occurrence of *Psammornis* on the Atlantic coast of Mauritania in a semi-popular short paper on eggshell fragments he had discovered at the "root" of the Cap Blanc (Râs Nouâdibhou) peninsula, in the northwestern corner of the country. The specimens were communicated to Schönwetter, who noted that they were 3 to 3,4 mm thick, and provided estimates for the dimensions of the complete shell (28 × 21 cm) and the weight of the egg (7 kg)—all of these figures being much larger than those for the living ostrich (as expressed by a comparative drawing). Monod gave no information about the geological provenance of the eggshell fragments but jokingly mentioned the omelets and water containers the Neolithic people of the Sahara could have made with such eggs, showing that he thought that they were of a comparatively recent date. In a more formal paper, including a better comparative drawing (Figure 4), Monod [52] quoted Schönwetter again; the German oologist had no doubt that the fragments belonged to *Psammornis rothschildi*, the thickness being the same, although he remarked that the disposition of the pores was not exactly similar.

The geological provenance of the Mauritanian *Psammornis* was made clear in 1971, when Tessier et al. [53] showed that the eggshells came from Aguerguerian deposits (the Aguerguerian is a local stage of the Pleistocene). They were found in calcareous and clayey sandstones, which, as in Tunisia, also yielded helicid gastropods. Seven localities (see map in [9]) were mentioned where *Psammornis* eggshell fragments were found in situ, with sharp, unworn edges (unlike the surface-collected specimens showing clear signs of aeolian erosion). Voisin [54] described the surface features (Figure 3B) and microstructure of eggshells collected by Hébrard at three Mauritanian localities, noting that they were extremely abundant at one of them, which seemed to correspond to a nesting site. She found a complete correspondence with *Psammornis rothschildi* and noted great similarities with ostrich eggs in the surface features of the shell. No association with human artefacts was noted. Dughi and Sirugue [9] also gave a description of Mauritanian eggshells, which they attributed to *Psammornis*, noting a certain variability in the eggshell surface.

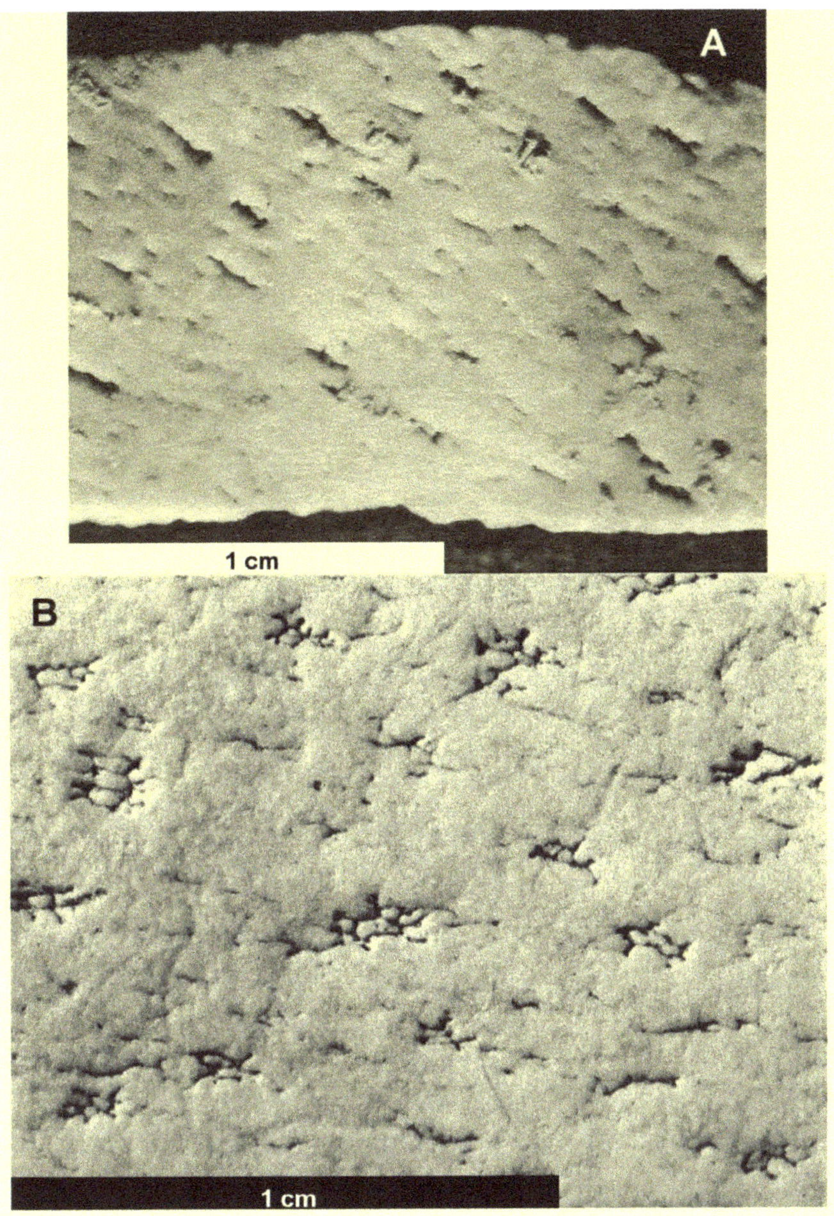

Figure 3. Pores on the outer surface of *Psammornis* eggshell fragments from Tunisia, (**A**) modified after Dughi and Sirugue [8]) and Mauritanaia (**B**), modified after Voisin [54].

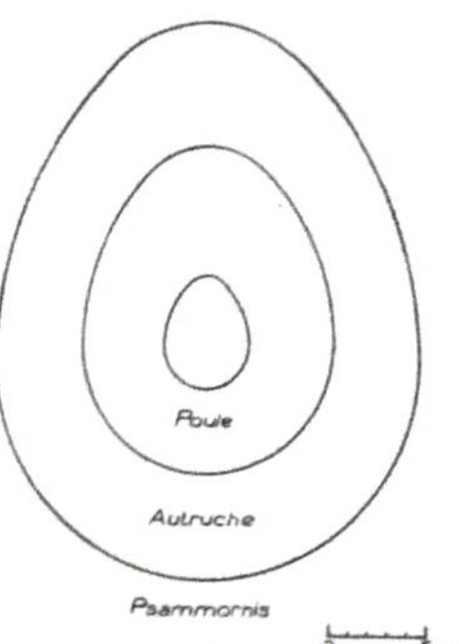

Figure 4. Outline reconstruction of a *Psammornis* egg by Monod on the basis of Schönwetter's size estimates, with outlines of the eggs of an ostich and a hen for comparison. After Monod [51].

The specimens from Mauritania, like those from the Chebket Safra, are of especial importance because they were found in situ in a stratigraphic context, and therefore, provide evidence about the geological age of *Psammornis*. As noted by Tessier et al. [53], Ortlieb [55] and Hébrard [56], they occur in the Aguerguer sandstones, attributed to the Aguerguerian stage. A section of one of the localities was published by Hébrard [56] (Figure 5). As noted by Hébrard et al., [57], the Aguerguerian is, in fact, a regressive facies of the Middle Pleistocene Aioujian stage, and the clayey sandstones containing helicid gastropods and *Psammornis* correspond to paleosols.

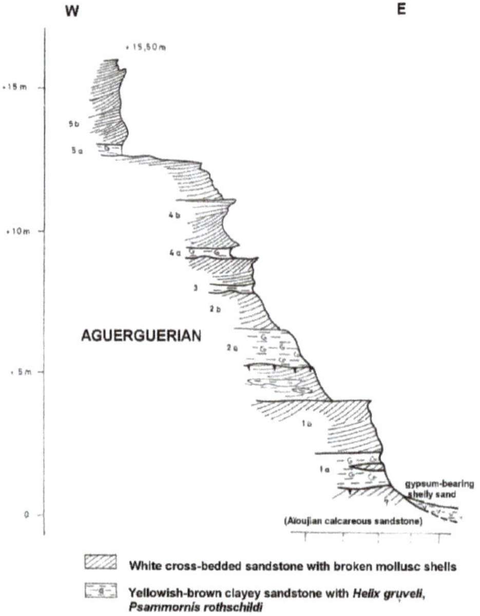

Figure 5. Section showing Aguerguerian deposits at 'point kilométrique 52' at the north-eastern end of Lévrier Bay, north-western Mauritania. White cross-bedded sandstones alternate with yellowish-brown clayey sandstones containing the snail *Helix gruveli* and in situ *Psammornis* eggshells. The main *Psammornis*-bearing horizon is at the bottom of the section. Modified after Hébrard [56].

Although the Aguerguerian has long been considered as Middle Pleistocene, its exact age remained uncertain until radiometric dates became available. C14 ages of more than 39,900 years were obtained for *Psammornis* eggshell fragments [55], but this was a minimum age. Giresse et al. [58] published U/Th dates from fossil mollusc shells from geological formations corresponding to several Pleistocene local stages along the Mauritanian coast. No ages were provided for the Aguerguerian; however, ages ranging from 241,000 to 258,000 years were obtained for the underlying Aioujian, and of up to 111,000 years for the overlying Inchirian. Since the Aguerguerian is considered a facies of the Aioujian, an age of about 200,000 years seems likely (Middle Pleistocene, Chibanian). As noted by Hébrard [56], *Psammornis* does not occur in the Inchirian deposits.

It is worth noting that engraved ostrich eggshells and ostrich eggshell beads are found in some abundance at prehistoric (Epipalaeolithic and Neolithic) sites in Mauritania [59]. Vernet et al. [59] mentioned that at the Cansado prehistoric site, many *Psammornis* eggshell fragments were collected in addition to fragments of *Struthio camelus* eggs; they noted that *Psammornis* had disappeared well before the occupation of the site by humans. More generally, there does not seem to be any record of the use of *Psammornis* eggs (for ornamentation or other purposes) by prehistoric humans, in Mauritania (R. Vernet, pers. com.) or elsewhere. Nevertheless, the presence of *Psammornis* eggs in the Middle Pleistocene Aguerguerian deposits of coastal Mauritania implies that this giant bird must have been contemporaneous with early humans, even though no early Palaeolithic industries have been found in Aïoujian and Aguerguerian sediments [56,60]. This absence may possibly be linked to unfavourable climatic conditions, since the Aguerguerian climate appears to show an evolution towards greater aridity [56,61].

8. The Stratigraphic Record of *Psammornis* and its Implications

(1) Most reported occurrences of *Psammornis* lack a reliable stratigraphic control. This applies to all records from the Algerian Sahara, including the type material from Touggourt. They correspond to surface finds, often in a sand dune environment, and the geological origin of the fragments could not be ascertained. This also applies to finds from Saudi Arabia (which may belong to *Psammornis*) and Iran (unlikely to be *Psammornis* in view of the thinness of the shell).

(2) The attribution to *Psammornis* of various eggshell fragments is dubious. This applies mainly to fragments which are significantly thinner than the original specimens from Touggourt and do not seem to be much thicker than normal *Struthio camelus* eggshells. A case in point is the material of *"Psammornis libycus"* from Libya which, in all likelihood, can be attributed to *Struthio*. As pointed out by Dughi and Sirugue [8], the unusual thickness of *Psammornis* eggshells was one of the main defining characteristics used by Andrews [1] to establish the taxon *Psammornis rothschildi*. However, among thick eggshells, differences in the pore system have been noted. This is reflected by the divergences between the interpretations of Andrews [1], Sauer [5] and Voisin [54], who find great similarities between *Psammornis* and *Struthio*, and that of Dughi and Sirugue [8,9], who conclude that *Psammornis* is related to Rheiformes, Casuariformes, Aepyornithiformes and Dinornithiformes. A revision of the material from Chebket Safra, on which Dughi and Sirugue's conclusions were largely based, would help to clarify the question.

(3) Only two groups of localities seem to provide reliable stratigraphic evidence concerning the geological age of *Psammornis*:

- The Segui Formation of south-western Tunisia, where *Psammornis* 'nests' have been discovered in situ in red sandy clays apparently containing paleosols. Although some uncertainties remain, the likeliest age for the Segui Formation seems to be latest Miocene (about 6 My ago);
- The Aguerguerian of the Mauritanian coast, especially in its northern part, where what are apparently *Psammornis* nesting sites have been discovered in situ in sandstone formations showing evidence of paleosols. The Aguerguerian is placed in the Middle Pleistocene and may be about 200,000 years old.

These two relatively well dated *Psammornis* records are rather far apart in time, implying that, if the specimens from Chebket Safra do belong to this oogenus, *Psammornis* was present in North Africa over a long time span, covering at least 6 million years. This in itself is not unlikely for a genus of giant bird. The genus *Struthio* has a record extending from the Early Miocene (*S. coppensi*) to the present. It seems less likely that the Miocene and Pleistocene specimens of *Psammornis* should belong to a single species (*P. rothschildi*). It should be remembered, however, that *Psammornis* is an ootaxon, the evolution of which probably cannot be followed over time with the same precision as that of taxa based on skeletal material. In fact, the distribution of *Psammornis* eggshells mainly shows that very large ratite birds were present in North Africa over a fairly long period of time, from the Late Miocene to the Middle Pleistocene. There is no consensus about the systematic position of these birds, although the idea that they may have been related to the Aepyornithiformes of Madagascar has never been well-supported. Dughi and Sirugue [8,9], on the basis of microstructural characteristics, suggested relationships with rheas and moas. However, following Sauer's revision [5], most authors [3,10,11,62] seem to agree that *Psammornis* was closely related to *Struthio*—if not, in fact, a member of that genus—and this would lead to the possible consideration of *Psammornis* as a junior synonym of the oogenus *Struthiolithus*. Further research is needed to decide whether the specimens from Chebket Safra can, indeed, safely be referred to *Psammornis*.

It may be worth noting that both in Tunisia and in Mauritania the deposits that yield *Psammornis* eggshells apparently show signs of aridity. The giant ostrich may have been adapted to dry environments, and this may, in turn, explain its large size. Size increase in the Australian dromornithids has been interpreted as an adaptation to an increasingly arid environment with decreasing food resources [63,64], and a similar process has been proposed for some of the giant ostriches of Eurasia [65].

(4) When *Psammornis* became extinct remains uncertain. The Mauritanian record indicates that it survived until the Middle Pleistocene, roughly 200,000 years ago; this implies that it was contemporaneous with early humans, although evidence of interactions between hominins and these giant birds has never been reported. Attributions of *Psammornis* to the Holocene (e.g., [4]) are not supported by the available evidence. A point worth noting, as mentioned above, is that apparently no engraved or otherwise transformed (for instance for bead-making) *Psammornis* eggshells have been reported, whereas decorated, cut or perforated *Struthio* eggshells are very common at Epipalaeolithic and Neolithic sites in North Africa [66,67]. This strongly suggests that *Psammornis* was no longer in existence when these cultures flourished in the Late Pleistocene and Holocene.

(5) Although very few *Psammornis* localities are well dated, the available reliable ages are important in terms of the evolution of giant ratites in North Africa. It should of course be borne in mind that ootaxa are more difficult to interpret in terms of evolutionary history than skeletal remains, and that no reliable association between *Psammornis*-type eggshells and fossil bones has been reported. Large ostriches have been reported on the basis of skeletal material from North Africa. They include *Struthio barbarus*, from the 'Villafranchian' (early Pleistocene) of Algeria [68], which is not associated with egg remains. At Ahl al Oughlam (late Pliocene or earliest Pleistocene of Morocco), a few bones of a large ostrich have been erroneously [3] assigned to *Struthio asiaticus* by Moure-Chauviré and Geraads [69]; they are accompanied by eggshell fragments described as being different from *Psammornis*. As long as no clear association between skeletal remains and *Psammornis* eggs is reported, it will remain difficult to interpret the *Psammornis* record in evolutionary terms. However, the stratigraphic evidence yielded by the Tunisian and Mauritanian records leads us to question some assumptions. For instance, Mikhailov and Zelenkov [3] assumed that *Psammornis rothschildi* is late Early Pleistocene to Middle Pleistocene in age and suggested that the *Psammornis* lineage evolved in more northern territories (perhaps from *Struthio chersonensis* from eastern Europe) and then dispersed to North Africa. However, if they actually belong to *Psammornis*, the latest Miocene specimens from Tunisia are not in agreement with this hypothesis. They indicate the presence in North Africa of a giant ostrich at a much earlier date and may even suggest an African origin for the large ostriches of late Neogene Europe, although the direction of dispersal remains uncertain. The whole picture is made even more complicated by the struthionid record

from southern and eastern Africa, consisting of both bone and egg remains (but apparently not including *Psammornis*-type eggs), which is beyond the scope of the present paper and has led to speculations about the evolutionary history of the ostriches [70].

9. Conclusions

Ever since the initial description by Andrews in 1911, the egg-based taxon *Psammornis* has been the subject of widely divergent interpretations in terms of both zoological affinities and stratigraphic distribution. There now seems to be a measure of consensus about the fact that the large thick-shelled *Psammornis* eggs were produced by giant ostriches, although the lack of association with skeletal material makes an interpretation in evolutionary terms difficult. The stratigraphic conundrum which began at the time of the initial description, when an Eocene age was suggested without any solid evidence, has proved even more difficult to resolve, because many of the *Psammornis* finds (even when doubtful records are eliminated), consisting of surface finds, lack any solid stratigraphic context. However, a review of the literature, including various important papers which, for some reason, have been ignored by most authors dealing with the topic, shows that eggs attributed to *Psammornis* have been discovered in situ in relatively well-dated formations at two geographically widely separated groups of localities, in south-western Tunisia and on the Mauritanian coast. Interestingly, these sites are also quite distinct in geological age: the Tunisian finds are apparently latest Miocene in age, whereas the Mauritanian occurrences are in Middle Pleistocene deposits. Whether the Tunisian and Mauritanian *Psammornis* really belong to a single taxon is a moot point, but the occurrence in North Africa of large eggs probably produced by giant ostriches at two periods widely distant in time may suggest a long-lasting lineage of large ratites in that region. This should be taken into consideration when trying to reconstruct ostrich evolutionary history. In particular, it seems difficult to accept that giant ostriches dispersed from Eurasia to Africa as late as the Pleistocene, since they appear to have been present on the African continent as early as the late Miocene. The large eggs from Tunisia may even suggest dispersal in the opposite direction, from Africa to Eurasia. A detailed revision of the Tunisian specimens would, therefore, be very welcome.

Funding: This research received no external funding.

Acknowledgments: Thanks to Jihed Dridi and Beya Mannaï-Tayech for the information about the geology of south-western Tunisia, to Robert Vernet for the information about *Psammornis* in Mauritania, and to the three anonymous reviewers for their helpful comments.

Conflicts of Interest: The author declares no conflict of interest.

References

1. Andrews, C.W. Note on some fragments of the fossil egg-shell of a large struthious bird from southern Algeria, with some remarks on some pieces of the egg-shell of an ostrich from northern India. In *Verhandlungen des V. Internationalen Ornithologen-Kongresses in Berlin 1911*; Schalow, H., Ed.; Deutsche Ornithologische Gesellschaft: Berlin, Germany, 1911; pp. 169–174.
2. Rothschild, W.; Hartert, E. Ornithological explorations in Algeria. *Novit. Zool.* **1911**, *18*, 456–550. [CrossRef]
3. Mikhailov, K.E.; Zelenkov, N. The late Cenozoic history of the ostriches (Aves: Struthionidae), as revealed by fossil eggshell and bone remains. *Earth Sci. Rev.* **2020**, *208*, 103270. [CrossRef]
4. Wiemann, J.; Fabbri, M.; Yang, T.R.; Stein, K.; Sander, P.M.; Norell, M.A.; Brigg, D.E.G. Fossilization transforms vertebrate hard tissue proteins into N-heterocyclic polymers. *Nat. Com.* **2018**, *9*, 1234567890. [CrossRef] [PubMed]
5. Sauer, E.G.F. Taxonomic evidence and evolutionary interpretation of *Psammornis*. *Bonn. Zool. Beitr.* **1969**, *20*, 290–310.
6. Lambrecht, K. *Fossilium Catalogus I Animalia*; Backhuys Publishers: Leiden, The Netherlands, 2007.
7. Lambrecht, K. *Handbuch der Palaeornithologie*; Borntrager: Berlin, Germany, 1933.
8. Dughi, R.; Sirugue, F. Sur la structure des œufs des Sauropsidés vivants ou fossiles; le genre *Psammornis* Andrews. *Bull. Soc. géol. Fr.* **1964**, *6*, 240–252. [CrossRef]
9. Dughi, R.; Sirugue, F. Sur le *Psammornis rotschildi* [sic]Andrews. *Bull. Inst. Fr. Afr.Noire* **1978**, *40*, 6–27.
10. Hartert, E. *Die Vögel der Paläarktischen Fauna. Band III*; Friedländer & Sohn: Berlin, Germany, 1922.
11 Knox, A.G.; Walters, M.P. Extinct and endangered birds in the collections of the Natural History Museum. *Brit. Ornith. Club Occas. Pub.* **1994**, *1*, 1–292.

12. Hartert, E. Expedition to the central western Sahara. IV. Birds. *Novit. Zool.* **1913**, *20*, 37–76.

13. Rothschild, W. On the former and present distribution ot the so called Ratitae or ostrich-like birds with certain deductions and a description of a new form by C.W. Andrews. In *Verhandlungen des V. Internationalen Ornithologen-Kongresses in Berlin 1911*; Schalow, H., Ed.; Deutsche Ornithologische Gesellschaft: Berlin, Germany, 1911; pp. 144–169.

14. Andrews, C.W. On the pelvis and hind-limb of *Mullerornis betsilei* M.-Edw. & Grand.; with a note on the occurrence of a ratite bird in the upper Eocene beds of the Fayum, Egypt. *Proc. Zool. Soc.London* **1904**, *1*, 163–171.

15. Rasmussen, D.T.; Simons, E.L.; Hertel, F.; Judd, A. hindlimb of a giant terrestrial bird from the Upper Eocene, Fayum, Egypt. *Palaeontology* **2001**, *44*, 325–337. [CrossRef]

16. Geyr von Schweppenburg, H. Ins Land der Tuareg. *J. Ornithol.* **1918**, *66*, 121–176. [CrossRef]

17. Abel, O. *Die Stämme der Wirbeltiere*; Walter de Gruyter & Co.: Berlin, Germany; Leipzig, Germany, 1919.

18. Piveteau, J. Oiseaux. Aves Linné. In *Traité de Paléontologie*; Tome, V., Piveteau, J., Eds.; Masson: Paris, France, 1955; pp. 994–1091.

19. Brodkorb, P. Catalogue of fossil birds. Part I. (Archaeopterygiformes through Ardeiformes). *Bull. Florida State Mus.* **1963**, *7*, 179–293.

20. Dementiev, G.P. Class Aves. Birds. In *Fundamentals of Palaeontology, Amphibians, Reptiles and Birds*; Rozhdestvenskii, A.K., Tatarinov, L.P., Eds.; Nauka: Moscow, Russia, 1964; pp. 660–699. (In Russian)

21. Swinton, W.E. *Fossil Birds*; British Museum (Natural History): London, UK, 1965.

22. Lowe, P.R. On Some Struthious Remains:-1. Description of some pelvic remains of a large fossil ostrich, *Struthio oldawayi*, sp. n., from the Lower Pleistocene of Oldaway (Tanganyika Territory); 2. Egg-shell fragments referable to *Psammornis* and other Struthiones collected by Mr. St. John Philby in southern Arabia. *Ibis* **1933**, *75*, 652–658.

23. Bibi, F.; Shabel, A.B.; Kraatz, B.P.; Stidham, T.A. New fossil ratite (Aves: Palaeognathae) eggshell discoveries from the late Miocene Baynunah Formation of the United Arab Emirates, Arabian Peninsula. *Palaeont. Electr.* **2006**, *9*, 1–13.

24. Lowe, P.R. Struthious remains from Northern China and Mongolia, with descriptions of *Struthio wimani*, *Struthio anderssoni* and *Struthio mongolicus*, spp. nov. *Palaeont. Sin. C* **1931**, *6*, 1–47.

25. Buffetaut, E. 'Dragon's eggs' from the 'Yellow Earth': The discovery of the fossil ostriches of China. *Historia Natural* **2021**, *11*, 47–63.

26. Buffetaut, E.; Angst, D. A giant ostrich from the Lower Pleistocene Nihewan Formation of North China, with a review of the fossil ostriches of China. *Diversity* **2021**, *13*, 47. [CrossRef]

27. Schönwetter, M. Fossile Vogelei-Schalen. *Novit. Zool.* **1929**, *35*, 192–203.

28. Schönwetter, M. *Handbuch der Oologie. Lieferung 1*; Akademie Verlag: Berlin, Germany, 1960.

29. Heim de Balsac, H. Premières données sur les oiseaux du Sahara occidental. *Alauda* **1930**, *2*, 451–463.

30. Moltoni, E. Risultati zoologici della missione inviata dalla R. Società Geografica Italiana per l'esplorazione dell'oasi de Giarabub (1926–1927). Uccelli. *Ann. Mus. Civ. Stor. Nat. G. Doria* **1928**, *52*, 387–401.

31. Arambourg, C. & Magnier, P. Gisements de vertébrés tertiaires dans le bassin de Syrte, Libye. *C. R. Acad. Sci. Paris* **1961**, *252*, 1181–1183.

32. Arambourg, C. Note préliminaire sur quelques vertébrés nouveaux du Burdigalien de Libye. *C. R. somm. Séances Soc. Géol. Fr.* **1961**, *4*, 107–109.

33. Arambourg, C. Continental vertebrate faunas of the Tertiary of North Africa. In *African Ecology and Human Evolution*; Howell, F.C., Bourlière, F., Eds.; Aldine Publishing Company: Chicago, IL, USA, 1967; pp. 55–64.

34. Savage, R.J.G.; Hamilton, W.R. Introduction to the Miocene mammal faunas of Gebel Zelten, Libya. *Bull. Brit. Mus. (Nat. Hist.) Geol.* **1973**, *22*, 513–527.

35. Vickers-Rich, P. Significance of the Tertiary avifaunas from Africa (with emphasis on a mid to late Miocene avifauna from southern Tunisia). *Ann. Geol. Surv. Egypt* **1974**, *4*, 167–210.

36. Mlíkovky, J. Early Miocene birds of Djebel Zelten, Libya. *Čas. Národ. Muz., Řad. Přírod.* **2003**, *172*, 114–120.

37. Bédé, P. Bericht über den v. intern orn. Kongress 1911. Psammornis Rothschildi, nov. Gen., nov. sp., by C.-W. Andrews. *Rev. crit. Paléozool.* **1919**, *23*, 56–57.

38. Choumowitch, W. Sur l'Autruche géante de la Koudiat Safra. *C.R. 70e Congr. Ass. Fr. Avanc. Sci. Tunis* **1951**, *1*, 172–175.

39. Solignac, M. Le Pontien dans le Sud tunisien. *Ann. Univ. Lyon N.S. I. Sci. Med.* **1931**, *48*, 1–29.

40. Rybkina, A.I.; Rostovtseva, Y.V. New evidence of the age of the Black Sea Pontian substage. *Russ. J. Earth Sci.* **2017**, *17*, ES5004. [CrossRef]

41. Roux, H.; Douvillé, H. La géologie des environs de Redeyef (Tunisie). *Bull. Soc. géol. Fr.* **1911**, *10*, 646–660.

42. Robinson, P.; Wiman, S.K. A revision of the stratigraphic subdivision of the Miocene rocks of sub-dorsale Tunisia. *Notes Serv. Géol. Tunisie* **1976**, *42*, 71–86.

43. Regaya, K.; Laatar, S.; Chaouachi, A. Metlaoui. *Carte géologique de la Tunisie 1/100,000* **1991**, 65.

44. Werdelin, L. Chronology of Neogene mammal localities. In *Cenozoic Mammals of Africa*; Werdelin, L., Sanders, W.J., Eds.; University of Californa Press: Berkeley, CA, USA; London, UK, 2010; pp. 27–43.

45. Vickers-Rich, P. A fossil avifauna from the Upper Miocene Beglia Formation of Tunisia. *Notes Serv. Géol. Tunisie* **1972**, *35*, 29–66.

46. Biely, A.; Rakus, M.; Robinson, P.; Salaj, J. Essai de corrélation des formations miocènes au Sud de la Dorsale tunisienne. *Notes Serv. Géol. Tunisie* **1972**, *38*, 73–92.

47. Biely, A.; Rakus, M.; Robinson, P. Le Miocène de la Tunisie. *Ann. Geol. Surv. Egypt* **1974**, *4*, 307–318.

48. Mannaï-Tayech, B. Les séries silico-clastiques miocènes du Nord-Est au Sud-Ouest de la Tunisie: Une mise au point. *Geobios* **2006**, *39*, 71–84. [CrossRef]

49. Mannaï-Tayech, B. The lithostratigraphy of Miocene series from Tunisia, revisited. *J. Afr; Earth Sci.* **2009**, *54*, 53–61. [CrossRef]

50. Arambourg, C. *La Paléontologie des Vertébrés en Afrique du Nord Française*; XIXe Congrès Géologique International, Monographies régionales: Alger, Algeria, 1952.

51. Monod, T. Nous aussi. *Notes afr. Instit. Fr. Afr. Noire* **1939**, *4*, 47–48.

52. Monod, T. Sur la présence d'un *Psammornis* en Afrique Occidentale Française. *C.R. 1e Conf. Intern. Africanistes Ouest* **1951**, *1*, 202–203.

53. Tessier, F.; Hébrard, L.; Lappartient, J.R. Découverte de fragments d'oeufs de *Psammornis* et de *Struthio* dans le Quaternaire de la presqu'île du Cap Blanc (République Islamique de Mauritanie). *C.R. Acad. Sci. Paris D* **1971**, *273*, 2418–2421.

54. Voisin, C. Etude de la structure de fragments de coquilles d'oeufs de *Psammornis rothschildi* Andrews provenant de Mauritanie. *L'Oiseau Rev. Franç. Ornith.* **1971**, *41*, 245–256.

55. Ortlieb, L. Recherches sur les formations plio-quaternaires du littoral ouest-saharien (28°30'- 20°40' lat. N). *Trav. Doc. O.R.S.T.O.M.* **1975**, *48*, 1–267.

56. Hébrard, L. Contribution à l'étude géologique du Quaternaire du littoral mauritanien entre Nouakchott et Nouadhibou, 18°—21° latitude Nord. Participation à l'étude des désertifications du Sahara. *Doc. Lab. Géol. Facult. Sci. Lyon* **1978**, *71*, 1–210.

57. Hébrard, L.; Elouard, P.; Faure, H. La synthèse stratigraphique du Quaternaire du littoral mauritanien entre Nouakchott et Nouadhibou. In *Lexique Stratigraphique International. Nouvelle Série n°1. Afrique de l'Ouest. Introduction Géologque et Termes Stratigraphiques*; Fabre, J., Ed.; Pergamon Press: Oxford, UK, 1983; pp. 158–170.

58. Giresse, P.; Barrusseau, J.P.; Causse, C.; Diouf, M. Successions of sea-level changes during the Pleistocene in Mauritania and Senegal distinguished by sedimentary facies study and U/Th dating. *Mar. Geol.* **2000**, *170*, 123–139. [CrossRef]

59. Vernet, R.; Rodrigue, A.; Tous, P. Les tests d'œuf d'autruche gravés du littoral atlantique saharien du nord du Banc d'Arguin à l'oued Draa. *Sahara* **2006**, *17*, 59–72.

60. Vernet, R. *Préhistoire de la Mauritanie*; Centre Culturel Français de Nouakchott—Sépia: Nouakchott, Mauritania, 1993.

61. Vernet, R. *Le Golfe d'Arguin de la Préhistoire à l'Histoire: Littoral et plaines intérieures*; Parc National du Banc d'Arguin: Nouakchott, Mauritania, 2007.

62. Mikhailov, K.E. Fossil and recent eggshell in amniotic vertebrates: Fine structure, comparative morphology and classification. *Spec. Pap. Palaeont.* **1997**, *56*, 1–80.

63. Murray, P.F.; Vickers-Rich, P. *Magnificent Mihirungs: The Colossal Flightless Birds of the Australian Dreamtime*; Indiana University Press: Bloomington, IN, USA, 2004.

64. Angst, D.; Buffetaut, E. Paleobiology of Giant Flightless Birds. Elsevier & ISTE Press: Oxford, UK; London, UK, 2017.

65. Zelenkov, N.V.; Lavrov, A.V.; Startsev, D.B.; Vislobokova, I.A.; Lopatin, A.V. A giant early Pleistocene bird from eastern Europe: Unexpected component of terrestrial faunas at the time of early Homo arrival. *J. Vert. Paleont.* **2019**, *39*, e1605521. [CrossRef]

66. Camps-Fabrer, H. *La disparition de l'autruche en Afrique du Nord*; Institut Français des Sciences Humaines en Algérie: Alger, Algeria, 1963.

67. Hodos, T.; Cartwright, C.R.; Montgomery, J.; Nowell, G.; Crowder, K.; Fletcher, A.C.; Gönster, Y. The origins of decorated ostrich eggs in the ancient Mediterranean and Middle East. *Antiquity* **2020**, 1–20. [CrossRef]

68. Arambourg, C. *Vertébrés villafranchiens d'Afrique du Nord (Artiodactyles, Carnivores, Primates, Reptiles, Oiseaux)*; Fondation Singer-Polignac: Paris, France, 1979.

69. Mourer-Chauviré, C.; Geraads, D. The Struthionidae and Pelagornithidae (Aves: Struthioniformes, Odontopterygiformes) from the Late Pliocene of Ahl Al Oughlam, Morocco. *Oryctos* **2008**, *7*, 169–194.

70. Mourer-Chauviré, C.; Senut, B.; Pickford, M.; Mein, P.; Dauphin, Y. Ostrich legs, eggs and phylogenies. *S. Afr. J. Sci.* **1996**, *92*, 492–495.

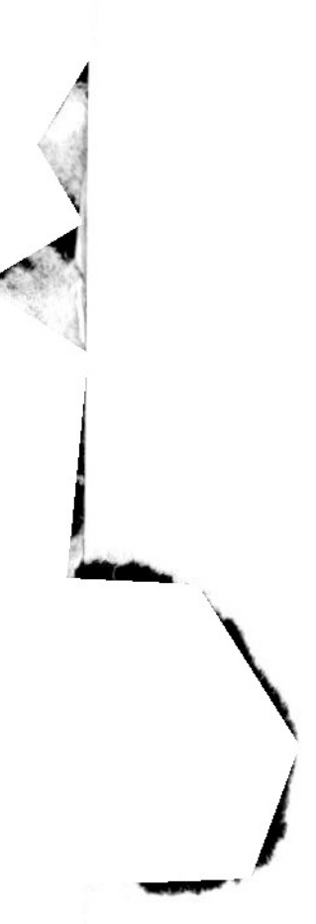

MDPI
St. Alban-Anlage 66
4052 Basel
Switzerland
Tel. +41 61 683 77 34
Fax +41 61 302 89 18
www.mdpi.com

Diversity Editorial Office
E-mail: diversity@mdpi.com
www.mdpi.com/journal/diversity